Switching Theory

Switching Theory
Architectures and Performance in Broadband ATM Networks

Achille Pattavina
Politecnico di Milano, Italy

JOHN WILEY & SONS
Chichester • New York • Weinheim • Brisbane • Singapore • Toronto

Other Wiley Editorial Offices

John Wiley & Sons, Inc., 605 Third Avenue,
New York, NY 10158–0012, USA

Weinheim • Brisbane • Singapore • Toronto

British Library Cataloguing in Publication Data

A catalogue record for this book is available from the British Library

ISBN 0 471 96338 0

Typeset in 10/12pt Monotype Bembo from the author's disks by WordMongers Ltd, Treen
Printed and bound in Great Britain by Biddles Ltd, Guildford and King's Lynn
This book is printed on acid-free paper responsibly manufactured from sustainable forestry, in which at least two trees are planted for each one used for paper production.

"Behold, I will send my angel who shall go before thee, keep thee in the journey and bring thee into the place that I have prepared."

(The Holy Bible, *Exodus* 23, 20)

to Chiara
    Matteo, Luca, Sara, Maria

"…………

d'i nostri sensi ch'è del rimanente
non vogliate negar l'esperienza,
di retro al sol, del mondo senza gente.

Considerate la vostra semenza:
fatti non foste a viver come bruti,
ma per seguir virtute e conoscenza."

(Dante, *Inferno*, Canto XXVI)

"…………

of your senses that remains, experience
of the unpeopled world behind the Sun.

Consider your origin: ye were not
formed to live like brutes, but to follow
virtue and knowledge."

(Dante, *Inferno*, Canto XXVI)

Contents

Preface

Broadband networks based on the Asynchronous Transfer Mode (ATM) standard are becoming more and more popular worldwide for their flexibility in providing an integrated transport of heterogeneous kinds of communication services. The book is intended to provide the state of the art in the field of switching for broadband ATM networks by covering three different areas: the theory of switching in interconnection networks, the architectures of ATM switching fabrics and the traffic performance of these switching fabrics.

A full coverage of switching theory is provided starting from the very first steps taken in this area about forty years ago to describe three-stage interconnection networks, either non-blocking or rearrangeable. It is pointed out how this classical theory is no longer effective to describe switching environments when hundreds of million packets per second must be switched. The brand new theory of multistage interconnection networks that has emerged from the studies of the last ten years is described and made homogeneous. The key role played by sorting and banyan networks of this new theory within the area of broadband ATM networking is highlighted.

The different types of ATM switching architectures are classified according to their fundamental parameters, related to the properties of their interconnection network and to the type of cell queueing adopted in the switching fabric. ATM switching fabrics are characterized by enormous amounts of carried traffic if compared to the operations of classical circuit-switching or packet-switching fabrics. Thus the type of banyan network classes that can be effectively used in these switching fabrics is shown.

Each class of switching architecture is evaluated in terms of its traffic performance, that is switch capacity, packet delay and loss probability, when a random traffic is offered to the switch. Analytical models, as well as computer simulation, are used for the numerical evaluation of the traffic parameters by studying the effect of the different network parameters.

Putting together the material for this book has required roughly ten years of studies and research. The active contribution across these years of many students of the Engineering Faculty of the Politecnico di Milano, who have prepared their theses, has made possible the collection of the material used here. Therefore I am grateful to all of them. I am deeply indebted to Professor Maurizio Decina for inspiring interest and passion in the study of switching theory and for his continual confidence in me. Without him this book could never have been completed. Last but not least, my family deserves the same gratitude, since writing this book has been possible only by stealing a lot of free time that I should have spent with my wife and children.

Achille Pattavina
Milan, 1997

Chapter 1 *Broadband Integrated Services Digital Network*

A broad overview on the Broadband Integrated Services Digital Network (B-ISDN) is here given. The key issues of the communication environment are first outlined (Section 1.1). Then the main steps leading to the evolution to the B-ISDN are described (Section 1.2), by also discussing issues related to the transfer mode and to the congestion control of the B-ISDN (Section 1.3). The main features of the B-ISDN in terms of transmission systems that are based on the SDH standard (Section 1.4) and of communication protocols that are based on the ATM standard (Section 1.5) are also presented.

1.1. Current Networking Scenario

The key features of the current communication environment are now briefly discussed, namely the characterization of the communication services to be provided as well as the features and properties of the underlying communication network that is supposed to support the previous services.

1.1.1. Communication services

The key parameters of a telecommunication service cannot be easily identified, owing to the very different nature of the various services that can be envisioned. The reason is the rapidly changing technological environment taking place in the eighties. In fact, a person living in the sixties, who faced the only provision of the basic telephone service and the first low-speed data services, could rather easily classify the basic parameters of these two services. The tremendous push in the potential provision of telecommunication services enabled by the current networking capability makes such classification harder year after year. In fact, not only are new services being thought and network-engineered in a span of a few years, but also the tremendous progress in VLSI technology makes it very difficult to foresee the new network capabilities that the end-users will be able to exploit even in the very near future.

A feature that can be always defined for a communication service provided within a set of *n* end-users irrespective of the supporting network is the service direction. A service is *unidirectional* if only one of the *n* end-users is the source of information, the others being the sink; a typical example of unidirectional service is broadcast television. A service is *multidirectional* if at least one of the *n* end-users is both a source and a sink of information. For decades a multidirectional telecommunication service involved only two end-users, thus configuring a bidirectional communication service. Only in the seventies and eighties did the interest in providing communication service within a set of more than two users grow; consider, e.g., the electronic-mail service, videoconferencing, etc. Apparently, multidirectional communication services, much more than unidirectional services, raise the most complete set of issues related to the engineering of a telecommunication network.

It is widely agreed that telecommunications services can be divided into three broad classes, that is *sound*, *data* and *image* services. These three classes have been developed and gradually enriched during the years as more powerful telecommunication and computing devices were made available. Sound services, such as the basic telephone service (today referred to as *plain old telephone service* - POTS), have been provided first with basically unchanged service characteristics for decades. Data services have started to be provided in the sixties with the early development of computers, with tremendous service upgrades in the seventies and eighties in terms of amounts of information transported per second and features of the data service. For about three decades the image services, such as broadcast television, have been provided only as unidirectional. Only in the last decade have the multidirectional services, such as video on demand, videotelephony, been made affordable to the potential users.

Communication services could be initially classified based on their information capacity, which corresponds to the typical rate (bit/s) at which the information is required to be carried by the network from the source to the destination(s). This parameter depends on technical issues such as the recommendations from the international standard bodies, the features of the communication network, the required network performance, etc. A rough indication of the information capacity characterizing some of the communication services is given in Table 1.1, where three classes have been identified: *low-speed services* with rates up to 100 kbit/s, *medium-speed services* with rates between 0.1 and 10 Mbit/s, and *high-speed services* with rates above 10 Mbit/s. Examples of low-speed services are voice (PCM or compressed), telemetry, terminal-to-host interaction, slow-scan video surveillance, videotelephony, credit-card authorization at point of sales (POS). HI-FI sound, host-to-host interaction in a LAN and videoconferencing represent samples of medium-speed services. Among data applications characterized by a high speed we can mention high-speed LANs or MANs, data exchange in an environment of supercomputers. However, most of the applications in the area of high speed are image services. These services range from compressed television to conventional uncompressed television, with bit rates in the range 1–500 Mbit/s. Nevertheless, note that these indicative bit rates change significantly when we take into account that coding techniques are progressing so rapidly that the above rates about video services can be reduced by one order of magnitude or even more.

Table 1.1. Service capacities

Class	Mbit/s	Service
Low speed	0.0001–0.001	Telemetry/POS
	0.005–0.1	Voice
	0.001–0.1	Data/images
Medium speed	0.1–1	HI-FI sound
	0.1–1	Videoconference
	0.1–10	Data/images
High speed	10–50	Compressed TV
	100–500	Uncompressed TV
	10–1000	Data/images

Some of the above services can be further classified as *real-time services*, meaning that a timing relationship exists between the end-users of the communication service. Real-time services are those sound and image services involving the interactions between two or more people: the typical example is the basic telephone service where the information has to be transferred from one person to the other within a time frame not exceeding a certain threshold (e.g., 500 ms), otherwise a satisfactory interaction between the two users would become impossible. On the other hand, data services as well as unidirectional sound or image services are not real-time services, since even a high delay incurred by the information units in the transport network does not impair the service itself, rather it somewhat degrades its quality.

A very important factor to characterize a service when supported by a communication channel with a given peak rate (bit/s) is its *burstiness factor*, defined as the ratio between the average information rate of the service and the channel peak rate. Apparently, the service burstiness decreases as the channel peak rate grows. Given a channel rate per service direction, users cooperating within the same service can well have very different burstiness factors: for example an interactive information retrieval service providing images (e.g. a video library) involves two information sources, one with rather high burstiness (the service center), the other with a very low burstiness (the user).

Figure 1.1 shows the typical burstiness factors of various services as a function of the channel peak rate. Low-speed data sources are characterized by a very wide range of burstiness and are in general supported by low-speed channels (less that 10^4 bit/s or so). Channels with rates of 10^4–10^5 bit/s generally support either voice or interactive low-speed data services, such the terminal-to-host communications. However, these two services are characterized by a very different burstiness factor: packetized voice with silence suppression is well known to have a very high burstiness (talkspurts are generated for about 30% of the time), whereas an interactive terminal-to-host session uses the channel for less than 1% of the time. Channel rates in the range 10^6–10^8 bit/s are used in data networks such as local area networks (LAN) or metropolitan area networks (MAN) with a burstiness factor seldom higher than 0.1. Image services are in general supported by channels with peak rates above 10^6 bit/s and can be both low-burstiness services, such as the interactive video services, and high-burstiness services as the unidirectional broadcasting TV (either conventional or high quality). However the mentioned progress in coding techniques can significantly modify the burstiness factor of an image information source for a given channel rate enabling its reduction by more than one order of magnitude.

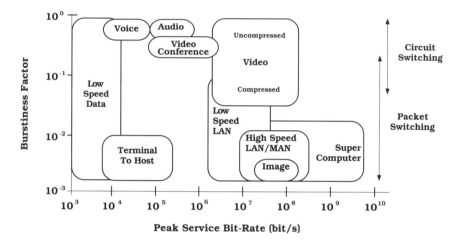

Figure 1.1. Service burstiness factor

Two features of a communication service are felt as becoming more and more important to the user, that is the *multipoint* and *multimedia* capability of a communication service. A multipoint service, representing the evolution of the basic point-to-point service, enables more than two users to be involved in the same communication. Also a multimedia service can be seen as the evolution of the "single-medium" service; a multimedia service consists in transporting different types of information between the end-users by keeping a time relation in the transport of the different information types, for example voice and data, or images coupled with sounds and texts. Both multipoint and multimedia communication services are likely to play a very important role in the social and business community. In fact a business meeting to be joined by people from different cities or even different countries can be accomplished by means of videoconferencing by keeping each partner in his own office. University lectures could be delivered from a central university to distributed faculty locations spread over the country by means of a multipoint multimedia channel conveying not only the speaker's image and voice, as well as the students' questions, but also texts and other information.

1.1.2. Networking issues

A parallel evolution of two different network types has taken place in the last decades: networks for the provision of the basic voice service on the one hand, and networks for the support of data services on the other hand. Voice signals were the first type of information to be transported by a communication network several decades ago based on the *circuit-switching* transfer mode: a physical channel crossing one or more switching nodes was made available exclusively to two end-users to be used for the information transfer between them. The set-up and release of the channel was carried out by means of a signalling phase taking place immediately before and after the information transfer.

Fast development of data networks took place only after the breakthroughs in the microelectronics technology of the sixties that made possible the manufacture of large computers (mainframes) to be shared by several users (either local or remote). In the seventies and eighties data networks had a tremendous penetration into the business and residential community owing to the progress in communication and computer technologies. Data networks are based on the *packet-switching* transfer mode: the information to be transported by the network is fragmented, if necessary, into small pieces of information, called packets, each carrying the information needed to identify its destination. Unlike circuit-switching networks, the nodes of a packet-switching network are called "store-and-forward", since they are provided with a storage capability for the packets whose requested outgoing path is momentarily busy. The availability of queueing in the switching nodes means that statistical multiplexing of the packets to be transported is accomplished on the communication links between nodes.

The key role of the burstiness factor of the information source now becomes clear. A service with high burstiness factor (in the range 0.1–1.0) is typically better provided by a circuit-switching network (see Figure 1.1), since the advantage of statistically sharing transmission and switching resources by different sources is rather limited and performing such resource sharing has a cost. If the burstiness factor of a source is quite small, e.g. less than 10^{-2}, supporting the service by means of circuit-switching becomes rather expensive: the connection would be idle for at least 99% of the time. This is why packet-switching is typically employed for the support of services with low burstiness factor (see again Figure 1.1).

Even if the transport capability of voice and data networks in the seventies was limited to narrowband (or low-speed) services, both networks were gradually upgraded to provide upgraded service features and expanded network capabilities. Consider for example the new voice service features nowadays available in the POTS network such as call waiting, call forwarding, three-party calls etc. Other services have been supported as well by the POTS network using the voice bandwidth to transmit data and attaching ad hoc terminals to the connection edges: consider for example the facsimile service. Progress witnessed in data networks is virtually uncountable, if we only consider that thousands of data networks more or less interconnected have been deployed all over the world. Local area networks (LAN), which provide the information transport capability in small areas (with radius less than 1 km), are based on the distributed access to a common shared medium, typically a bus or a ring. Metropolitan area networks (MAN), also based on a shared medium but with different access techniques, play the same role as LANs in larger urban areas. Data networks spanning over wider areas fully exploit the store-and-forward technique of switching nodes to provide a long-distance data communication network. A typical example is the ARPANET network that was originally conceived in the early seventies to connect the major research and manufacturing centers in the US. Now the INTERNET network interconnects tens of thousand networks in more than fifty countries, thus enabling communication among millions of hosts. The set of communication services supported by INTERNET seems to grow without apparent limitations. These services span from the simplest electronic mail (e-mail) to interactive access to servers spread all over the world holding any type of information (scientific, commercial, legal, etc.).

Voice and data networks have evolved based on two antithetical views of a communication service. A voice service between two end-users is provided only after the booking of the required transmission and switching resources that are hence used exclusively by that communication. Since noise on the transmission links generally does not affect the service effectiveness, the quality of service in POTS networks can be expressed as the probability of call acceptance. A data service between two-end-users exploits the store-and-forward capability of the switching nodes; a statistical sharing of the transmission resources among packets belonging to an unlimited number of end-users is also accomplished. Therefore, there is in principle no guarantee that the communication resources will be available at the right moment so as to provide a prescribed quality of service. Owing to the information transfer mode in a packet-switching network that implies a statistical allocation of the communication resources, two basic parameters are used to qualify a data communication service, that is the average packet delay and the probability of packet loss. Moreover in this case even a few transmission errors can degrade significantly the quality of transmission.

1.2. The Path to Broadband Networking

Communication networks have evolved during the last decades depending on the progress achieved in different fields, such as transmission technology, switching technology, application features, communication service requirements, etc. A very quick review of the milestones along this evolution is now provided, with specific emphasis on the protocol reference model that has completely revolutionized the approach to the communication world.

1.2.1. Network evolution through ISDN to B-ISDN

An aspect deeply affecting the evolution of telecommunication networks, especially telephone networks, is the progress in digital technology. Both transmission and switching equipment of a telephone network were initially analogue. Transmission systems, such as the multiplexers designed to share the same transmission medium by tens or hundreds of channels, were largely based on the use of frequency division multiplexing (FDM), in which the different channels occupy non-overlapping frequencies bands. Switching systems, on which the multiplexers were terminated, were based on space division switching (SDS), meaning that different voice channels were physically separated on different wires: their basic technology was initially mechanical and later electromechanical. The use of analogue telecommunication equipment started to be reduced in favor of digital system when the progressing digital technology enabled a saving in terms of installation and management cost of the equipment. Digital transmission systems based on time division multiplexing (TDM), in which the digital signal belonging to the different channels are time-interleaved on the same medium, are now widespread and analogue systems are being completely replaced. After an intermediate step based on semi-electronic components, nowadays switching systems have become completely electronic and thus capable of operating a time division switching (TDS) of the received channels, all of them carrying digital information interleaved on the same physical support in the time domain. Such combined evolution of transmission and switching equipment of a telecommunication network into a full digital scenario has represented the advent of the *integrated digital network* (IDN) in which both time division techniques TDM and TDS are used for the transport of the user information through the network. The IDN offers the advantage of keeping the (digital) user signals unchanged while passing through a series of transmission and switching equipment, whereas previously signals transmitted by FDM systems had to be taken back to their original baseband range to be switched by SDS equipment.

Following an approach similar to that used in [Hui89], the most important steps of network evolution can be focused by looking first at the narrowband network and then to the broadband network. Different and separated communication networks have been developed in the (narrowband) network according to the principle of traffic *segregated transport* (Figure 1.2a). Circuit-switching networks were developed to support voice-only services, whereas data services, generally characterized by low speeds, were provided by packet-switching networks. Dedicated networks completely disjoint from the previous two networks have been developed as well to support other services, such as video or specialized data services.

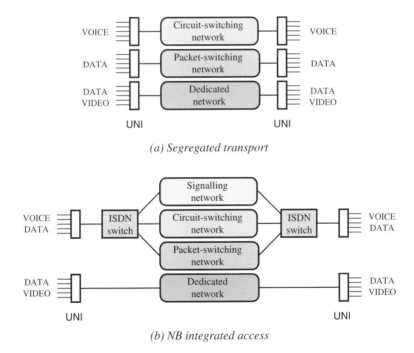

(a) Segregated transport

(b) NB integrated access

Figure 1.2. Narrowband network evolution

The industrial and scientific community soon realized that *service integration* in one network is a target to reach in order to better exploit the communication resources. The IDN then evolved into the *integrated services digital network* (ISDN) whose scope [I.120] was to provide a unique user-network interface (UNI) for the support of the basic set of narrowband (NB) services, that is voice and low-speed data, thus providing a *narrowband integrated access*. The ISDN is characterized by the following main features:

- standard user–network interface (UNI) on a worldwide basis, so that interconnection between different equipment in different countries is made easier;
- integrated digital transport, with full digital access, inter-node signalling based on packet-switching and end-to-end digital connections with bandwidth up to 144 kbit/s;
- service integration, since both voice and low-speed non-voice services are supported with multiple connections active at the same time at each network termination;
- intelligent network services, that is flexibility and customization in service provision is assured by the ISDN beyond the basic end-to-end connectivity.

The transition from the existing POTS and low-speed-data networks will be gradual, so that interworking of the ISDN with existing networks must be provided. The ISDN is thought of as a unified access to a set of existing networking facilities, such as the POTS network, public and private data networks, etc. ISDN has been defined to provide both circuit-switched and packet-switched connections at a rate of 64 kbit/s. Such choice is clearly dependent on the PCM voice-encoded bit rate. Channels at rates lower than 64 kbit/s cannot be set up. Therefore, for example, smarter coding techniques such as ADPCM generating a 32 kbit/s digital voice signal cannot be fully exploited, since a 64 kbit/s channel has always to be used.

Three types of channels, B, D and H, have been defined by ITU-T as the transmission structure to be provided at the UNI of an ISDN. The *B channel* [I.420] is a 64 kbit/s channel designed to carry data, or encoded voice. The *D channel* [I.420] has a rate of 16 kbit/s or 64 kbit/s and operates on a packet-switching basis. It carries the control information (signalling) of the B channels supported at the same UNI and also low-rate packet-switched information, as well as telemetry information. The *H channel* is [I.421] designed to provide a high-speed digital pipe to the end-user: the channel H_0 carries 384 kbit/s, i.e. the equivalent of 6 B channels; the channels H_{11} and H_{12} carry 1536 and 1920 kbit/s, respectively. These two channel structures are justified by the availability of multiplexing equipment operating at 1.544 Mbit/s in North America/Japan and at 2.048 Mbit/s in Europe, whose "payloads" are the H_{11} and H_{12} rates, respectively.

It is then possible to provide a narrowband network scenario for long-distance interconnection: two distant ISDN local exchanges are interconnected by means of three network types: a circuit-switching network, a packet-switching network and a signalling network (see Figure 1.2b). This last network, which handles all the user-to-node and node-to-node signalling information, plays a key role in the provision of advanced networking services. In fact such a network is developed as completely independent from the controlled circuit-switching network and thus is given the flexibility required to enhance the overall networking capabilities. This handling of signalling information accomplishes what is known as *common-channel signalling* (CCS), in which the signalling relevant to a given circuit is not transferred in the same band as the voice channel (*in-band associated signalling*). The signalling system number 7 (SS7) [Q.700] defines the signalling network features and the protocol architecture of the common-channel signalling used in the ISDN. The CCS network, which is a fully digital network based on packet-switching, represents the "core" of a communication network: it is used not only to manage the set-up and release of circuit-switched connections, but also to control and manage the overall communication network. It follows that the "network intelligence" needed to provide any service other than the basic connectivity between end-users resides in the CCS network. In this scenario (Figure 1.2b) the ISDN switching node is used to access the still existing narrowband dedicated networks and all the control functions of the ISDN network are handled through a specialized signalling network. Specialized services, such as data or video services with more or less large bandwidth requirements, continue to be supported by separate dedicated networks.

The enormous progress in optical technologies, both in light source/detectors and in optical fibers, has made it possible optical transmission systems with huge capacities (from hundreds of Mbit/s to a few Gbit/s and even more). Therefore the next step in the evolution of network architectures is represented by the integration of the transmission systems of all the different networks, either narrowband (NB) or broadband (BB), thus configuring the first step of the broadband integrated network. Such a step requires that the switching nodes of the different networks are co-located so as to configure a multifunctional switch, in which each type of traffic (e.g., circuit, packet, etc.) is handled by its own switching module. Multifunctional switches are then connected by means of *broadband integrated transmission* systems terminated onto network–node interfaces (NNI) (Figure 1.3a). Therefore in this networking scenario broadband integrated transmission is accomplished with partially integrated access but with segregated switching.

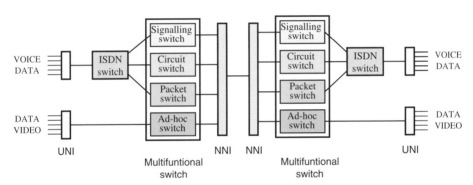

(a) NB-integrated access and BB-integrated transmission

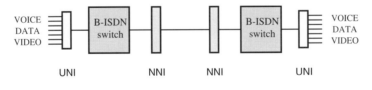

(b) BB-integrated transport

Figure 1.3. Broadband network evolution

The narrowband ISDN, although providing some nice features, such as standard access and network integration, has some inherent limitations: it is built assuming a basic channel rate of 64 kbit/s and, in any case, it cannot support services requiring large bandwidth (typically the video services). The approach taken of moving from ISDN to *broadband integrated services digital network* (B-ISDN) is to escape as much as possible from the limiting aspects of the narrowband environment. Therefore the ISDN rigid channel structure based on a few basic channels with a given rate has been removed in the B-ISDN whose transfer mode is called *asynchronous transfer mode* (ATM).

The ATM-based B-ISDN is a connection-oriented structure where data transfer between end-users requires a preliminary set-up of a virtual connection between them. ATM is a packet-switching technique for the transport of user information where the packet, called a *cell*, has a fixed size. An ATM cell includes a payload field carrying the user data, whose length is 48 bytes, and a header composed of 5 bytes. This format is independent from any service requirement, meaning that an ATM network is in principle capable of transporting all the existing telecommunications services, as well as future services with arbitrary requirements. The objective is to deploy a communication network based on a single transport mode (packet-switching) that interfaces all users with the same access structure by which any kind of communication service can be provided.

The last evolution step of network architectures has been thus achieved by the *broadband integrated transport*, that is a network configuration provided with broadband transport capabilities and with a unique interface for the support of both narrowband (sound and low-speed data) and broadband (image and high-speed data) services (Figure 1.3b). Therefore an end-to-end digital broadband integrated transport is performed. It is worth noting that choosing the packet-switching technique for the B-ISDN that supports also broadband services means also assuming the availability of ATM nodes capable of switching hundreds of millions of packets per second. In this scenario also all the packet-switching networks dedicated to medium and long-distance data services should migrate to incorporate the ATM standard and thus become part of a unique worldwide network. Therefore brand new switching techniques are needed to accomplish this task, as the classical solutions based on a single processor in the node become absolutely inadequate.

1.2.2. The protocol reference model

The interaction between two or more entities by the exchange of information through a communication network is a very complex process that involves communication protocols of very different nature between the end-users. The International Standards Organization (ISO) has developed a layered structure known as Open Systems Interconnection (OSI) [ISO84] that identified a set of layers (or levels) hierarchically structured, each performing a well-defined function. Apparently the number of layers must be a trade-off between a too detailed process description and the minimum grouping of homogeneous functions. The objective is to define a set of hierarchical layers with a well-defined and simple interface between adjacent layers, so that each layer can be implemented independently of the others by simply complying with the interfaces to the adjacent layers.

The OSI model includes seven layers: the three bottom layers providing the network services and the four upper layers being associated with the end-user. The physical layer (layer 1) provides a raw bit-stream service to the data-link layer by hiding the physical attributes of the underlying transmission medium. The data-link layer (layer 2) provides an error-free communication link between two network nodes or between an end-user and a network node, for the exchange of data-link units, often called frames. The function of the network layer (layer 3) is to route the data units, called packets, to the required downstream node, so as to reach the final end-user. The functions of these three lower layers identify the tasks of each node of a communication network. The transport layer (layer 4) ensures an in-sequence, loss- and duplicate-free exchange of information between end-users through the underlying communication network. Session (layer 5), presentation (layer 6) and application (layer 7) layers are solely related to the end-user characteristics and have nothing to do with networking issues.

Two transport layer entities exchange transport protocol data units (T-PDU) with each other (Figure 1.4), which carry the user information together with other control information added by the presentation and session layers. A T-PDU is carried as the payload at the lower layer within a network protocol data unit (N-PDU), which is also provided with a network header and trailer to perform the network layer functions. The N-PDU is the payload of a data-link protocol data unit (DL-PDU), which is preceded and followed by a data-link header and trailer that accomplish the data-link layer functions. An example of standard for the physical layer is X.21 [X.21], whereas the High-Level Data-link Control (HDLC) [Car80] represents the typical data-link layer protocol. Two representative network layer protocols are the level 3 of [X.25] and the Internet Protocol (IP) [DAR83], which provide two completely different network services to the transport layer entities. The X.25 protocol provides a *connection-oriented* service in that the packet transfer between transport entities is always preceded by the set-up of a virtual connection along which all the packets belonging to the connection will be transported. The IP protocol is *connectionless* since a network path is not set up prior to the transfer of datagrams carrying the user information. Therefore, a connection-oriented network service preserves packet sequence integrity, whereas a connectionless one does not, owing to the independent network routing of the different datagrams.

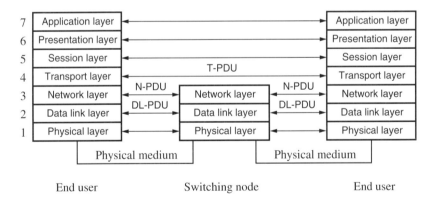

Figure 1.4. Interaction between end-users through a packet-switched network

We have seen how communication between two systems takes place by means of a proper exchange of information units at different layers of the protocol architecture. Figure 1.5 shows formally how information units are exchanged with reference to the generic layers N and $N+1$. The functionality of layer N in a system is performed by the N-entity which provides service to the $(N+1)$-entity at the N-SAP (service access point) and receives service from the

(N–1)-entity at the (N–1)-SAP. The (N+1)-entities of the two communicating systems exchange information units of layer (N+1), i.e. (N+1)-PDUs (protocol data units). This process requires that the (N+1)-PDU of each system is delivered at its N-SAP thus becoming an N-SDU (service data units). The N-entity treats the N-SDU as the payload of its N-PDU, whose control information, provided by the N-entity, is the N-PCI (protocol control information). N-PDUs are then exchanged by means of the service provided by the underlying (N–1)-layer at the (N–1)-SAP and so on.

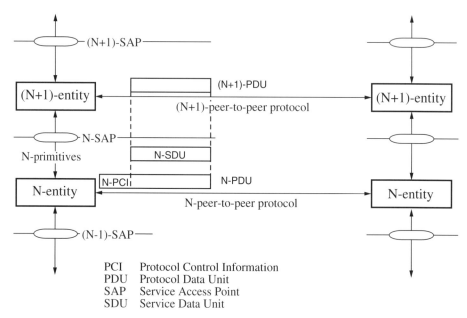

PCI   Protocol Control Information
PDU   Protocol Data Unit
SAP   Service Access Point
SDU   Service Data Unit

Figure 1.5. Interaction between systems according to the OSI model

According to the OSI layered architecture each node of the communication network is required to perform layer 1 to 3 functions, such as interfacing the transmission medium at layer 1, frame delimitation, sequence control and error detection at layer 2, routing and multiplexing at layer 3. A full error recovery procedure is typically performed at layer 2, whereas flow control can be carried out both at layer 3 (on the packet flow of each virtual circuit) and at layer 2 (on the frame flow) (Figure 1.6a). This operating mode, referred to as *packet-switching*, was mainly due to the assumption of a quite unreliable communication system, so that transmission errors or failures in the switching node operations could be recovered. Moreover, these strict coordinated operations of any two communicating switching nodes can severely limit the network throughput.

Progress in microelectronics technology and the need to carry more traffic by each node suggested the simplification of the protocol functionalities at the lower layers. A new simpler transfer mode was then defined for a connection–oriented network, termed a *frame relay*, according to which some of the functions at layer 2 and 3, such as error recovery and flow control are moved to the network edges, so that the functions to be performed at each switch-

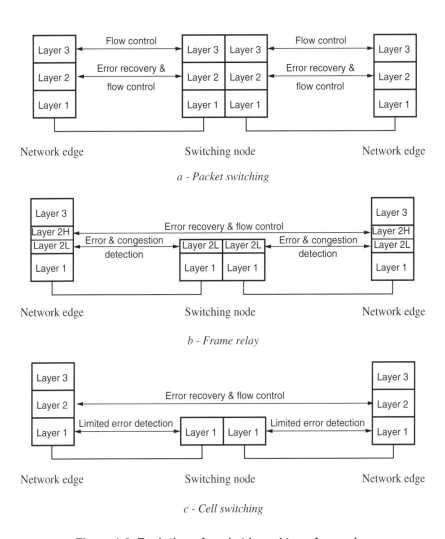

Figure 1.6. Evolution of packet-based transfer modes

ing node are substantially reduced [I.122]. In particular the protocol architecture of the lower layers can be represented as in Figure 1.6b. In the switching node the routing function is just a table look-up operation, since the network path is already set-up. The data-link (DL) layer can be split into two sublayers: a DL-core sublayer (Layer 2L) and a DL-control sublayer (Layer 2H). Error detection, just for discarding errored frames, and a very simple congestion control can be performed at Layer 2L in each network node, whereas Layer 2H would perform full error recovery and flow control but only at the network edges. Only the packet-switching protocol architecture was initially recommended in the ISDN for the packet base operations, whereas frame mode has been lately included as another alternative.

The final stack of this protocol architecture is set by the recommendations on the B-ISDN, where the basic information to be switched is a small fixed-size packet called a *cell*. With the *cell-switching* mode, each switching node is required to carry throughputs on the order of millions of cells per second on each interfaced digital link, so that the cell-switching functionality must be reduced as much as possible. Therefore the switching node will only perform functions that can be considered basically equivalent to the OSI layer 1 functions, by simply performing table look-up for cell routing and an error detection limited to the cell control fields. All other flow control and error recovery procedures are performed end-to-end in the network (Figure 1.6c).

1.3. Transfer Mode and Control of the B-ISDN

For the broadband network B-ISDN the packet-switching has been chosen as the only technique to switch information units in a switching node. Among the two well-known modes to operate packet-switching, i.e. datagram and virtual circuit, the latter approach, also referred to as connection-oriented, has been selected for the B-ISDN. In other words, in the B-ISDN network any communication process is always composed of three phases: virtual call set-up, information transfer, virtual call tear-down. During the set-up phase a sequence of virtual circuits from the calling to the called party is selected; this path is used during the information transfer phase and is released at the end of the communication service. The term asynchronous transfer mode (ATM) is associated with these choices for the B-ISDN. A natural consequence of this scenario is that the ATM network must be able to accommodate those services previously (or even better) provided by other switching techniques, such as circuit-switching, or by other transfer modes, e.g. datagram.

Migration to a unique transfer mode is not free, especially for those services better supported by other kinds of networks. Consider for example the voice service: a packetization process for the digital voice signal must be performed which implies introducing overhead in voice information transfer and meeting proper requirements on packet average delay and jitter. Again, short data transactions that would be best accomplished by a connectionless operation, as in a datagram network, must be preceded and followed by a call set-up and release of a virtual connection. Apparently, data services with a larger amount of information exchanged would be best supported by such broadband ATM network.

1.3.1. Asynchronous time division multiplexing

The *asynchronous transfer mode* (ATM) adopted in the B-ISDN fully exploits the principle of statistical multiplexing typical of packet-switching: bandwidth available in transmission and switching resources is not preallocated to the single sources, rather it is assigned on demand to the virtual connections requiring bandwidth. Since the digital transmission technique is fully synchronous, the term "asynchronous" in the acronym ATM refers to the absence of any TDM preallocation of the transmission bandwidth (time intervals) to the supported connections.

It is interesting to better explain the difference between ATM, sometimes called *asynchronous time division multiplexing* (ATDM), and pure time division multiplexing, also referred to as *synchronous transfer mode* or STM. Figure 1.7 compares the operation of an STM multiplexer and an ATM multiplexer. Transmission bandwidth is organized into periodic frames in STM with a proper pattern identifying the start of each frame. Each of the n inlets of the STM multiplexer is given a slot of bandwidth in each frame thus resulting in a deterministic allocation of the available bandwidth. Note that an idle inlet leaves the corresponding slot idle, thus wasting bandwidth in STM. The link bandwidth is allocated on demand to the n inlets of the ATM multiplexer, thus determining a better utilization of the link bandwidth. Note that each information unit (an ATM cell) must be now accompanied by a proper header specifying the "ownership" of the ATM cell (the virtual channel it belongs to). It follows that, unlike STM, now a periodic frame structure is no longer defined and queueing must be provided in the multiplexer owing to the statistical sharing of the transmission bandwidth. Cells can be transmitted empty (idle cells) if none of the inlets has a cell to transmit and the multiplexer queue is empty.

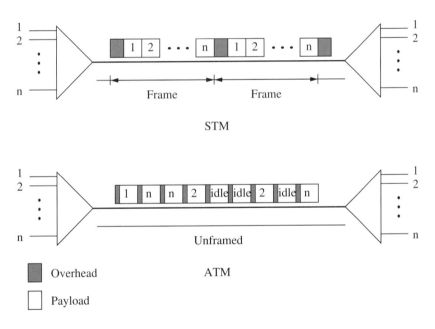

Figure 1.7. STM versus ATM

The ATM cell has been defined as including a payload of 48 bytes and a header of 5 bytes. We have already mentioned that ATM has been defined as a worldwide transport technique for existing and future communication services. We would like to point out now that the choice of a fixed packet size is functional to this objective: all information units, independent of the specific service they support, must be fragmented (if larger than an ATM cell payload) so as to fit into a sequence of ATM cells. Therefore the format for the transport of user information is not affected by the service to be supported. Nevertheless, the network transport requirements vary from service to service; thus a proper adaptation protocol must be performed that adapts the indistinguishable ATM transport mode to the specific service. Some classes of these protocols have been identified and will be later described. Note that owing to the absence of any rigid preallocation of services to channels of a given rate, what distinguishes a low-speed service from a high-speed service is simply the rate at which cells are generated for the two services.

A few words should also be spent on the rationale behind the ATM cell size. The cell header size is 5 bytes, as it is intended to carry the identifier of the virtual circuit for the cell and a few other items of information, such as type of cell, a control code, etc. The cell payload size is a much more critical parameter. In fact larger cell payloads would reduce the cell overhead, which results in bandwidth wastage, but would determine a larger number of partially filled cells, especially for services with short information units. On the other hand real-time services, for which a bounded network delay must be ensured, call for small cell payloads owing to the fixed delay determined by the packetization process. The objective of also supporting voice services in the ATM network together with data and image services, suggested that the cell payload should be limited to 32 bytes, which implies a packetization delay of 4 ms for a 64 kbit/s voice source. In fact in order to avoid the use of echo cancellers in the analogue subscriber loop of a POTS network interworking with an ATM network the one-way delay, including packetization and propagation delay, should not exceed a given threshold, say 25 ms. As a compromise between a request for a cell payload of 64 bytes, thought to better accommodate larger information units, and 32 bytes, arising from voice traffic needs, the payload size of 48 bytes has been selected as standard by the international bodies.

1.3.2. Congestion control issues

The pervasive exploitation of the principle of statistical multiplexing in the ATM network implies that guaranteeing a given quality of service (QOS) becomes a non-trivial task. In fact, let us assume that the traffic sources can be described rather accurately in terms of some traffic parameters, such as the peak bit rate, the long-term average rate, the maximum burst size, etc. Then the target of achieving a high average occupancy of the communication links implies that large buffers are required at the network nodes in order to guarantee a low cell loss probability. Therefore there is a trade-off between link occupancy and cell loss performance that can be obtained by a certain queueing capacity. The picture becomes even more complicated if we take into account that the statistical characterization of a voice source is well established, unlike what happens for data and, more importantly, for video sources.

In order to achieve a high link utilization without sacrificing the performance figures, a partition of the ATM traffic into service classes has been devised at the ATM Forum[1], by specifying each class with its peculiar performance targets. Four service classes have been identified by the ATM Forum [Jai96]:

- constant bit rate (CBR): used to provide circuit-emulation services. The corresponding bandwidth is allocated on the peak of the traffic sources so that a virtually loss-free communication service is obtained with prescribed targets of cell transfer delay (CTD) and cell delay variation (CDV), that is the variance of CTD;

---

1. The ATM Forum is a consortium among computer and communications companies formed to agree on *de facto* standards on ATM networking issues more rapidly than within ITU-T.

- variable bit rate (VBR): used to support sources generating traffic at a variable rate with specified long-term average rate (sustained cell rate) and maximum burst size at the peak rate (burst tolerance). Bandwidth for this service is allocated statistically, so as to achieve a high link utilization while guaranteeing a maximum cell loss ratio (CLR), e.g. CLR $\leq 10^{-7}$, and a maximum CTD, e.g. CTD $\leq$ 10 ms. The CDV target is specified only for real-time VBR sources;

- available bit rate (ABR): used to support data traffic sources. In this class a minimum bandwidth can be required by the source that is guaranteed by the network. The service is supported without any guarantee of CLR or CTD, even if the network makes any efforts to minimize these two parameters;

- unspecified bit rate (UBR): used to support data sources willing to use just the capacity left available by all the other classes without any objective on CLR and CTD; network access to traffic in this class is not restricted, since the corresponding cells are the first to be discarded upon congestion.

The statistical multiplexing of the ATM traffic sources onto the ATM links coupled with the very high speed of digital links makes the procedures for congestion control much more critical than in classical packet switched networks. In fact classical procedures for congestion prevention/control based on capacity planning or dynamic routing do not work in a network with very high amounts of data transported in which an overload condition requires very fast actions to prevent buffer overflows. It seems that congestion control in an ATM network should rely on various mechanisms acting at different levels of the network [Jai96]: at the UNI, both at the call set-up and during the data transfer, and also between network nodes.

Some forms of admission control should be exercised on the new virtual connection requests, based on suitable schemes that, given the traffic description of a call, accepts or refuses the new call depending on the current network load. Upon virtual connection acceptance, the network controls the offered traffic on that connection to verify that it is conforming to the agreed parameters. Traffic in excess of the declared one should either be discarded or accepted with a proper marking (see the usage of field CLP in the ATM cell header described in Section 1.5.3) so as to be thrown away first by a switching node experiencing congestion. Congestion inside the network should be controlled by feedback mechanisms that by proper upstream signalling could make the sources causing congestion decrease their bit rate. Two basic feedback approaches can be identified: the *credit-based* and the *rate-based* approach. In the former case each node performs a continuous control of the traffic it can accept on each virtual connection and authorizes the upstream node to send only the specific amount of cells (the credit) it can store in the queue for that connection. In the latter case an end-to-end rate control is accomplished using one bit in the ATM cell header to signal the occurrences of congestions. Credit-based schemes allow one to guarantee avoidance of cell loss, at least for those service classes for which it is exercised (for example CBR); in fact the hop-by-hop cell exchange based on the availability of buffers to hold cells accomplishes a serial back-pressure that eventually slows down the rate of the traffic sources themselves. Rate-based schemes cannot guarantee cell loss values even if large buffers in the nodes are likely to provide very low loss performance values. Credit-based schemes need in general smaller buffers, since the buffer requirements is proportional both to the link rate (equal for both schemes) and to the propagation delay along the controlled connection (hop by hop for credit-based, end-to-end for rate-based). In spite of these disadvantages, rate-based schemes are being preferred to window-based schemes due the higher complexity required by the latter in the switching node to track the credit status in the queue associated with each single virtual connection.

1.4. Synchronous Digital Transmission

The capacity of transmission systems has gradually enlarged in the last decades as the need for the transfer of larger amounts of information grew. At the same time the frequency division multiplexing (FDM) technique started to be gradually replaced by the time division multiplexing (TDM) technique. The reason is twofold: first digital multiplexing techniques have become cheaper and cheaper and therefore more convenient than analogue techniques for virtually all the transmission scenarios. Second, the need for transporting inherently digital information as in the case of data services and partly of video services has grown substantially in the last two decades. Therefore also researches on digital coding techniques of analogue signals have been pushed significantly so as to fully exploit a targeted all-digital transmission network for the transport of all kinds of information.

The evolution of the digital transmission network in the two most developed world regions, that is North America and Europe, followed different paths, leading to the deployment of transmission equipment and networks that were mutually incompatible, being based on different standards. These networks are based on the so called *plesiochronous digital hierarchy* (PDH), whose basic purpose was to develop a step-by-step hierarchical multiplexing in which higher rate multiplexing levels were added as the need for them arose. This kind of development without long-term visibility has led to a transmission network environment completely lacking flexibility and interoperability capabilities among different world regions. Even more important, the need for potential transport of broadband signals of hundreds of Mbit/s, in addition to the narrowband voice and data signals transported today, has pointed out the shortcomings of the PDH networks, thus suggesting the development of a brand new standard digital transmission systems able to easily provide broadband transmission capabilities for the B-ISDN.

The new digital transmission standard is based on synchronous rather than plesiochronous multiplexing and is called *synchronous digital hierarchy* (SDH). SDH was standardized in the late eighties by ITU-T [G.707] by reaching an agreement on a worldwide standard for the digital transmission network that could be as much as possible future-proof and at the same coexist by gradually replacing the existing PDH networks. Four bit rate levels have been defined for the synchronous digital hierarchy, shown in Table 1.2 [G.707]. The basic SDH transmission signal is the STM-1, whose bit rate is 155.520 Mbit/s; higher rate interfaces called STM-n have also been defined as n times the basic STM-1 interface $(n = 4, 16, 64)$. The SDH standard was not built from scratch, as it was largely affected by the SONET (synchronous optical network) standard of the ANSI-T1 committee, originally proposed in the early eighties as an optical communication interface standard by Bellcore [Bel92]. The SONET standard evolved as well in the eighties so as to become as much as possible compatible with the future synchronous digital network. The basic building block of the SONET interface is the signal STS-1 whose bit rate of 51.840 Mbit/s is exactly one third of the STM-1 rate. It will be shown later how the two standards relate to each other.

Table 1.2. Bit rates of the SDH levels

SDH level	Bit rate (Mbit/s)
1	155.520
4	622.080
16	2488.320
64	9953.280

The SDH interface STM-1 represents the basic network node interface (NNI) of the B-ISDN. It also affected the choice of the user network interface (UNI), since the basic UNI has exactly the same rate of 155.520 Mbit/s. It will be shown how the B-ISDN packets, called cells, can be mapped onto the STM-1 signal.

The basic features of SDH are now described by discussing at the same time how the drawbacks of the existing digital hierarchy have been overcome. The SDH multiplexing structure and the signal elements on which it is built are then described. Since SDH has the target of accommodating most of the current digital signals (plesiochronous signals) whose rates vary in a wide range, it is shown how the various multiplexing elements are mapped one into the other so as to generate the final STM-n signal.

1.4.1. SDH basic features

The plesiochronous multiplexing of existing digital networks relies on the concept that tributaries to be multiplexed are generated by using clocks with the same nominal bit rate and a given tolerance. Two different PDH structures are used in current networks, one in North America and one in Europe[1]. In North America the first PDH levels are denoted as DS-1 (1.544 Mbit/s), DS-1C (3.152 Mbit/s), DS-2 (6.312 Mbit/s), DS-3 (44.736 Mbit/s), whereas in Europe they are DS-1E (2.048 Mbit/s), DS-2E (8.448 Mbit/s), DS-3E (34.368 Mbit/s), DS-4E (139.264 Mbit/s). Plesiochronous multiplexing is achieved layer by layer by bit stuffing with justification to allow the alignment of the tributary digital signals generated by means of a clock with a certain tolerance.

It should be noted that a DS-1E tributary can be extracted from a DS-4E signal only by demultiplexing such a 139.264 Mbit/s signal three times by thus extracting all its 64 DS-1E tributaries. Moreover a tributary of level i can be carried only by a multiplex signal of level $i + 1$, not by higher levels directly. SDH has been developed in order to overcome such limits and to provide a flexible digital multiplexing scheme of synchronous signals. The basic features of SDH are:

---

1. The PDH hierarchy in Japan is close to the North American one, since they share the first two levels of the plesiochronous hierarchy (DS-1 and DS-2), the following two levels being characterized by the bit rates 32.064 and 97.728 Mbit/s.

- provision of a single worldwide transmission network with very high capacity capable of accommodating digital signals of arbitrary rate and of coexisting with current digital networks in order to gradually replace them;

- easy multiplexing/demultiplexing of lower rate tributaries in synchronous digital flow without needing to extract all the other tributaries at the same or higher level;

- flexibility in adapting the internal signal structure according to future needs and in accommodating other tributaries with rates higher than currently foreseen;

- effective provision of operation and maintenance functions with easy tasks performed at each transmission equipment.

As will be shown later, all digital signals defined in the plesiochronous hierarchy can be transported into an SDH signal. Single tributaries are directly multiplexed onto the final higher rate SDH signals without needing to pass through intermediate multiplexing steps. Therefore direct insertion and extraction of single tributaries by means of add/drop multiplexers is a simple and straightforward operation. Advanced network management and maintenance capabilities, as required in a flexible network, can be provided owing to the large amount of bandwidth in the SDH frame reserved for this purpose (about four percent of the overall link capacity).

The basic SDH digital signal is called STM-1 and its rate is 155.520 Mbit/s. Its structure is such that it can accommodate all the North American (except for DS-1C) and European DS digital signals in one-step multiplexing. Higher-rate SDH signals have also been defined and are referred to as STM-n with $n = 4, 16, 64$; these signals are given by properly byte interleaving lower-level SDH multiplexing elements.

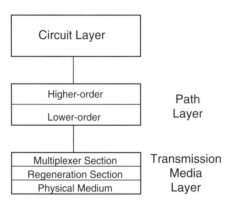

Figure 1.8. SDH layers

SDH relies on the layered architecture shown in Figure 1.8, which includes top to bottom the *circuit layer*, the *path layer* and the *transmission layer* [G.803]. Their basic functions are now described.

- Transmission media layer: this provides the means to transport synchronous digital signals between SDH devices through a physical medium. The transmission layer can be subdivided into *section layer* and *physical medium layer* and the former can be further subdivided into the *regenerator section* and *multiplexer section*. These two sections are distinguished since, as will be seen later, both regenerators and multiplexers perform their functions based on their dedicated overhead carried by the STM-*n* signal. The physical medium layer describes the physical device transporting the information, typically optical fibers or radio links, and masks the device characteristics to the section layer.

- Path layer: this provides the means to transport digital signals between network devices where a tributary enters and exits a SDH multiplexed signal through the transmission layer. The path layer can be subdivided into a *lower-order path layer* and *higher-order path layer* depending on the information transport capacity of the path. Also in this case an overhead is associated with the signal to perform all the functions needed to guarantee the integrity of the transported information.

- Circuit layer: this provides the means to transport digital information between users through the path layer, so as to provide a communication service, based, e.g., on circuit-switching, packet-switching, or leased lines.

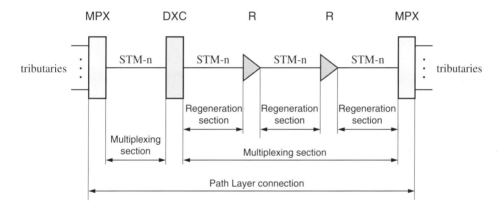

Figure 1.9. SDH layers in a transmission system example

An example of SDH sublayer occurrences in a unidirectional transmission system including multiplexers (MPX), digital cross-connect (DXC) and regenerators (R) is shown in Figure 1.9.

1.4.2. SDH multiplexing structure

Unlike the plesiochronous hierarchy, the SDH digital signals are organized in frames each lasting 125 μs. The reason behind this choice is the need for an easy access in a multiplexed high-rate signal to low-rate signals with capacity 64 kbit/s that are so widespread in plesiochronous digital networks. The final product of the SDH multiplexing scheme is the STM-*n* signal shown in Figure 1.10 [G.707]: a 125 μs frame includes $270 \times n$ columns each composed of 9 bytes. The first $9 \times n$ columns represent the STM-*n frame header* and the other $261 \times n$ columns the STM-*n frame payload*. For simplicity, reference will be now made to the lowest rate SDH signal, that is the STM-1. The frame header includes 3×9 bytes of *regeneration section overhead* (RSOH), 5×9 bytes of *multiplexer section overhead* (MSOH) and 1×9 bytes of pointer information for the carried payload (AU-PTR). RSOH and MSOH, which together represent the *section overhead*, perform the functions required at the regeneration section and multiplexer section of the layered SDH structure described in Section 1.4.1. The header needed to perform the functions at the path layer, that is the *path overhead* (POH), is carried by the STM-1 payload and is located in the first payload column in the case of a higher-order path, while it is embedded within the payload with lower-order paths.

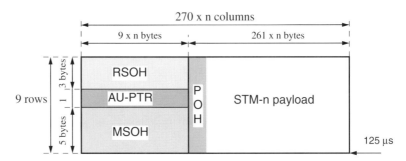

Figure 1.10. STM-*n* frame structure

Field AU-PTR performs the payload synchronization function for higher-order path connections, that is it identifies the starting byte of each frame payload frame by frame, as shown in Figure 1.11 (an analogous function is required for lower-order path connections). This operation is needed when the tributary to be multiplexed has a rate only nominally equal to that corresponding to its allocated capacity within its multiplexed signal, as in the case of plesiochronous tributaries. Therefore the starting position of the payload frame must be able to float within the STM payload and AU-PTR provides the synchronization information.

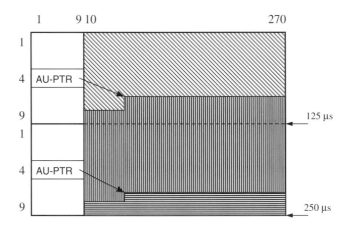

Figure 1.11. Pointer action in AU-3/AU-4

Given the described sizes of STM-1 headers and payloads and a frame duration of 125 μs, it follows that a STM-1 signal has a capacity of 155.520 Mbit/s, whereas its payload for higher-order paths is 149.760 Mbit/s. The analogous rates of the STM-*n* digital signals are simply *n* times those of the STM-1 signal. Since *n* can assume the value $n = 1, 4, 16, 64$, the rates are 622.080 Mbit/s (STM-4), 2,488.320 Mbit/s (STM-16) and 9,953.280 Mbit/s (STM-64) according to the SDH bit rates specified in Table 1.2.

Figure 1.12 shows the structure of STM-1 headers. The first three (last five) rows of the section overhead are used only at the end-points of a regeneration (multiplexer) section. A1 and A2 are used for alignment of the STM-n frame, which is entirely scrambled before transmission except for bytes A1 and A2 to avoid long sequences of 0s and 1s. J0 is used as a regenerator section trace so that the a section receiver can verify its continued connection to the intended transmitter. B1 is used for a parity check and is computed over all the bits of the previous frame after scrambling. Bytes X are left for national use, whereas bytes Δ carry media-dependent information. D1, D2, D3, E1, and F1 are used for operation and maintenance in the regeneration section. In particular D1–D3 form a 192 kbit/s data communication channel at the regenerator section, E1 and F1 provide two 64 kbit/s channels usable, e.g., for voice communications. Fields B2, D4–D12, E2 play in the multiplexer section header a role analogous to the equal-lettered fields of the regeneration section header. In particular B2 is computed over all the bits of the previous frame excluding RSOH, whereas D4–D12 form a 576 kbit/s channel available at the multiplexer section. K1 and K2 are used for protection switching and the synchronization status byte S1 indicates the type of clock generating the synchronization signal. Byte M1 is used to convey the number of errors detected at the multiplexer section by means of the B2 bytes. Blank fields in Figure 1.12 are still unspecified in the standard. The meaning of fields H1–H3 of the pointer will be explained in Section 1.4.3.

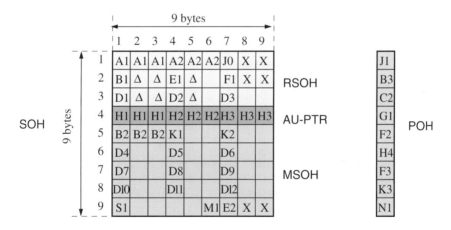

Figure 1.12. STM-1 headers

In order to explain the meaning of the path overhead fields with reference to the specific transported payload, the SDH multiplexing elements and their mutual relationship are described first. These elements are [G.707]: the *container* (C), the *virtual container* (VC), the *tributary unit* (TU), the *tributary unit group* (TUG), the *administrative unit* (AU), the *administrative unit group* (AUG) and finally the *synchronous transport module* (STM). Their functions and their hierarchical composition are now briefly described (see also Figure 1.13).

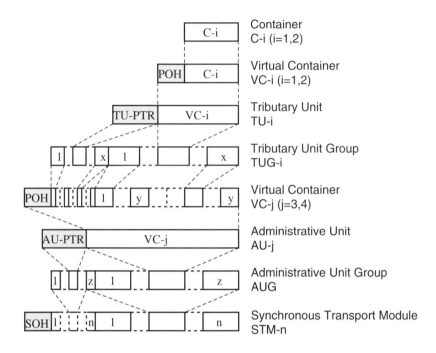

C-i	Container C-i (i=1,2)
POH C-i	Virtual Container VC-i (i=1,2)
TU-PTR VC-i	Tributary Unit TU-i
1 x 1 x	Tributary Unit Group TUG-i
POH 1 y y	Virtual Container VC-j (j=3,4)
AU-PTR VC-j	Administrative Unit AU-j
1 z 1 z	Administrative Unit Group AUG
SOH 1 n 1 n	Synchronous Transport Module STM-n

Figure 1.13. Relation between multiplexing elements

- Container: this is the basic building element of the whole SDH structure. Four classes of containers have been defined, namely C-1, C-2, C-3 and C-4 to accommodate easily most of the plesiochronous signals built according to both the North American and European standards. In particular containers C-11 and C-12 accommodate the digital signals DS-1 and DS-1E, respectively, whereas signals DS-2, DS-3 and DS-3E, DS-4E fit into containers C-2, C-3, C-4, respectively

- Virtual container: this is the information transport unit exchanged on a path layer connection. A virtual container includes a container and the corresponding path overhead processed in the SDH multiplexers. According to the container classes, four classes of virtual containers have been defined, namely VC-1, VC-2, VC-3, VC-4, each carrying as the payload the equal-numbered container. By referring to the layered SDH architecture that subdivides the path layer into two sublayers, VC-1 and VC-2 are lower-order VCs, whereas VC-3 and VC-4 are higher-order VCs. The frame repetition period depends on the VC class: VC-1 and VC-2 frames last 500 μs, while VC-3 and VC-4 frames last 125 μs.

- Tributary unit: this is the multiplexing element that enables a VC to be transported by a higher-class VC by keeping the direct accessibility to each of the transported lower-class VC. A TU includes a VC and a pointer (TU-PTR). TU-PTR indicates the starting byte of a VC, which can then float within the TU payload, by thus performing the same synchronization function carried out by the pointer AU-PTR for the STM payload. Three classes of TU have been defined, namely TU-1, TU-2 and TU-3, where the last two accommodate the virtual containers VC-2 and VC-3, respectively. TU-1 is further split into TU-11 and TU-12 to accommodate VC-11 and VC-12.

- Tributary unit group: this performs the function of assembling together several TUs without further overhead, so that the several TUGs can be byte interleaved into the payload of a higher-order VC. The two classes TUG-2 and TUG-3 have been defined, the former to accommodate lower-class TUs, that is four TU-11s, three TU-12s or just one TU-2, the latter to assemble seven TUG-2s or just one TU-3. A VC-3 can accommodate seven TUG-2s and a VC-4 three TUG-3s.

- Administrative unit: this is the multiplexing element that makes higher-order virtual containers transportable on a multiplexer section layer. An AU includes a higher-order VC and a pointer (AU-PTR) that permits the VC floating within the AU payload by specifying the starting position of each VC frame. The two administrative units AU-3 and AU-4 carry one VC-3 and VC-4 each, respectively.

- Administrative unit group: this performs the function of assembling together several AUs without further overhead, so that the several AUs can be properly byte interleaved into an STM-n signal. VCs and AU-PTRs will be interleaved separately, since the former will be located into the STM-n payload, whereas the latter will be placed into the fourth row of the STM-n header. An AUG can accommodate either three AU-3s or just one AU-4.

- Synchronous transport module: this is the largest SDH multiplexing elements and is the unit physically transmitted onto the physical medium. A STM-n is the assembling of n AUGs ($n = 1, 4, 16, 64$) and n SOHs properly interleaved.

An overall representation of the synchronous multiplexing (SM) structure is given in Figure 1.14, which also shows how the plesiochronous multiplexing (PM) signals are related one another and where they access the SDH multiplexing scheme.

By going back to Figure 1.12, the path overhead there represented refers only to higher-order paths, that is to signals VC-3 and VC-4 and occupies the first column of the corresponding virtual container. Note that this column is actually the STM-1 payload first column (column 10 of STM-1) only for VC-4, which is 261 columns long, since three VC-3s, each 85 columns long, are byte interleaved (and hence also their POHs) in the STM-1 payload. The byte B3 is used for parity check and is computed based on all the bits of the previous VC, whereas C2 indicates the type of load carried by the virtual container. H4 acts as a position indicator of the payload, such that the starting position of a multiframe when lower-order VCs are carried. J1 is used as a path trace so that the a path-receiving terminal can verify its continued connection to the intended transmitter. K3 performs the function of automatic protection switching at the VC-3/4 level. G1, F2, F3 and N1 are used for miscellaneous operations. In the case of lower-order paths, VC-11, VC-12 and VC-2 carry their POH as the their first byte.

As already mentioned the SONET standard is closely related to SDH since the latter has been significantly affected by the former; also SONET has been modified to increase compatibility to SDH. The basic SONET signal is called synchronous transport signal STS-1: it is a frame structure with repetition period 125 μs consisting of 90 columns of 9 bytes each. The first three columns are the STS-1 overhead so that the STS-1 payload includes 87×9 bytes and the STS-1 signal has a rate of 51.840 Mbit/s with a payload capacity of 50.112 Mbit/s. Higher layer interfaces are obtained as integer multiples of the basic building block. The relation

Plesiochronous Multiplexing

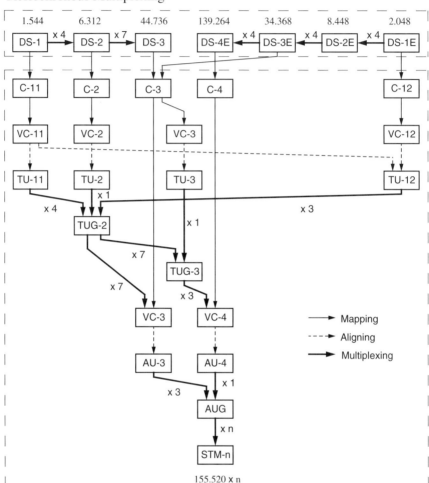

Synchronous Multiplexing

Figure 1.14. SDH multiplexing structure

between SONET and SDH is immediately seen. The 155.520 Mbit/s interface includes 3 basic blocks and is called STS-3 in SONET: it corresponds to STM-1 in SDH. The 622.080 Mbit/s interface includes 12 basic blocks in SONET and is called STS-12: it corresponds to the SDH STM-3 interface.

The SONET multiplexing elements are very similar to those of SDH. In fact signals DS-1, DS-1E and DS-2 are carried by the virtual tributaries (VT) VT1.5, VT2, VT6; unlike SDH also the signal DS-1C is now carried in the tributary VT3. These VTs, as well as the signals DS-3, are mapped into the STS-1 payload analogously to the mapping of VCs and C-3 into VC-3 in SDH.

1.4.3. Synchronization by pointers

We already described how the virtual containers are allowed to float within the STM-1 payload, by thus crossing the STM-1 frame boundaries. In the particular case of the higher-order VCs, namely VC-3 and VC-4, the pointer AU-PTR carries the information about the byte position where the virtual container starts. This information is called the *offset* with respect to a predetermined frame position, which for VC-4 immediately follows the last byte of AU-PTR. We will see how the pointer works for all the multiplexing elements that include this field, that is TU-11, TU-12, TU-2, TU-3, VC-3 and VC-4. The VC floating occurs every time the rate of the VC signal to be multiplexed differs from the rate used to generate TUs or AUs.

In the case of AU-4, AU-3 and TU-3 the pointer is represented by the fields H1, H2, H3, whose logical arrangement is shown in Figure 1.15. These bytes are carried by the field AU-PTR of the STM-1 header for AU-4 and AU-3, while they occupy the first three positions of the first column in TU-3. H1 and H2 include one byte each, whereas H3 has a length dependent on the VC type. H1 and H2 carry the pointer information to the start of the VC (the bits actually used are the last two of H1 and all eight of H2); conventionally the byte with address 0 is the one immediately following H3. Negative and positive justification, to be explained later and signalled through bits D and I respectively, are carried out using the byte(s) H3 and the byte(s) immediately following H3. Bits N (new data flag) are inverted to signal an arbitrary change in the pointer value due to a change in the payload.

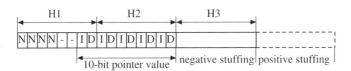

Figure 1.15. Pointer coding with AU-4/AU-3 and TU-3

The AU-4 PTR is represented by the first byte of H1 and H2, as well as by the three H3 bytes, in the STM-1 header (see Figure 1.12). The VC-4 frame occupies the entire STM-1 payload and thus includes 261×9 bytes numbered three by three from 0 to 782. Therefore 10 bits of the couple H1-H2 are enough to address any of these three-cell sets. The AU-4 format is represented in Figure 1.16.

If the VC-4 data rate is higher than the AU-4 rate, the received data will eventually accumulate a (positive) offset three bytes long. When this event occurs a negative justification takes place: the three H3 bytes carry VC-4 information and the pointer value is decreased by one to keep the same VC-4 frame content. Nevertheless, before decreasing the pointer value, the justification occurrence is signalled by inverting only once the bits D (decrement bits) of bytes H1 and H2. According to the standard the pointer can be adjusted for positive or negative justification only every 4 frames (500 μs). An example of negative justification is shown in Figure 1.17 (the pointer value decreases from n to $n-1$).

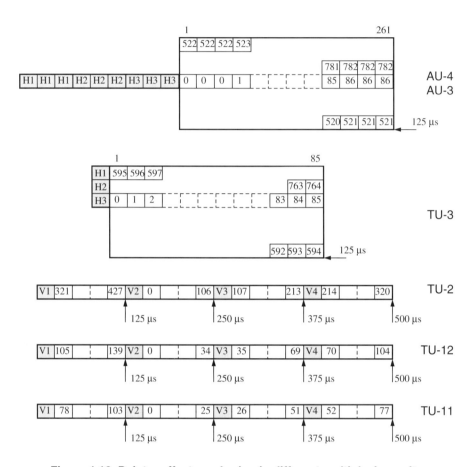

Figure 1.16. Pointer offset numbering in different multiplexing units

On the other hand, when the VC-4 data rate is lower than the AU-4 rate, the received data accumulate a negative offset compared to the AU-4 transmitted information. When the accumulated offset sums up to three bytes, a positive justification takes place: the three bytes immediately following H3 carry no information and the pointer value is increased by one. As before the occurrence of justification is signalled by inverting once the bits I (increment bits) of bytes H1 and H2. Figure 1.18 shows an example of positive justification (the pointer value increases from n to $n + 1$).

The pointer AU-3 PTR of a VC-3 includes exactly three bytes, that is H1, H2 and H3 (Figure 1.16). Compared to AU-4 PTR, one byte rather than three is used for justification, since the capacity of a VC-3 frame (87×9 bytes) is one third of the VC-4 frame. Now the nine AU-PTR bytes of STM-1 header are all used, as each triplet H1, H2, H3 is the pointer AU-3 PTR of a different AU-3 carried by STM-1. The VC-3 bytes are singly numbered 0 through 782 with offset 0 again assigned to the VC-3 byte immediately following the byte H3 of AU-3. Negative and positive justification are operated as in AU-4, by filling byte H3 and

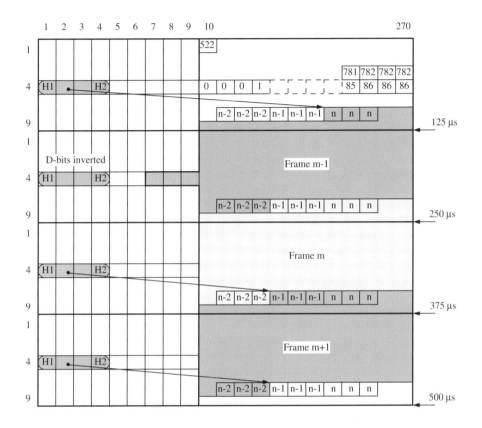

Figure 1.17. Example of negative justification with AU-4

emptying byte 0 of AU-3, respectively and adjusting correspondingly the pointer value in H1–H2. Apparently justification occurs when the negative or positive accumulated offset of the received data compared to the transmitted data equals one byte. If we look at the overall STM-1 frame, now byte 0 used for positive justification does not follow the H3 byte, since the AU-3s are byte interleaved in the STM-1 frame (see Section 1.4.4).

The pointer of TU-3, TU-3 PTR, includes three bytes H1, H2, H3 which act as in AU-3 PTR. These three bytes are now included in the first column of TU-3, whose frame is composed of 86×9 bytes (its payload VC-3 occupies the last 85 columns). Therefore the VC-3 bytes are numbered 0 through 764 and the byte 0 follows immediately H3 (see again Figure 1.16).

The pointer information to synchronize lower-order VCs, that is VC-11, VC-12 and VC-2, operates a little differently from the previous pointers. Now the bytes carrying the pointer information are called V1, V2 and V3 and they play the same role as H1, H2 and H3, respectively. A 500 μs multiframe including 4 basic frames is defined, which includes 104 (4×26), 140 (4×35) and 428 (4×107) bytes for VC-11, VC-12 and VC-2, respectively. A TU is built from these multiframe VCs by adding one V byte per frame, so that the four bytes V1, V2,

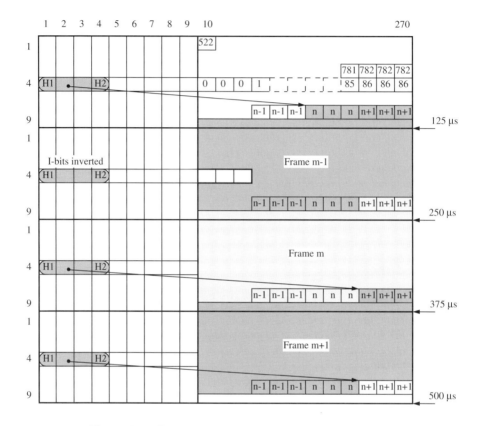

Figure 1.18. Example of positive justification with AU-4

V3,V4 are added, as shown in Figure 1.16.V1,V2 and V3 represent the pointers TU-11 PTR, TU-12 PTR, TU-2 PTR in their respective multiframe, whereas V4 is left for future usage. Bytes are numbered 0 through 103 (TU-11), 0 through 139 (TU-12), 0 through 427 (TU-2) with byte 0 conventionally allocated to the byte following V2. So V1 and V2 carry the offset to indicate the current starting byte of the multiframe, whereas V3 and the byte following V3 are used for negative and positive justification. Since also the multiframe needs to be synchronized to properly extract the pointer bytes, all TUs are multiplexed so have to same phase in the multiframe and the phase alignment information is carried by byte H4 of POH in the higher-order VC carrying the TUs.

Using the pointer information allows a VC to float within its TU, which is called the *floating mode* of multiplexing. There are some cases where this floating is not needed, namely when a lower-order VC-*i* ($i = 11, 12, 2$) is mapped onto the a higher-order VC-*j* ($j = 3, 4$), the two VC signals being synchronous. In this situation, called *locked mode*, VC-*i* will keep always the same position within its TU-*i*, thus making useless the pointer TU-*i* PTR. Therefore the 500 µs multiframe is not required and the basic 125 µs frame is used for these signals.

1.4.4. Mapping of SDH elements

How the mappings between multiplexing elements are accomplished is now described [G.707]. Figure 1.19 shows how a TUG-2 signal is built starting from the elements VC-11, VC-12 and VC-2. These virtual containers are obtained by adding a POH byte to their respective container C-11, C-12 and C-2 with capacity 25, 34 and 106 bytes (all these capacities are referred to a 125 μs period). By adding a pointer byte V to each of these VCs (recall that these VCs are structured as 500 μs multiframe signals) the corresponding TUs are obtained with capacities 27, 36 and 108 bytes, which are all divisible by the STM-1 column size (nine). TU-2 fits directly into a TUG-2 signal whose frame has a size 12×9 bytes, whereas TU-11 and TU-12 are byte interleaved into TUG-2. Recall that since the alignment information of the multiframe carrying lower-order VCs is contained in the field H4 of a higher-order VC, this pointer is implicitly unique for all the VCs. Therefore all the multiframe VCs must be phase aligned within a TUG-2.

A single VC-3 whose size is 85×9 bytes is mapped directly onto a TU-3 by adding the three bytes H1-H3 of the TU-3 PTR in the very first column (see Figure 1.20). This signal becomes a TUG-3 by simply filling the last six bytes of the first column with stuff bytes. A TUG-3 can also be obtained by interleaving byte-by-byte seven TUG-2s and at the same time filling the first two columns of TUG-3, since the last 84 columns are enough to carry all the TUG-2 data (see Figure 1.20). Compared to the previous mapping by VC-3, now the pointer information is not present in the first column since we are not assembling floating VCs. This absence of pointers will be properly signalled by a specific bit configuration in the H1–H3 positions of TU-3.

TUG-2s can also be interleaved byte-by-byte seven by seven so as to fill completely a VC-3, whose first column carries the POH (see Figure 1.21). Alternatively a VC-3 can also carry a C-3 whose capacity is 84×9 bytes.

A VC-4, which occupies the whole STM-1 payload, can carry 3-byte interleaved TUG-3s each with a capacity 86×9 bytes so that the first two columns after the POH are filled with stuff bytes (see Figure 1.22). Analogously to VC-3, VC-4 can carry directly a C-4 signal whose size is 260×9 bytes.

An AUG is obtained straightforwardly from a VC-4 by adding the AU-PTR, which gives the AU-4, and the AU-4 is identical to AUG. Figure 1.23 shows the mapping of VC-3s into an AU-3. Since a VC-3 is 85 columns long, two stuff columns must be added to fill completely the 261 columns of AUG, which are specifically placed after column 29 and after column 57 of VC-3. Adding AU-3 PTR to this modified VC-3 gives AU-3. Three AU-3s are then byte interleaved to provide an AUG. The STM-1 signal is finally given by adding RSOH (3×9 bytes) and MSOH (5×9 bytes). The byte interleaving of n AUGs with the addition of SOH in the proper positions of the first $9 \times n$ columns gives the signal STM-n.

SDH enables also signals with rate higher than the payload capacity of a VC-4 to be transported by the synchronous network, by means of the *concatenation*. A set of x AU-4s can be concatenated into an AU-4-xc, which is carried by an STM-n signal. Since only one pointer is needed in the concatenated signal only the first occurrence of H1–H2 is actually used and the other $x - 1$ bytes H1–H2 are filled with a null value. Analogously only the first AU-4 carries the POH header in its first column, whereas the same column in the other AU-4s is filled with

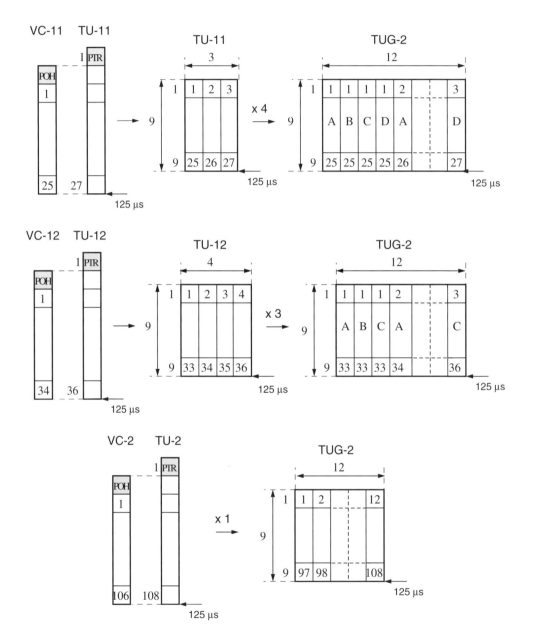

Figure 1.19. Mappings into TUG-2

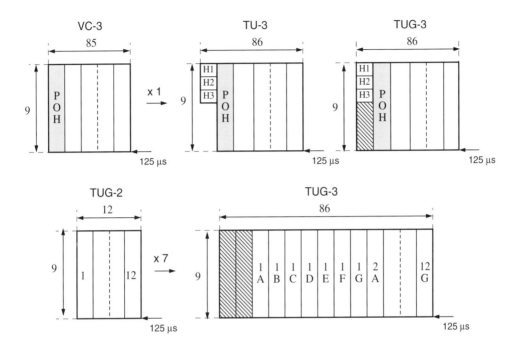

Figure 1.20. Mappings into TUG-3

null information. Therefore a signal AUG-n, which is the byte interleaving of n unrelated AUGs, and an AUG-nc, although differing in the AU-PTR and POH information, carry exactly the same payload.

A few words should be dedicated to explaining how the SOH of Figure 1.12 is modified passing from STM-1 to STM-n [G.707]. In particular only fields A1, A2, B2 are actually expanded to size n, whereas the others remain one-byte fields (therefore null information will be inserted in the $n-1$ byte positions following each of these fields). More bytes A1–A2 satisfy the need for a fast acquisition of STM-n frame alignment. Using more B2 bytes improves the error control function at the multiplexer level.

1.5. The ATM Standard

A brief review is now given for the ITU-T specification of the B-ISDN environment that relies on the ATM concept. The protocol reference model is first summarized and then the lower protocol layers are explained in detail, that is the physical layer, the ATM layer and the ATM adaptation layer.

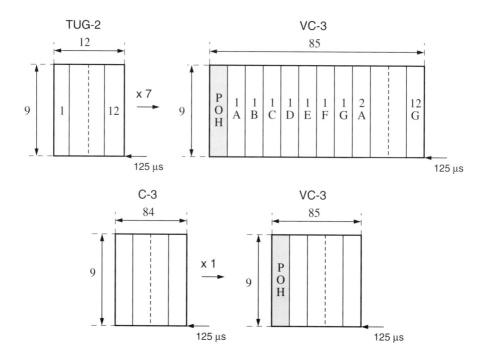

Figure 1.21. Mappings into VC-3

1.5.1. Protocol reference model

The protocol architecture of the ATM B-ISDN, as defined by ITU-T [I.321], refers to three different planes where the functions performed for the provision of a telecommunication service take place (Figure 1.24):

- *user plane*, whose function is to provide the transfer of user information and to apply all the relative control functions, such as flow and error control;
- *control plane*, whose task is to set up, monitor and release the virtual connections, as well as to control their correct operations by means of proper exchange of signalling messages;
- *management plane*, which performs management functions both for the whole system (plane management) by coordinating the operations on the control and user planes, and for the single layers (layer management) by controlling their parameters and resources.

The user and control planes include three layers dealing with the data transport through the network and a set of higher layers [I.321]. The three lower layers are the *physical layer*, the *ATM layer* and the *ATM adaptation layer* (AAL) (see Figure 1.24). Signalling functions, which are performed by the control plane, are separate from the functions related to the transport of user information, handled by the user plane. So, a sort of out-of-band signalling is accomplished; in fact only the control plane is active during the set-up of virtual connections, whereas only the user plane is active during the transport of user information.

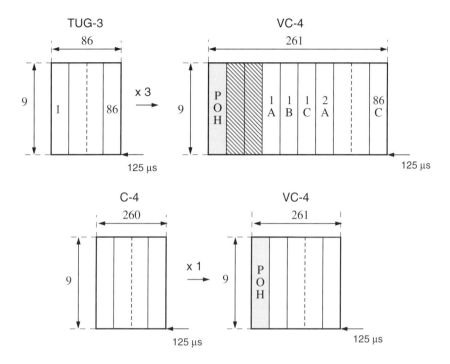

Figure 1.22. Mappings into VC-4

Sublayering is further used within a layer to identify sets of homogeneous functions that can be grouped and implemented separately from other sets within the same layer. In particular the physical layer is subdivided into two sublayers: the *physical medium sublayer* and the *transmission convergence sublayer*. The former sublayer has to deal mainly with the physical characteristics of the transmission medium, including the function of bit timing, whereas the latter sublayer performs hardware functions related to aggregates of bits, the ATM cells. Also the ATM adaptation layer is subdivided into two sublayers: the *segmentation and reassembly sublayer* and the *convergence sublayer*. The former sublayer performs the function of interfacing entities handling variable-length packets with an underlying (ATM) network transporting fixed-size packets (the cells). The latter sublayer plays the key role of adapting the indistinguishable transport service provided by the ATM layer to the specific service characteristics of the supported application. The functions performed at each (sub)layer are summarized in Table 1.3 and will be described in the following sections.

After defining layers (and sublayers) for the protocol architecture of an ATM network, a better understanding of the overall operations in transferring user data from one end-system to another is obtained by looking at how information units are handled while going through the overall protocol stack (see Figure 1.25). A user information unit is passed from a higher layer entity to the AAL-SAP to be delivered to a given end-user. This information unit, which represents the service data unit (SDU) of the AAL layer (AAL-SDU) is encapsulated within a protocol data unit (PDU) of the CS sublayer of AAL. The protocol control information (PCI)

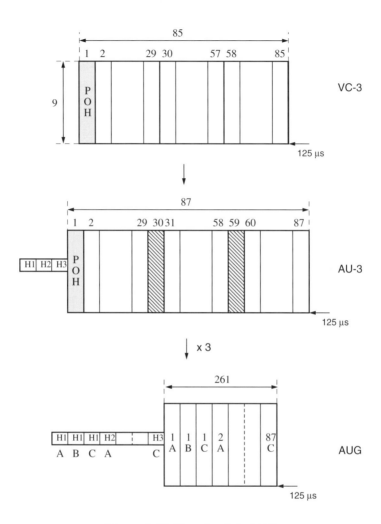

Figure 1.23. Mapping into AUG

(see Section 1.2.2 for the definition of PCI) is here represented by a proper header (H) and trailer (T) appended to the AAL-SDU. This new information unit AAL_CS-PDU, if larger than 48 bytes, is segmented into several fixed size units, each provided with an appropriate header and trailer at the SAR sublayer. Each of these new information units, called AAL_SAR-PDU, having a length of 48 bytes, becomes the ATM-SDU and is transferred through the ATM-SAP. The ATM layer adds a header (the ATM PCI) of 5 bytes to each ATM-SDU, thus producing the ATM-PDU, that is the ATM cell of 53 bytes. The ATM cells are passed through the physical layer SAP (PHY-SAP) to be transferred through the ATM network. All these operations take place in reverse order at the receiving end-user to finally deliver the user data.

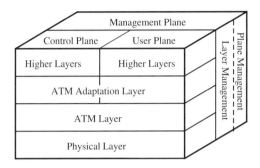

Figure 1.24. ATM protocol reference model

Table 1.3. Functions performed at each layer of the B-ISDN protocol reference model

		Higher Layers	Higher Layer Functions
Layer Management	ATM Adaptation Layer (AAL)	Convergence Sublayer (CS)	Service Specific (SS)
			Common Part (CP)
		Segmentation and Reassembly Sublayer (SAR)	Segmentation and reassembly
	ATM Layer		Generic flow control Cell header generation/extraction Cell VPI/VCI translation Cell multiplexing/demultiplexing
	Physical Layer	Transmission Convergence Sublayer (TC)	Cell rate decoupling HEC sequence generation/verification Cell delineation Transmission frame adaptation Transmission frame generation/recovery
		Physical Medium (PM)	Bit timing Physical medium

Different types of cells have been defined [I.321]:

- *idle cell* (physical layer): a cell that is inserted and extracted at the physical layer to match the ATM cell rate available at the ATM layer with the transmission speed made available at the physical layer, which depends on the specific transmission system used;
- *valid cell* (physical layer): a cell with no errors in the header that is not modified by the header error control (HEC) verification;
- *invalid cell* (physical layer): a cell with errors in the header that is not modified by the HEC verification;
- *assigned cell* (ATM layer): a cell carrying valid information for a higher layer entity using the ATM layer service;
- *unassigned cell* (ATM layer): an ATM layer cell which is not an assigned cell

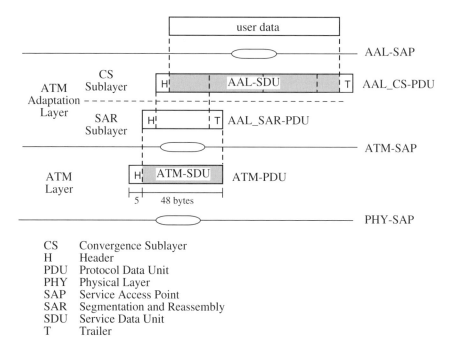

CS   Convergence Sublayer
H   Header
PDU   Protocol Data Unit
PHY   Physical Layer
SAP   Service Access Point
SAR   Segmentation and Reassembly
SDU   Service Data Unit
T   Trailer

Figure 1.25. Nesting of data units in the ATM protocol reference model

Note that only assigned and unassigned cells are exchanged between the physical and the ATM layer through the PHY-SAP. All the other cells have a meaning limited to the physical layer.

Since the information units switched by the ATM network are the ATM cells, it follows that all the layers above the ATM layer are end-to-end. This configuration is compliant with the overall network scenario of doing most of the operations related to specific service at the end-user sites, so that the network can transfer enormous amounts of data with a minimal processing functionality within the network itself. Therefore the protocol stack shown in Figure 1.4 for a generic packet switched network of the old generation becomes the one shown in Figure 1.26 for an ATM network.

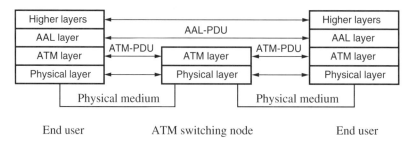

Figure 1.26. Interaction between end-users through an ATM network

Establishing a mapping between ATM layers and OSI layers is significant in understanding the evolution of processing and transmission technologies in the decade that followed the definition of the OSI model. The functions of the physical layer in an ATM network are a subset of the OSI physical layer (layer 1). From the ATM layer upwards the mapping to OSI layers is not so straightforward. The ATM layer could be classified as performing OSI layer 1 functions, since the error-free communication typical of OSI layer 2 is made available only end-to-end by the AAL layer, which thus performs OSI layer 2 functions. According to a different view, the ATM layer functions could be classified as belonging both to the OSI physical layer (layer 1) and to OSI data-link layer (layer 2). In fact the error-free communication link made available at the OSI layer 2 can be seen as available partly between ATM layer entities, which perform a limited error detection on the cells, and partly between AAL layer entities (end-to-end), where the integrity of the user message can be checked. Furthermore any flow control action is performed at the AAL layer. Therefore it could be stated that the ATM layer functions can be mapped onto both OSI layers 1 and 2, whereas the AAL layer functions belong to the OSI layer 2. As a proof that this mapping is far from being univocal, consider also that the handling at the ATM layer of the virtual circuit identifier by the switch configures a routing function typical of the OSI network layer (layer 3). The layers above the AAL can be well considered equivalent to OSI layers 3-7. Interestingly enough, the ATM switching nodes, which perform only physical and ATM layer functions, accomplish mainly hardware-intensive tasks (typically associated with the lower layers of the OSI protocol architecture), whereas the software-intensive functions (related to the higher OSI layers) have been moved outside the network, that is in the end-systems. This picture is consistent with the target of switching enormous amount of data in each ATM node, which requires the exploitation of mainly very fast hardware devices.

1.5.2. The physical layer

The physical layer [I.432] includes two sublayers: the physical medium sublayer, performing medium-dependent functions such as the provision of the timing in association with the digital channel, the adoption of a suitable line coding technique, etc., and the transmission convergence sublayer, which handles the transport of ATM cells in the underlying flow of bits.

At the physical medium sublayer, the physical interfaces are specified, that is the digital capacity available at the interface together with the means to make that capacity available on a specific physical medium. ITU-T has defined two user-network interfaces (UNI) at rates 155.520 Mbit/s and 622.080 Mbit/s[1]. These rates have been clearly selected to exploit the availability of digital links compliant with the SDH standard. The former interface can be either electrical or optical, whereas the latter is only optical. The 155.520 interface is defined as symmetrical (the same rate in both directions user-to-network and network-to-user); the

---

1. During the transition to the B-ISDN, other transport modes of ATM cells have been defined that exploit existing transmission systems. In particular ITU-T specifies how ATM cells can be accommodated into the digital flows at PDH bit rates DS-1E (2.048 Mbit/s), DS-3E (34.368 Mbit/s), DS-4E (139.264 Mbit/s), DS-1 (1.544 Mbit/s), DS-2 (6.312 Mbit/s), DS-3 (44.736 Mbit/s) [G.804].

622.080 interface can be either symmetrical or asymmetrical (155.520 Mbit/s in one direction and 622.080 Mbit/s in the opposite direction).

Two basic framing structures at the physical layer have been defined for the B-ISDN: an SDH-based structure and a cell-based structure [I.432]. In the SDH-based solution the cell flow is mapped onto the VC-4 payload, whose size is 260×9 bytes. Therefore the capacity of the ATM flow for an interface at 155.520 Mbit/s is 149.760 Mbit/s. An integer number of cells does not fill completely the VC-4 payload, since 2340 is not an integer multiple of $48 + 5 = 53$. Therefore the ATM cell flow floats naturally within the VC-4, even if the ATM cell boundaries are aligned with the byte boundaries of the SDH frame. Figure 1.27 shows how the ATM cells are placed within a VC-4 and VC-4-4c for the SDH interfaces STM-1 at 155.520 Mbit/s and STM-4 at 622.080 Mbit/s, respectively. Note that the payload C-4-4c in the latter case is exactly four times the payload of the interface STM-1, that is $149.760 \times 4 = 599.040$. This choice requires three columns to be filled with stuffing bytes, since the POH information in STM-4 requires just one column (nine bytes) as in the STM-1 interface.

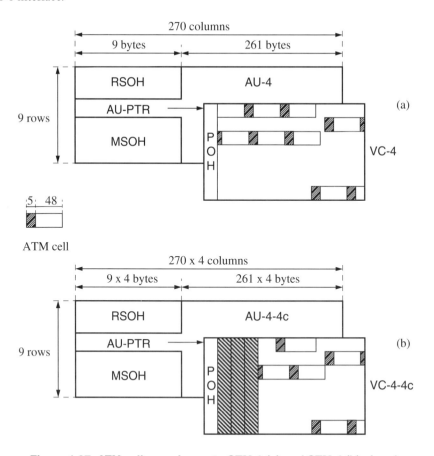

Figure 1.27. ATM cell mapping onto STM-1 (a) and STM-4 (b) signals

With a cell-based approach, cells are simply transmitted on the transmission link without relying on any specific framing format. On the transmission link other cells will be transmitted too: idle cells in absence of ATM cells carrying information, cells for operation and maintenance (OAM) and any other cell needed to make the transmission link operational and reliable. It is worth noting that for an interface at 155.520 Mbit/s after 26 contiguous cells generated by the ATM layer one idle or OAM cell is always transmitted: only in this way the actual payload available for the ATM layer on the cell-based interface is exactly the same as in the STM-1 interface, whose payload for ATM layer cells is 260 columns out of 270 of the whole frame.

The functions performed at the transmission convergence (TC) sublayer are

- transmission frame generation/recovery
- transmission frame adaptation
- cell delineation
- HEC header sequence generation/verification
- cell rate decoupling

The first two functions are performed to allocate the cell flow onto the effective framing structure used in the underlying transmission system (cell-based or SDH-based). Cell rate decoupling consists in inserting (removing) at the transmission (reception) side idle cells when no ATM layer cells are available, so that the cell rate of the ATM layer is independent from the payload capacity of the transmission system.

The HEC header sequence generation/verification consists in a procedure that protects the information carried by the ATM cell header, to be described in the next section, by a header error control (HEC) field included in the header itself. The HEC field is one byte long and therefore protects the other four bytes of the header. The thirty-first degree polynomial obtained from these four bytes multiplied by x^8 and divided by the generator polynomial $x^8 + x^2 + x + 1$ gives a remainder that is used as an HEC byte at the transmission side. The HEC procedure is capable of correcting single-bit errors and detecting multiple-bit errors. The receiver of the ATM cell can be in one of two states: correction mode and detection mode (see Figure 1.28). It passes from correction mode to detection mode upon single-bit error (valid cell with header correction) and multiple-bit error (invalid cell with cell discarding); a state transition in the reverse direction takes place upon receiving a cell without errors. Cells with error detected that are received in the detection mode are discarded, whereas cells without errors received in the correction mode are valid cells.

The last function performed by the TC sublayer is cell delineation, which allows at the receiving side the identification of the cell boundaries out of the flow of bits represented by the sequence of ATM cells generated by the ATM layer entity at the transmission side. Cell delineation is accomplished without relying on other "out-of-band" signals such as additional special bit patterns. In fact it exploits the correlation existing between four bytes of the ATM cell header and the HEC fifth byte that occupies a specific position in the header. The state diagram of the receiver referred to cell delineation is shown in Figure 1.29. The receiver can be in one of three states: hunt, presynch, synch. In the hunt state a bit-by-bit search of the header into the incoming flow is accomplished. As soon as the header is identified, the receiver passes to the presynch state where the search for the correct HEC is done cell-by-cell. A tran-

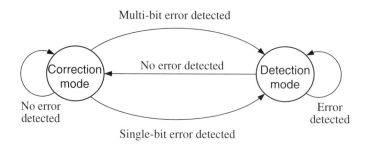

Figure 1.28. State diagram of the receiver in the HEC processing

sition occurs from the presynch state to the synch state if δ consecutive correct HECs are detected or to the hunt state if an incorrect HEC is detected[1]. Upon receiving α consecutive cells with incorrect HEC, the receiver moves from the synch state to the hunt state. Suggested values for the two parameters are $\alpha = 7$ and $\delta = 6$ for SDH-based interfaces, $\alpha = 7$ and $\delta = 8$ for cell-based interfaces.

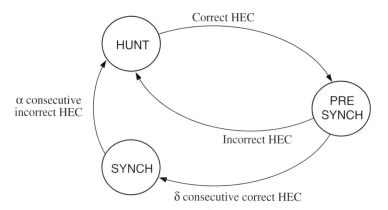

Figure 1.29. State diagram of the receiver during cell delineation

1.5.3. The ATM layer

The task of the ATM layer [I.361] is to provide virtual connections between two or more network nodes or between a network node and an end-user by relying on the transmission paths made available by the physical layer protocol. A hierarchy of two virtual connections are provided by the ATM layer: *virtual paths* and *virtual channels*. A virtual path is simply a bundle of virtual channels that thus terminate on the same ATM layer protocol entities. The difference between the two just lies in the fact that a virtual path is a set of virtual channels sharing a common identifier value. Virtual paths provide several advantages in the management of the

---

1. The hunt state becomes useless if the information about the byte boundaries is available, as in the case of the SDH-based interface.

virtual connections through the network, such as reduced processing for the set up of a new virtual channel once the corresponding virtual path is already set-up, functional separation of the tasks related to the handling of virtual paths and virtual channels, etc. The PDU of the ATM layer is the ATM cell [I.361]: it includes a cell payload of 48 bytes and a cell header of 5 bytes (see Figure 1.30).

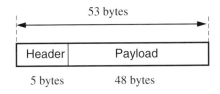

Figure 1.30. ATM cell format

The functions performed at the ATM layer are

- *cell multiplexing/demultiplexing*: cells belonging to different virtual channels or virtual paths are multiplexed/demultiplexed onto/from the same cell stream,
- *cell VPI/VCI translation*: the routing function is performed by mapping the virtual path identifier/virtual channel identifier (VPI/VCI) of each cell received on an input link onto a new VCI/VPI and an output link defining where to send the cell,
- *cell header generation/extraction*: the header is generated (extracted) when a cell is received from (delivered to) the AAL layer,
- *generic flow control*: a flow control information can be coded into the cell header at the UNI.

The cell header, shown in Figure 1.31 for the user network interface (UNI) and for the network node interface (NNI), includes

- the *generic flow control* (GFC), defined only for the UNI to provide access flow control functions,
- the *virtual path identifier* (VPI) and *virtual channel identifier* (VCI), whose concatenation represents the cell addressing information,
- the *payload type* (PT), which specifies the cell type,
- the *cell loss priority* (CLP), which provides information about cell discarding options,
- the *header error control* (HEC), which protects the other four header bytes.

The GFC field, which includes four bits, is used to control the traffic flow entering the network (upstream) onto different ATM connections. This field can be used to alleviate short-term overload conditions that may occur in the customer premises network. For example it can be used to control the upstream traffic flow from different terminals sharing the same UNI.

The addressing information VPI/VCI includes 24 bits for the UNI and 28 bits for the NNI, thus allowing an enlarged routing capability within the network. Some VPI/VCI codes cannot be used for ATM connections as being *a priori* reserved for other functions such as signalling, OAM, unassigned cells, physical layer cells, etc.

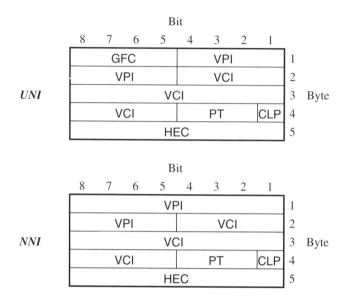

Figure 1.31. ATM cell header format

The PT field includes three bits: one is used to carry higher layer information in conjunction with one type of AAL protocol (AAL Type 5, to be described in Section 1.5.4.4), another indicates upstream congestion to the receiving user, the last discriminates between cell types. The different codings of field PT are described in Table 1.4. The least significant bit indicates the last cell of a higher layer multicell PDU and thus carry *ATM-layer-user-to-ATM-layer-user* (AUU) indication. The intermediate bit, which is referred to as *explicit forward congestion indication* (EFCI), can be set by any network element crossed by the cell to indicate congestion to the end-user. The most significant bit discriminates between cells carrying user data and cells performing other functions, such as control, management, etc.

Table 1.4. Coding of field PT in the ATM cell header

PT	Meaning
000	Data cell, EFCI=0, AUU=0
001	Data cell, EFCI=0, AUU=1
010	Data cell, EFCI=1, AUU=0
011	Data cell, EFCI=1, AUU=1
100	OAM cell
101	OAM cell
110	Resource management cell
111	Reserved for future functions

The one-bit field CLP is used to discriminate between high-priority cells (CLP=0) and low-priority cells (CLP=1), so that in case of network congestion a switching node can discard first the low-priority cells. The CLP bit can be set either by the originating user device, or by any network element. The former case refers to those situations in which the user declares which cells are more important (consider for example a coding scheme in which certain parts of the message carry more information than others and the former cells are thus coded as high priority). The latter case occurs for example at the UNI when the user is sending cells in violation of a contract and the cells in excess of the agreed amount are marked by the network as low-priority as a consequence of a traffic policing action.

The HEC field is an eight-bit code used to protect the other four bytes of the cell header. Its operation has been already described in Section 1.5.2. Note that at the ATM layer only the information needed to route or anyway handle the ATM cell are protected by a control code; the cell payload is not protected in the same way. This is consistent with the overall view of the ATM network which performs the key networking functionalities at each switching node and leaves to the end-users (that is to the layers above the ATM, e.g. to the AAL and above) the task of eventually protecting the user information by a proper procedure.

1.5.4. The ATM adaptation layer

The ATM adaptation layer is used to match the requirements and characteristics of the user information transport to the features of the ATM network. Since the ATM layer provides an indistinguishable service, the ATM adaptation layer is capable of providing different service classes [I.362]. These classes are defined on the basis of three service aspects: the need for a timing relationship between source and destination of the information, the source bit rate that can be either constant (constant bit rate - CBR) or variable (variable bit rate - VBR), and the type of connection supporting the service, that is connection-oriented or connectionless. Four classes have thus been identified (see Figure 1.32). A time relation between source and destination exists in Classes A and B, both being connection oriented, while Class A is the only one to support a constant bit-rate service. A service of circuit emulation is the typical example of Class A, whereas Class B is represented by a packet video service with variable bit rate. No timing information is transferred between source and destination in Classes C and D, the former providing connection-oriented services and the latter connectionless services. These two classes have been defined for the provision of data services for which the set-up of connection may (Class C) or may not (Class D) be required prior to the user information transfer. Examples of services provided by Class C are X.25 [X.25] or Frame Relay [I.233], whereas the Internet Protocol (IP) [DAR83] and the Switched Multimegabit Data Service (SMDS) [Bel91] are typical services supportable by Class D.

The AAL is subdivided into two sublayers [I.363]: the segmentation and reassembly (SAR) sublayer and the convergence (CS) sublayer. The SAR sublayer performs the segmentation (reassembly) of the variable length user information into (from) the set of fixed-size ATM cell payloads required to transport the user data through the ATM network. The CS sublayer maps the specific user requirements onto the ATM transport network. The CS sublayer can be thought of as including two hierarchical parts: the *common part convergence sublayer* (CPCS), which is common to all users of AAL services, and the *service specific convergence sublayer* (SSCS), which is dependent only on the characteristics of the end-user. Figure 1.33 shows how the

	Class A	Class B	Class C	Class D
Timing relation between source and destination	Required		Not required	
Bit rate	Constant	Variable		
Connection mode	Connection-oriented			Connection less

Figure 1.32. Features of service classes

sublayers SAR, CPCS and SSCS are related to each other. The SSCS sublayer can be null, meaning that it need not be implemented, whereas the CPCS sublayer is always present. It is apparent that the protocol functionalities performed at the SAR and CPCS are common to all AAL users.

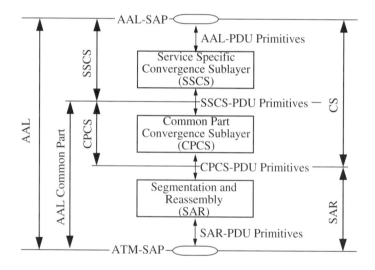

Figure 1.33. Protocol structure for the AAL

Four AAL protocols (Type 1, Type 2, Type 3/4 and Type 5) have been defined each to support one or more service class: Type 1 for Class A, Type 2 for Class B, Types 3/4 and 5 for Class C or Class D. Originally two different AAL protocols were defined, that is Type 3 and Type 4, to support respectively services of Class C and Class D. Later it was realized that the two protocols were quite similar and were thus merged into the Type 3/4 envisioned to support both connection-oriented (Class C) and connectionless (Class D) services. A new protocol was then defined suggested by the computer industry under the initial name of SEAL, meaning "simple efficient adaptation protocol" [Lyo91], later standardized as AAL Type 5. The main rationale of this protocol is to reduce as much as possible the functionalities performed at the AAL layer and thus the induced overhead.

1.5.4.1. AAL Type 1

The AAL Type 1 protocol is used to support CBR services belonging to three specific service classes: circuit transport (also known as circuit emulation), video signal transport and voice-band signal transport. Therefore the functions performed at the CS sublayer differ for each of these services, whereas the SAR sublayer provides the same function to all these services.

At the CS sublayer, 47 bytes are accumulated at the transmission side and are passed to the SAR sublayer together with a 3-bit sequence count and 1-bit convergence sublayer indication (CSI), which perform different functions. These two fields providing the sequence number (SN) field of the SAR-PDU together with the 4-bit sequence number protection (SNP) represent the header of the SAR-PDU (see Figure 1.34). The SAR sublayer computes a cyclic redundancy check (CRC) to protect the field SN and an even parity bit to protect the seven bits of fields SN and CRC. Such a 4-bit SNP field is capable of correcting single-bit errors and of detecting multiple-bit errors. At the receiving side the SNP is first processed to detect and possibly correct errors. If the SAR-PDU is free from errors or an error has been corrected, the SAR-PDU payload is passed to the upper CS sublayer together with the associated sequence count. Therefore losses or misinsertions of cells can be detected and eventually recovered at the CS sublayer, depending on the service being supported.

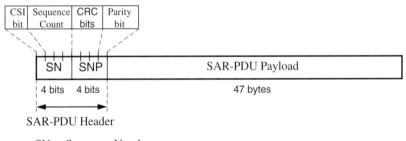

SN   Sequence Number
SNP   Sequence Number Protection
CSI   Convergence Sublayer Indication
CRC   Cyclic Redundancy Check

Figure 1.34. AAL1 SAR-PDU format

The CS is capable of recovering the source clock at the receiver by using the *synchronous residual time stamp* (SRTS) approach. With the SRTS mode an accurate reference network clock is supposed to be available at both ends of the connection, so that information can be conveyed by the CSI bit about the difference between the source clock rate and the network rate (the residual time stamp - RTS). The RTS is a four-bit information transmitted using CSI of the SAR-PDU with odd sequence count (1,3,5,7). The receiving side can thus regenerate with a given accuracy the source clock rate by using field CSI.

The CS is also able to transfer between source and destination a structured data set, such as one of $n \times 64$ kbit/s, by means of the *structured data transfer* (SDT) mode. The information about the data structure is carried by a pointer which is placed as the first byte of the 47-byte payload, which thus actually carries just 46 bytes of real payload information. The pointer is carried by even-numbered (0,2,4,6) SAR-PDU, in which CSI is set to 1 (CSI is set to 0 in odd SAR-PDUs). Since the pointer is transferred every two SAR-PDUs, it must be able to address any byte of the payload in two adjacent PDUs, i.e. out of $46 + 47 = 93$ bytes. Therefore seven bits are used in the one-byte pointer to address the first byte of an $n \times 64$ kbit/s structure.

1.5.4.2. AAL Type 2

AAL Type 2 is used to support services with timing relation between source and destinations, but unlike the services supported by the AAL Type 1 now the source is VBR. Typical target applications are video and voice services with real-time characteristics. This AAL type is not yet well defined. Nevertheless, its functions include the recovery of source clock at the receiver, the handling of lost or misinserted cells, the detection and possible corrections of errors in user information transported by the SAR-PDUs.

1.5.4.3. AAL Type 3/4

AAL Type 3/4 is used to support VBR services for which a source-to-destination traffic requirement is not needed. It can be used both for Class C (connection-oriented) services, such as frame relay. and for Class D (connectionless) services, such as SMDS. In this latter case, the mapping functions between a connectionless user and an underlying connection-oriented network is provided by the service specific convergence sublayer (SSCS). The common part convergence sublayer (CPCS) plays the role of transporting variable-length information units through an ATM network through the SAR sublayer. The format of the CPCS PDU is shown in Figure 1.35. The CPCS-PDU header includes the fields CPI (common part identifier), BTA (beginning tag) and BAS (buffer allocation size), whereas the trailer includes the fields AL (alignment), ETA (ending tag) and LEN (length). CPI is used to interpret the subsequent fields in the CPCS-PDU header and trailer, for example the counting units of the subsequent fields BAS and LEN. BTA and ETA are equal in the same CPCS-PDU. Different octets are used in general for different CPCS-PDUs and the receiver checks the equality of BTA and ETA. BAS indicates to the receiver the number of bytes required to store the whole CPCS-PDU. AL is used to make the trailer a four-byte field and LEN indicates the actual content of the CPCS payload, whose length is up to 65,535 bytes. A padding field (PAD) is also used to make the payload an integral multiple of 4 bytes, which could simplify the receiver design. The current specification of CPI is limited to the interpretation just described for the BAS and LEN fields.

The CPCS-PDU is segmented at the SAR sublayer of the transmitter into fixed-size units to be inserted into the payload of the SAR-PDU, whose format is shown in Figure 1.36. Reassembly of the SAR-PDU payloads into the original CPCS-PDU is accomplished by the SAR sublayer at the receiver. The two-byte SAR-PDU header includes a segment type (ST), a sequence number (SN), a multiplexing identifier (MID); a length indicator (LI) and a cyclic redundancy check (CRC) constitute the two-byte SAR-PDU trailer. It follows that the SAR-PDU payload is 44 bytes long. ST indicates whether a cell carries the beginning, the continuation, the end of a CPCS-PDU or a single-cell CPCS-PDU. The actual length of the useful information within the SAR-PDU payload is carried by the field LI. Its content will be 44 bytes for the first two cases, any value in the range 4-44 and 8-44 bytes in the third and fourth case, respectively. SN numbers the SAR-PDUs sequentially and its value is checked by the

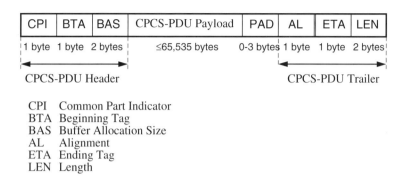

Figure 1.35. AAL3/4 CPCS-PDU format

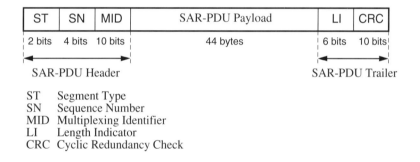

Figure 1.36. AAL3/4 SAR-PDU format

receiver. The field MID performs the important function of AAL Type 3/4 of allowing the multiplexing of different CPCS-PDUs onto the same virtual channel connection. The CRC field is a 10-bit sequence protecting all the other fields in the SAR-PDU.

1.5.4.4. AAL Type 5

AAL Type 5 supports the same service types as AAL Type 4 in a simpler manner with much less overhead but without the capability of multiplexing higher layer units onto the same ATM connection.

The CPCS-PDU format for AAL Type 5 is shown in Figure 1.37. Its payload has a variable length but now a PAD field makes the CPCS-PDU length an integral multiple of 48 bytes, so as to fill completely the SAR-PDU payloads. The CPCS-PDU overhead is only represented by a trailer including the field's CPCS user-to-user indication (UU), CPI, LEN, CRC. UU transparently carries CPCS user-to-user information. CPI makes the trailer 8 bytes long and its use is still undefined. Field LEN specifies the actual length of the CPCS-PDU payload, so as to identify the PAD size. CRC protects the remaining part of the CPCS-PDU content against errors.

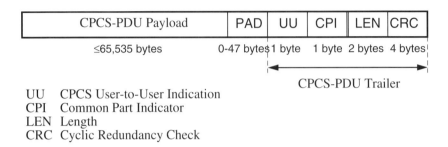

Figure 1.37. AAL5 CPCS-PDU format

The efficiency of the AAL Type 5 protocol lies in the fact that the whole ATM cell payload is taken by the SAR-PDU payload. Since information must be carried anyway to indicate whether the ATM cell payload contains the start, the continuation or the end of a SAR-SDU (i.e. of a CPCS_PDU) the bit AUU (ATM-layer-user-to-ATM-layer-user) of field PT carried by the ATM cell header is used for this purpose (see Figure 1.38). AUU=0 denotes the start and continuation of an SAR-SDU; AUU=1 means the end of an SAR-SDU and indicates that cell reassembly should begin. Note that such use of a field of the protocol control information (PCI) at the ATM layer to convey information related to the PDU of the upper AAL layer actually represents a violation of the OSI protocol reference model.

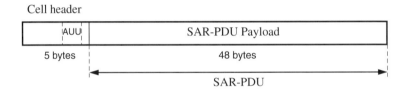

Figure 1.38. AAL5 SAR-PDU format

1.5.4.5. AAL payload capacity

After describing the features of the four AAL protocols, it is interesting to compare how much capacity each of them makes available to the SAR layer users. Let us assume a UNI physical interface at 155.520 Mbit/s and examine how much overhead is needed to carry the user information as provided by the user of the AAL layer as AAL-SDUs. Both the SDH-based and the cell-based interfaces use 1/27 of the physical bandwidth, as both of them provide the same ATM payload of 149.760 Mbit/s. The ATM layer uses 5/53 of that bandwidth, which thus reduces to 135.632 Mbit/s. Now, as shown in Figure 1.39, the actual link capacity made available by AAL Type 1 is only 132.806, as 1/47 of the bandwidth is taken by the SAR-PDU header. Even less capacity is available to the SAR layer user with AAL 3/4, that is 124.329, as the total SAR overhead sums up to four bytes. We note that the AAL Type 5 protocol makes available the same link capacity seen by the ATM cell payloads, that is 135.632, since its overhead is carried within the cell header.

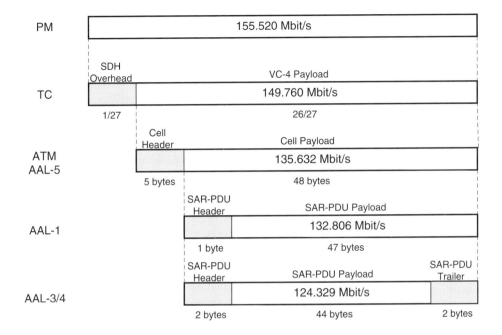

Figure 1.39. B-ISDN link capacity at various layers

1.6. References

[Bel91]   Bellcore, "Generic system requirements in support of Switched Multimegabit Data Ser-
           vice", Technical Advisory TA-TSV-000772, 1991.

[Bel92]   Bellcore, "Synchronous optical network (SONET) transport system: common generic cri-
           teria", TR–NWT-000253, 1992.

[Car80]   D.E. Carlson, "Bit-oriented data link control procedures", *IEEE Trans. on Commun.*, Vol. 28,
           No. 4, Apr. 1980, pp. 455-467.

[DAR83]   DARPA, *Internet protocol* (RFC 791), 1983.

[G.707]   ITU-T Recommendation G.707, "Network node interface for the synchronous digital
           hierarchy (SDH)", Geneva, 1996.

[G.803]   ITU-T Recommendation G.803, "Architectures of transport networks based on the syn-
           chronous digital hierarchy (SDH)", Geneva, 1993.

[G.804]   ITU-T Recommendation G.804, "ATM cell mapping into plesiochronous digital hierarchy
           (PDH)", Geneva, 1993.

[Hui89]   J.Y. Hui, "Network, transport, and switching integration for broadband communications",
           IEEE Network, Vol. 3, No. 2, pp.40–51.

[ISO84]   ISO, "Information processing system – Open systems interconnection (OSI) – Basic refer-
           ence model", American National Standards Association, New York, 1984.

[I.120]   ITU-T Recommendation I.120, "Integrated services digital networks", Geneva, 1993.

[I.122]   ITU-T Recommendation I.122, "Framework for providing additional packet mode bearer services", Geneva, 1993.

[I.233]   ITU-T Recommendation I.233, "Frame mode bearer services", Geneva, 1992.

[I.321]   ITU-T Recommendation I.321, "B-ISDN protocol reference model and its application", Geneva, 1991.

[I.361]   ITU-T Recommendation I.361, "B-ISDN ATM layer specification", Geneva, 1995.

[I.362]   ITU-T Recommendation I.362, "B-ISDN ATM adaptation layer (AAL) functional description", Geneva, 1993.

[I.363]   ITU-T Recommendation I.363, "B-ISDN ATM adaptation layer (AAL) specification", Geneva, 1993.

[I.420]   ITU-T Recommendation I.420, "Basic user-network interface", Geneva, 1989.

[I.421]   ITU-T Recommendation I.421, "Primary rate user-network interface", Geneva, 1989.

[I.432]   ITU-T Recommendation I.432, "B-ISDN user network interface physical layer specification", Geneva, 1993.

[Jai96]   R. Jain, "Congestion control and traffic management in ATM networks: recent advances and a survey", *Computer Networks and ISDN Systems*, Vol. 28, No. 13, Oct 1996, pp. 1723-1738.

[Lyo91]   T. Lyon, "Simple and efficient adaptation layer (SEAL)", ANSI T1S1.5/91-292, 1991.

[Q.700]   ITU-T Recommendation Q.700, "Introduction to CCITT Signalling System No. 7", Geneva, 1993.

[X.21]   ITU-T Recommendation X.21, "Interface between data terminal equipment and data circuit-terminating equipment for synchronous operation on public data networks", Geneva, 1992.

[X.25]   ITU-T Recommendation X.25, "Interface between data terminal equipment (DTE) and data circuit-terminating equipment (DCE) for terminals operating in the packet mode and connected to public data networks by dedicated circuit", Geneva, 1993.

1.7. Problems

1.1   Compute the maximum frequency deviation expressed in ppm (parts per million) between the clocks of two cascaded SDH multiplexers that can be accommodated by the pointer adjustment mechanism of an AU-4 (consider that the ITU-T standard sets this maximum tolerance as ± 4.6 ppm).

1.2   Repeat Problem 1.1 for a TU-3.

1.3   Repeat Problem 1.1 for a TU-2.

1.4   Compute the effective bandwidth or payload (bit/s) available at the AAL-SAP of a 155.520 Mbit/s interface, by thus taking into account also the operations of the CS sublayer, by using the AAL Type 1 with SDT mode.

1.5   Repeat Problem 1.4 for AAL Type 3/4 when the user information units are all 1 kbyte long.

Chapter 2   *Interconnection Networks*

This chapter is the first of three chapters devoted to the study of network theory. The basic concepts of the interconnection networks are briefly outlined here. The aim is to introduce the terminology and define the properties that characterize an interconnection network. These networks will be described independently from the context in which they could be used, that is either a circuit switch or a packet switch. The classes of rearrangeable networks investigated in Chapter 3 and that of non-blocking networks studied in Chapter 4 will complete the network theory.

The basic classification of interconnection network with respect to the blocking property is given in Section 2.1 where the basic crossbar network and EGS pattern are introduced before defining classes of equivalences between networks. Networks with full interstage connection patterns are briefly described in Section 2.2, whereas partially connected networks are investigated in Section 2.3. In this last section a detailed description is given for two classes of networks, namely banyan networks and sorting networks, that will play a very important role in the building of multistage networks having specific properties in terms of blocking. Section 2.4 reports the proofs of some properties of sorting networks exploited in Section 2.3.

2.1. Basic Network Concepts

The study of networks has been pursued in the last decades by researchers operating in two different fields: communication scientists and computer scientists. The former have been studying structures initially referred to as *connecting networks* for use in switching systems and thus characterized in general by a very large size, say with thousands of inlets and outlets. The latter have been considering structures called *interconnection networks* for use in multiprocessor systems for the mutual connection of memory and processing units and so characterized by a reasonably small number of inlets and outlets, say at most a few tens. In principle we could say that connecting networks are characterized by a centralized control that sets up the permutation required, whereas the interconnection networks have been conceived as based on a distributed processing capability enabling the set-up of the permutation in a distributed fashion. Interestingly enough the expertise of these two streams of studies have converged into a unique objective: the development of large interconnection networks for switching systems in which a distributed processing capability is available to set up the required permutations. The two main driving forces for this scenario have been the request for switching fabrics capable of carrying aggregate traffic on the order of hundreds of Gbit/s, as typical of a medium-size broadband packet switch, and the tremendous progress achieved in CMOS VLSI technology that makes the distributed processing of interconnection networks feasible also for very large networks.

The connection capability of a network is usually expressed by two indices referring to the absence or presence of traffic carried by the network: *accessibility* and *blocking*. A network has *full accessibility* when each inlet can be connected to each outlet when no other I/O connection is established in the network, whereas it has *limited accessibility* when such property does not hold. Full accessibility is a feature usually required today in all interconnection networks since electronic technology, unlike the old mechanical and electromechanical technology, makes it very easy to be accomplished. On the other hand, the blocking property refers to the network connection capability between *idle* inlets and outlets in a network with an arbitrary current permutation, that is when the other inlets and outlets are either *busy* or idle and arbitrarily connected to each other.

An interconnection network, whose taxonomy is shown in Table 2.1, is said to be:

- *Non-blocking*, if an I/O connection between an arbitrary idle inlet and an arbitrary idle outlet can be always established by the network independent of the network state at set-up time.
- *Blocking*, if at least one I/O connection between an arbitrary idle inlet and an arbitrary idle outlet cannot be established by the network owing to internal congestion due to the already established I/O connections.

Depending on the technique used by the network to set up connections, non-blocking networks can be of three different types:

- *Strict-sense non-blocking* (SNB), if the network can always connect each idle inlet to an arbitrary idle outlet independent of the current network permutation, that is independent of the already established set of I/O connections and of the policy of connection allocation.
- *Wide-sense non-blocking* (WNB), if the network can always connect each idle inlet to an arbitrary idle outlet by preventing blocking network states through a proper policy of allocating the connections.
- *Rearrangeable non-blocking* (RNB), if the network can always connect each idle inlet to an arbitrary idle outlet by applying, if necessary, a suitable internal rearrangement of the I/O connections already established.

Therefore, only SNB networks are free from blocking states, whereas WNB and RNB networks are not (see Table 2.1). Blocking states are never entered in WNB networks due to a suitable policy at connection set-up time. Blocking states can be encountered in RNB networks during the dynamic network evolution but new connections can always be established by possibly rearranging the connections already set up.

Table 2.1. Network taxonomy

Network class	Network type	Network states
Non-blocking	Strict-sense non-blocking	Without blocking states
	Wide-sense non-blocking	With blocking states
	Rearrangeable non-blocking	
Blocking	Others	

It is intuitively clear that a SNB network satisfies at the same time the definition of WNB and RNB networks, but not vice versa. We will not develop here the subject of WNB networks and will only focus on SNB networks, simply denoted in the following as *non-blocking networks*, and on RNB networks, referred to as *rearrangeable networks*. Note that an RNB network is also SNB if all the connections are set up and torn down at the same time.

We can intuitively assume that the above three non-blocking network types are characterized by a decreasing cost index, starting from strict-sense non-blocking and ending with rearrangeable non-blocking. Traditionally the cost index of the network has always assumed to be the number of crosspoints in the network, as reasonable in the space–division switching systems of the sixties and seventies. Nowadays such a performance index alone does not characterize the cost of an interconnection network for broadband applications, owing to the extreme degree of integration of the electronic components in a single chip enabled by VLSI technologies. Consider for example the other cost indices: the gates per chip, the chips per board, the crosspoints per board, etc. Nevertheless, since it is very hard to find a unique, even composite, cost index for a network, we will continue to refer to the number of crosspoints in the network as the cost index for the network, by always bearing in mind its limited significance.

The reference network is necessarily the well-known *crossbar* network $N \times M$ (Figure 2.1) with N inlets, M outlets. The cost index of such a network, that is the number of its crosspoints, referred to a squared structure $(N = M)$, is

$$C = N^2$$

Since each crosspoint is dedicated to a specific I/O connection, the crossbar network is implicitly non-blocking. A squared crossbar network $(N = M)$ is able to set up an arbitrary *network permutation*, that is an arbitrary set of N I/O connections; if P denotes the set of all the permutations set up by a generic network, in general $P \leq N!$, whereas $P = N!$ in a crossbar network.

Research activities have been undertaken for decades to identify network structures falling into one of the non-blocking classes, but cheaper than the crossbar network. The guideline for

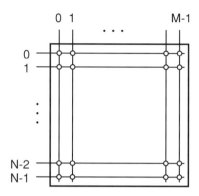

Figure 2.1. Crossbar network

this research is building *multistage networks*, with each stage including switching matrices each being a (non-blocking) crossbar network. The general model of an $N \times M$ multistage network includes s stages with r_i matrices $n_i \times m_i$ at stage i $(i = 1, ..., s)$, so that $N = n_1 r_1$, $M = m_s r_s$. The $n_i \times m_i$ matrix of the generic stage i, which is the basic building block of a multistage network, is assumed to be non-blocking (i.e. a crossbar network).

The key feature that enables us to classify multistage networks is the type of interconnection pattern between (adjacent) stages. The apparent condition $m_i r_i = n_{i+1} r_{i+1}$ $(0 \le i \le s - 1)$ always applies, that is the number of outlets of stage i equals the number of inlets of stage $i + 1$. As we will see later, a different type of interconnection pattern will be considered that cannot be classified according to a single taxonomy. Nevertheless, a specific class of connection pattern can be defined, the *extended generalized shuffle* (EGS) [Ric93], which includes as subcases a significant number of the patterns we will use in the following. Let the couple (j_i, k_i) represent the generic inlet (outlet) j of the matrix k of the generic stage i, with $j_i = 0, ..., n_i - 1$ $(j_i = 0, ..., m_i - 1)$ and $k_i = 1, ..., r_i$. The EGS pattern is such that the outlet j_i of matrix k_i, that is outlet (j_i, k_i) , is connected to inlet (j_{i+1}, k_{i+1}) with

$$j_{i+1} = \text{int} \left[\frac{m_i (k_i - 1) + j_i}{r_{i+1}} \right]$$

$$k_{i+1} = [m_i (k_i - 1) + j_i]_{\text{mod} r_{i+1}} + 1$$

In other words, we connect the $r_i m_i$ outlets of stage i starting from outlet $(0,1)$ sequentially to the inlets of stage $i + 1$ as

$$(0, 1), ..., (0, r_{i+1}), (1, 1), ..., (1, r_{i+1}), ..., (n_{i+1} - 1, 1), ..., (n_{i+1} - 1, r_{i+1})$$

An example is represented in Figure 2.2 for $m_i < r_{i+1}$. A network built out of stages interconnected by means of EGS patterns is said to be an *EGS network*.

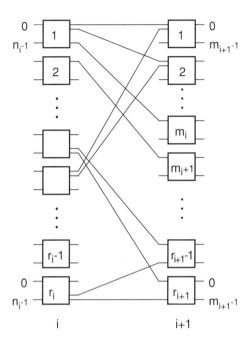

Figure 2.2. EGS interconnection pattern

Multistage networks including s stages will now be described according to the following type of interstage pattern configuration:

- *full connection* (FC), if each matrix in stage i $(i = 2, ..., s - 1)$ is connected to all the matrices in stages $i - 1$ and $i + 1$;

- *partial connection* (PC), if each matrix in stage i $(i = 2, ..., s - 1)$ is not connected to all the matrices in stages $i - 1$ and $i + 1$.

It is worth noting that an EGS network with $m_i \geq r_{i+1}$ $(i = 1, ..., s - 1)$ is a FC network, whereas it is a PC network if $m_i < r_{i+1}$ $(i = 1, ..., s - 1)$.

2.1.1. Equivalence between networks

We refer now to a squared $N \times N$ network and number the network inlets and outlets 0 through $N - 1$ sequentially from the top to the bottom. If the network includes more than one stage and the generic matrix includes b outlets, each outlet matrix being labelled from 0 to $b - 1$, an underlying graph can be identified for the network: each matrix is mapped onto a graph node and each interstage link onto a graph edge. Graph nodes and edges are not labelled, so that network inlets and outlets need not be mapped onto edges, since they carry no information. Therefore the most external graph elements are the nodes representing the matrices of the first and last network stage.

Two graphs A and B are said to be isomorphic if, after relabelling the nodes of graph A with the node labels of graph B, graph A can be made identical to graph B by moving its nodes and hence the attached edges. The mapping so established between nodes in the same position in the two original graphs expresses the "graph isomorphism". A network is a more complex structure than a graph, since an I/O path is in general described not only by a sequence of nodes (matrices) but also by means of a series of labels each identifying the outlet of a matrix (it will be clear in the following the importance of such a more complete path description for the routing of messages within the network). Therefore an isomorphism between networks can also be defined that now takes into account the output matrix labelling.

Two networks A ad B are said to be isomorphic if, after relabelling the inlets, outlets and the matrices of network A with the respective labels of network B, network A can be made identical to network B by moving its matrices, and correspondingly its attached links. It is worth noting that relabelling the inlets and outlets of network A means adding a proper inlet and outlet permutation to network A. Note that the network isomorphism requires the modified network A topology to have the same matrix output labels as network B for matrices in the same position and is therefore a label-preserving isomorphism. The mapping so established between inlets, outlets and matrices in these two networks expresses the "network isomorphism". In practice, since the external permutations to be added are arbitrary, network isomorphism can be proven by just moving the matrices, together with the attached links, so that the topologies of the two networks between the first and last stage are made identical.

By relying on the network properties defined in [Kru86], three kinds of relations between networks can now be stated:

- *Isomorphism*: two networks are isomorphic if a label-preserving isomorphism holds between them.
- *Topological equivalence*: two networks are topologically equivalent if an isomorphism holds between the underlying graphs of the two networks.
- *Functional equivalence*: two networks A and B are functionally equivalent if they perform the same permutations, that is if $P_A = P_B$.

Two isomorphic networks are also topologically equivalent, whereas the converse is not always true. In general two isomorphic or topologically equivalent networks are not functionally equivalent. Nevertheless, if the two networks (isomorphic or not, topologically equivalent or not) are also non-blocking, they must also be functionally equivalent since both of them perform the same permutations (all the $N!$ permutations). Note that the same number of network components are required in two isomorphic networks, not in two functionally equivalent networks.

For example, consider the two networks A and B of Figure 2.3: they are topologically equivalent since their underlying graph is the same (it is shown in the same Figure 2.3). Nevertheless, they are not isomorphic since the above-defined mapping showing a label-preserving isomorphism between A and B cannot be found. In fact, if matrices in network A are moved, the two networks A' and A" of Figure 2.3 can be obtained, which are close to B but not the same. A' has nodes with the same label but the links outgoing from H exchanged compared to the analogous outgoing from Y, whereas A" has the same topology as B but with

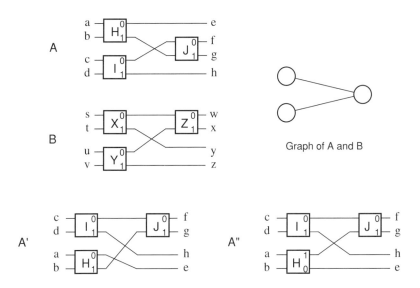

Figure 2.3. Example of topologically equivalent non-isomorphic networks

the labels in H exchanged compared to Y. On the other hand both networks A' and A" are isomorphic to network B in Figure 2.4 and the required mappings are also given in the figure.

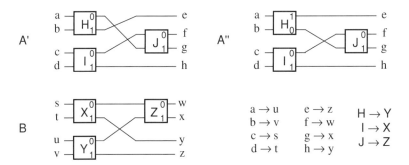

Figure 2.4. Example of isomorphic networks

Note that the two networks A and B of both examples above are not functionally equivalent, since, e.g., the two connections (0,0) and (1,1) can belong to the same permutation in network A, whereas this is not true for network B (remember that inlets and outlets are numbered 0 through 3 top to bottom of the network). The example of Figure 2.5 shows two isomorphic and functionally equivalent networks (it will be shown in Section 3.2.1.1 that both networks are rearrangeable).

A useful tool for the analysis of multistage networks is the *channel graph*. A channel graph is associated with each inlet/outlet pair and is given by the sequence of network elements that are crossed to reach the selected outlet from the selected input. In the channel graph, matrices

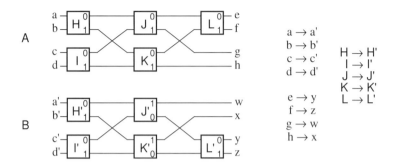

Figure 2.5. Example of isomorphic and functionally equivalent networks

are represented as nodes and interstage links as edges. Therefore the number of I/O paths in the channel graph represents the number of different modes in which the network outlet can be reached from the network inlet. Two I/O paths in the channel graph represent two I/O network connections differing in at least one of the crossed matrices. A network in which a single channel graph is associated with all the inlet/outlet pairs is a *regular network*. In a *regular channel graph* all the nodes belonging to the same stage have the same number of incoming edges and the same number of outgoing edges. Two isomorphic networks have the same channel graph.

   The channel graphs associated with the two isomorphic networks of Figure 2.4 are shown in Figure 2.6. In particular, the graph of Figure 2.6a is associated with the inlet/outlet pairs terminating on outlets e or f in network A (y or z in network B), whereas the graph of Figure 2.6b represents the I/O path leading to the outlets g or h in network A (w or x in network B). In fact three matrices are crossed in the former case engaging either of the two middle-stage matrices, while a single path connects the inlet to the outlet, crossing only two matrices in the latter case.

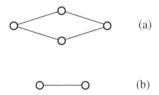

(a)

(b)

Figure 2.6. Channel graphs of the networks in Figure 2.4

2.1.2. Crossbar network based on splitters and combiners

In general it is worth examining how a crossbar network can be built by means of smaller building blocks relying on the use of special asymmetric connection elements called *splitters* and *combiners*, whose size is respectively $1 \times K$ and $K \times 1$ with a cost index $1 \cdot K = K \cdot 1 = K$. Note that a splitter, as well as a combiner, is able to set up one connection at a time. The *crossbar tree* network (Figure 2.7) is an interconnection network functionally equivalent to the crossbar network: it includes N splitters $1 \times N$ and N combiners $N \times 1$ interconnected by means of an EGS pattern and its cost is

$$C = 2N^2$$

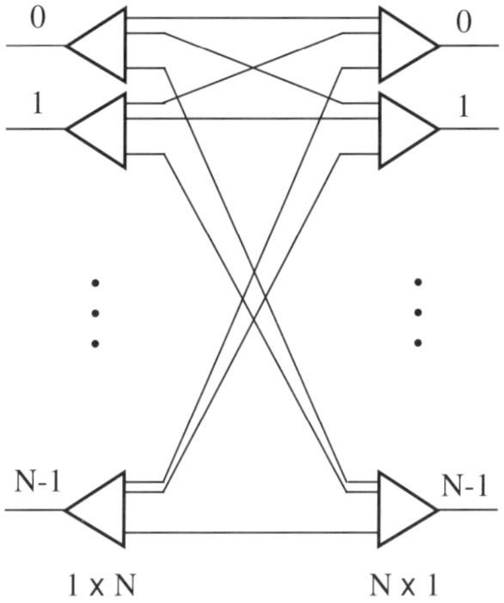

Figure 2.7. Crossbar tree

The crossbar tree can be built also by using multistage splitting and combining structures based on the use of elementary 1×2 splitters and 2×1 combiners, as shown in Figure 2.8 for $N = 8$. The cost index of such structure, referred to as a *crossbar binary tree* network, is

$$C = 4N \sum_{i=0}^{\log_2 N - 1} 2^i = 4N(N-1) = 4N^2 - 4N$$

It is interesting to note how the basic crossbar tree network of Figure 2.7 can be built using smaller splitters and combiners by means of a central switching stage. Our aim is a central stage with the smallest cost that still guarantees full input/output accessibility, thus suggesting a set of crossbar matrices each with the smallest possible size. In general if we have an expansion stage with size $1 \times K$ with $K = 2^k$ ($k = 1, ..., \log_2 N - 1$), each inlet has access to K switching matrices from which all the N outlets must be reached, so that each switching matrix must have a number of outlets (and inlets) at least equal to N/K (each matrix in the switching stage is connected to splitters and combiners with at most one link). By adopting for the two inter-stage connections the EGS pattern, which provides a cyclic connection to the elements of the following stage, it follows that such matrix size is sufficient to give full accessibility, as is shown in Figure 2.9 for $N = 8$ and $K = 2, 4$. Since the number of these matrices is

$$\frac{KN}{\frac{N}{K}} = K^2$$

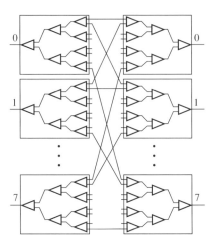

Figure 2.8. Crossbar binary tree

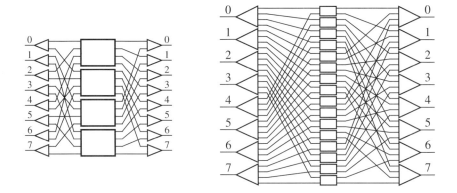

Figure 2.9. Crossbar tree with one switching stage

the cost function of such a network is given by

$$C = K^2 \left(\frac{N}{K} \right)^2 + 2NK = N^2 + 2NK$$

2.2. Full-connection Multistage Networks

The description of FC networks assumes as a general rule, unless stated otherwise, that matrices in adjacent stages are always connected by a single link. Thus the general model of an $N \times M$ full-connection (FC) multistage network is given by Figure 2.10, in which $n_i = r_{i-1}$, $m_{i-1} = r_i$ $(i = 2, \ldots, s)$ and the $n_i \times m_i$ matrix of the generic stage i is a crossbar network.

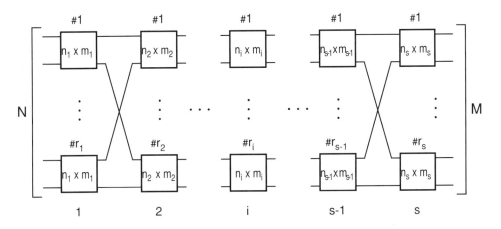

Figure 2.10. FC network with *s* stages

Note that such network is a subcase of an EGS network with $m_i = r_{i+1}$ $(i = 1, \ldots, s-1)$. We examine here the two cases of two- and three-stage networks $(s = 2, 3)$ as they provide the basic concepts for understanding the properties of full-connection multistage networks. In the following n and m will denote the inlets to each first-stage matrix and the outlets from each last-stage matrix $(n_1 = n, m_s = m)$.

The model of a two-stage FC network is represented in Figure 2.11. This network clearly has full accessibility, but is blocking at the same time. In fact, no more than one connection between two matrices of different stages can be established. Only the equipment of multiple links between each couple of matrices in different stages can provide absence of blocking.

The scheme of a three-stage FC network is given in Figure 2.12. Adopting three stages in a multistage network, compared to a two-stage arrangement, introduces a new important concept: different I/O paths are available between any couple of matrices in the first and third stage, each engaging a different matrix in the second stage. Full accessibility is implicitly guaranteed by the full connection feature of the two interstage patterns. Given N and M, the three-stage network has to be engineered so as to minimize its cost. In particular the number of second-stage matrices is the parameter determining the network non-blocking condition.

The control of multistage FC networks requires in general a centralized storage device which keeps the information about the busy/idle state of all the network terminations and interstage links. So a new connection between an idle inlet and an idle outlet of a non-block-

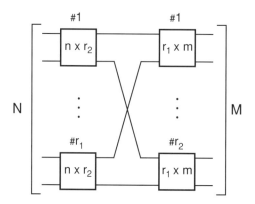

Figure 2.11. FC two-stage network

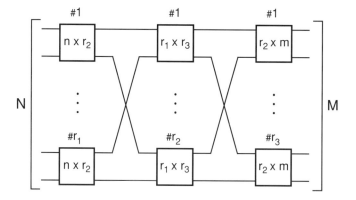

Figure 2.12. FC three-stage network

ing network can easily be found by properly visiting the network connection map, whereas more complex algorithms are in general required in rearrangeable networks to select the connections to be moved and their new path.

2.3. Partial-connection Multistage Networks

The class of partial-connection (PC) networks, in which each matrix of the intermediate stages is connected only to a subset of the matrices in the two adjacent stages, is becoming more and more important in today's high-speed communication networks based on packet switching. The interconnection networks of these scenarios are expected to carry very large amounts of traffic and the network permutation must be changed at a very high rate, e.g. the state duration is on the order of a few microseconds. Thus a mandatory requirement to build such interconnection networks with a significant size seems to be the availability of a high degree of parallel processing in the network. This result is accomplished by designing multi-stage PC networks in which the matrices of all stages are in general very small, usually all of the same size, and are provided with an autonomous processing capability. These matrices, referred to as *switching elements* (SEs), have in general sizes that are powers of two, that is $2^k \times 2^k$ with k typically in the range [1,5]. In the following, unless stated otherwise, we assume the SEs to have size 2×2.

By relying on its processing capability, each SE becomes capable of routing autonomously the received packets to their respective destinations. Such feature is known as *packet self-routing* property. The networks accomplishing this task are blocking structures referred to as *banyan networks*. These networks, which provide a single path per each inlet/outlet pair, can be suitably "upgraded" so as to obtain RNB and SNB networks. *Sorting networks* play a key role as well in high-speed packet switching, since the RNB network class includes also those structures obtained by cascading a sorting network and a banyan network. All these topics are investigated next.

2.3.1. Banyan networks

We first define four basic permutations, that is one-to-one mappings between inlets and outlets of a generic network $N \times N$, that will be used as the basic building blocks of self-routing networks. Let $a = a_{n-1} \ldots a_0$ $(n = \log_2 N)$ represent a generic address with base-2 digits a_i, where a_{n-1} is the most significant bit.

Four basic network permutations are now defined that will be needed in the definition of the basic banyan networks; the network outlet connected to the generic inlet a is specified by one of the following functions:

- $\sigma_h(a_{n-1} \ldots a_0) = a_{n-1} \ldots a_{h+1} a_{h-1} \ldots a_0 a_h$ $(0 \leq h \leq n-1)$
- $\sigma_h^{-1}(a_{n-1} \ldots a_0) = a_{n-1} \ldots a_{h+1} a_0 a_h \ldots a_1$ $(0 \leq h \leq n-1)$
- $\beta_h(a_{n-1} \ldots a_0) = a_{n-1} \ldots a_{h+1} a_0 a_{h-1} \ldots a_1 a_h$ $(0 \leq h \leq n-1)$
- $j(a_{n-1} \ldots a_0) = a_{n-1} \ldots a_0$

The permutations σ_3 and β_2 are represented in Figure 2.13 for $N = 16$. The permutations σ_h and σ_h^{-1} are called *h-shuffle* and *h-unshuffle*, respectively, one being the mirror image of the other (if the inlet a is connected to the outlet b in the shuffle, the inlet b is connected to the outlet a in the unshuffle). The *h*-shuffle (*h*-unshuffle) permutation consists in a circular left (right) shift by one position of the $h+1$ least significant bits of the inlet address. In the case of $h = n-1$ the circular shift is on the full inlet address and the two permutations are referred to as *perfect shuffle* (σ) and *perfect unshuffle* (σ^{-1}). Moreover, β is called the *butterfly* permutation and j the *identity* permutation. Note that $\sigma_0 = \sigma_0^{-1} = \beta_0 = j$. It can be verified that a permutation σ_h $(0 < h < n-1)$ corresponds to $k = 2^{n-h-1}$ perfect shuffle permutations each applied to N/k adjacent network inlets/outlets (only the $h+1$ least significant bits are rotated in σ_h). It is interesting to express the perfect shuffle and unshuffle permutations by using addresses in base 10. They are

- $\sigma(i) = (2i + \lfloor 2i/N \rfloor) \bmod N$
- $\sigma^{-1}(i) = \lfloor i/2 \rfloor + (i \bmod 2) N/2$

In the perfect shuffle the input address i is doubled modulo N, which corresponds to a left shift of the $n-1$ least significant bits, whereas the last term of $\sigma(i)$ accounts for the value of i_{n-1} (inlets larger than $N/2$ are mapped onto odd outlets so that a unity is added to the first term). Analogously, in the perfect unshuffle the input address i is halved and the integer part is taken, which corresponds to a right shift of the $n-1$ most significant bits, whereas the last term of $\sigma^{-1}(i)$ accounts for the value of i_0 (even and odd addresses are mapped onto the first and last $N/2$ outlets respectively).

Two other permutations are also defined which will be useful in stating relations between banyan networks

- $\delta(a_{n-1}\ldots a_0) = a_1 a_2 \ldots a_{n-1} a_0$
- $\rho(a_{n-1}\ldots a_0) = a_0 a_1 \ldots a_{n-2} a_{n-1}$

The permutation δ, called *bit switch*, and ρ, called *bit reversal*, are shown in Figure 2.13 for $N = 16$.

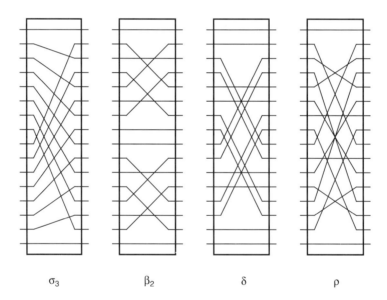

$$\sigma_3 \qquad\qquad \beta_2 \qquad\qquad \delta \qquad\qquad \rho$$

Figure 2.13. Examples of network permutations

2.3.1.1. Banyan network topologies

Banyan networks are multistage arrangements of very simple $b \times b$ switching elements whose inlets and outlets are labelled 0 through $b-1$. The interstage connection patterns are such that only one path exists between any network inlet and network outlet, and different I/O paths can share one or more interstage links. SEs in a banyan network are organized in $n = \log_b N$ stages each comprising N/b SEs. As is usually the case with banyan networks, we assume N, as well as b, to be a power of 2.

A general construction rule is now given to build a banyan network that, as explained later, provides the nice feature of a very simple distributed routing of packets through the network to their own destination. The rule consists in connecting each network outlet to all the inlets in such a way that at each step of this backward tree construction the inlets of the SEs of the stage i are connected to outlets of the same index in the SEs of stage $i-1$. In the case of $b = 2$ this corresponds to connecting the inlets of the SEs in a stage to all top or bottom outlets of upstream SEs. An example for a network with $N = 8$ is given in Figure 2.14. The building of the whole network is split into four steps, each devoted to the connection of a couple of network outlets, terminated onto the same SE of the last stage, to the eight network inlets. The process is started from a network without interstage links. The new interstage links added at each step to provide connection of a network outlet to all the network inlets are drawn in bold. Given the construction process being used, it follows that a single path descriptor including n outlet indices (one per stage) identifies all the I/O paths leading to the same network outlet. The class of banyan networks built using this rule is such that all the N paths leading to a given network outlet are characterized by the same path descriptor, given by the sequence of outlet indices selected in the path stage by stage. The banyan networks in which such a path descriptor is a permutation of the path network outlet are also called "delta" networks [Pat81]. Only this kind of banyan network will be considered in the following.

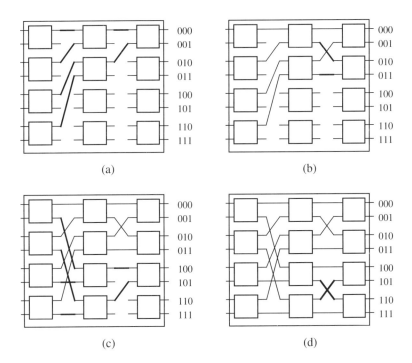

Figure 2.14. Construction of a banyan network

For simplicity we consider now 2×2 SEs, but the following description of banyan networks can be easily extended to $b \times b$ SEs $(b > 2)$. A 2×2 SE, with top and bottom inlets (and outlets) labelled 0 and 1, respectively, can assume only two states, *straight* giving the I/O paths 0-0 and 1-1 in the SE, and *cross* giving the I/O paths 0-1 and 1-0 (Figure 2.15).

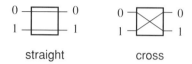

Figure 2.15. SE states

Several topologies of banyan networks have been described in the technical literature that differ in the way of interconnecting SEs in adjacent stages and network inlets (outlets) to the SEs of the first (last) stage. Figure 2.16 shows the structure of four of these networks for $N = 16$: *Omega* [Law75], *SW-banyan* [Gok73], *n-cube* [Sie81], *Baseline* [Wu80a]. The reverse topology of each of these networks is easily obtained as the mirror image of the network itself: the topology of a *reverse Baseline* network (or *Baseline^{-1}*) is shown at the end of this section when the routing property of a banyan network is described. The *reverse n-cube* is also known as *indirect binary n-cube* [Pea77]. Figure 2.16 also shows how the SW-banyan and Baseline network can be built by applying $\log_2 N - 1$ times a recursive construction. In fact an $N \times N$ Baseline network Φ_N includes a first stage of $N/2$ SEs 2×2 interconnected through a perfect unshuffle permutation σ_{n-1}^{-1} to two Baseline networks $\Phi_{N/2}$ of half size. An analogous recursive construction is applied in the SW-banyan network Σ_N: two SW-banyan networks $\Sigma_{N/2}$ are connected through a butterfly permutation β_{n-1} to a last stage of $N/2$ SEs 2×2.

In our representation stages are numbered 1 through n with stage 1 (n) interfacing the network inlets (outlets). The permutation set up by the N links following the switching stage h is denoted by $P(h)$, whereas $P(0)$ indicates the input permutation of the network (that is the mapping between network inlets and the inlets of the first stage).

Thus the four networks of Figure 2.16 and their reverse structures are formally described by the first three columns of Table 2.2. It is worth noting that the interstage pattern of the Omega network is a subcase of an EGS pattern with $n_i = m_i = 2$, $r_i = N/n_i = N/2$.

Functional equivalences between the different banyan topologies have been found [Wu80b], in the sense that one topology can be obtained from another by applying on the inlets and/or on the outlets one of the two permutations δ and ρ. Table 2.3 shows functional equivalence between networks starting from each the four basic topologies of Figure 2.16, the Omega (Ω), the SW-banyan (Σ), the n-cube (Γ) and the Baseline (Φ). A sequence of permutations $\alpha\beta\gamma$ means a sequence of networks applying on an input sequence the permutation α, followed by β and ending with γ. For example, a Baseline network preceded by a bit reversal permutation of the inlets is functionally equivalent to the Omega network. Note that the operation of network reversing does not affect the Baseline network as this network and its reverse perform the same permutations. Furthermore, since Figure 2.16 shows that a reverse n-cube is obtained by an SW-banyan followed by a perfect unshuffle permutation, it follows from the correspondence in the Table 2.3 that $\delta\rho = \sigma^{-1}$, that is a bit switch permutation followed by a bit reversal is equivalent to a perfect unshuffle (it can be immediately verified by simply applying the permutation definitions).

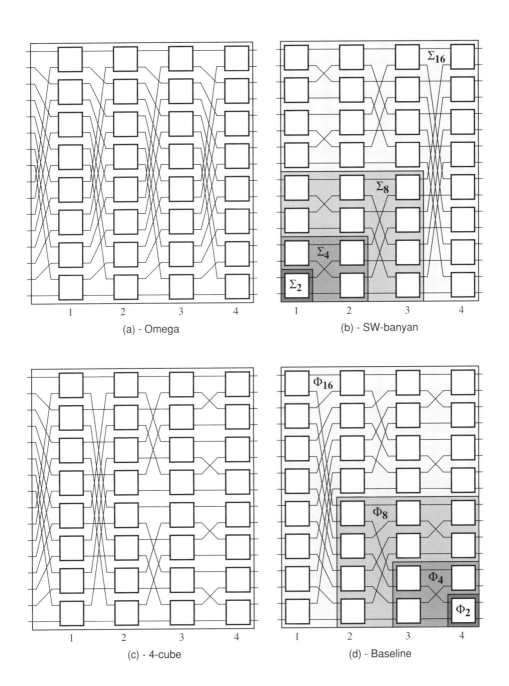

(a) - Omega

(b) - SW-banyan

(c) - 4-cube

(d) - Baseline

Table 2.2. Topology and routing rule in banyan networks

	$P(0)$	$P(h)$ $0 < h < n$	$P(n)$	Self-routing bit $I \to O$	Self-routing bit $O \to I$
Omega	σ_{n-1}	σ_{n-1}	j	d_{n-h}	d_{n-h}
Omega^{-1}	j	σ^{-1}_{n-1}	σ^{-1}_{n-1}	d_{h-1}	d_{h-1}
SW-banyan	j	β_h	j	$d_{h \bmod n}$	d_{h-1}
SW-banyan^{-1}	j	β_{n-h}	j	d_{n-h}	$d_{(n-h+1)\bmod n}$
n-cube	σ_{n-1}	β_{n-h}	j	d_{n-h}	d_{n-h}
n-cube^{-1}	j	β_h	σ^{-1}_{n-1}	d_{h-1}	d_{h-1}
Baseline	j	σ^{-1}_{n-1}	j	d_{n-h}	d_{h-1}
Baseline^{-1}	j	σ_h	j	d_{n-h}	d_{h-1}

Table 2.3. Functional equivalence between banyan networks

	Ω	Σ	Γ	Φ
Omega	Ω	$\rho\Sigma\delta$	Γ	$\rho\Phi$
Omega^{-1}	$\rho\Omega\rho$	$\Sigma\delta\rho$	$\rho\Gamma\rho$	$\Phi\rho$
SW-banyan	$\rho\Omega\delta$	Σ	$\rho\Gamma\delta$	$\Phi\delta$
SW-banyan^{-1}	$\delta\rho\Omega$	$\delta\Sigma\delta$	$\delta\rho\Gamma$	$\delta\Phi$
n-cube	Ω	$\rho\Sigma\delta$	Γ	$\rho\Phi$
n-cube^{-1}	$\rho\Omega\rho$	$\Sigma\delta\rho$	$\rho\Gamma\rho$	$\Phi\rho$
Baseline	$\rho\Omega$	$\Sigma\delta$	$\rho\Gamma$	Φ
Baseline^{-1}	$\rho\Omega$	$\Sigma\delta$	$\rho\Gamma$	Φ

According to Table 2.3, Figure 2.16 includes in reality only three basic networks that are not functionally equivalent, since the Omega and n-cube networks are functionally equivalent. In fact it can be easily verified that one topology can be obtained from the other by suitable position exchange of SEs in the intermediate stages.

2.3.1.2. Banyan network properties

All the four topologies of banyan networks defined here are based on interstage patterns satisfying two properties, called the *buddy property* [Dia81] and the *constrained reachability property*.

Buddy property. If SE j_i at stage i is connected to SEs l_{i+1} and m_{i+1}, then these two SEs are connected also to the same SE k_i in stage i.

In other words, switching elements in adjacent stages are always interconnected in couples to each other. By applying the buddy property across several contiguous stages, it turns out that certain subsets of SEs at stage i reach specific subsets of SEs at stage $i+k$ of the same size, as stated by the following property.

Constrained reachability property. The 2^k SEs reached at stage $i+k$ by an SE at stage i are also reached by exactly $2^k - 1$ other SEs at stage i.

The explanation of this property relies on the application of the buddy property stage by stage to find out the set of reciprocally connected SEs. An SE is selected in stage i as the root of a forward tree crossing 2 SEs in stage $i+1$, 4 SEs in stage $i+2$, ..., 2^{k-1} SEs in stage $i+k-1$ so that 2^k SEs are reached in stage $i+k$. By selecting any of these SEs as the root of a tree reaching backwards stage i, it is easily seen that exactly 2^k SEs in stage i are reached including the root of the previous forward tree. If all the forward and backward subtrees are traced starting from the SEs already reached in stages $i+1$, ..., $i+k-1$, exactly 2^k SEs per stage will have been crossed in total. Apparently, the buddy property will be verified as holding between $2^k/2$ couples of SEs in adjacent stages between i and $i+k$. An example of these two properties can be found in Figure 2.17, where two couples of buddy SEs in stages 1 and 2 are shown together with the constrained reachability between sets of 4 SEs in stages 1 through 3.

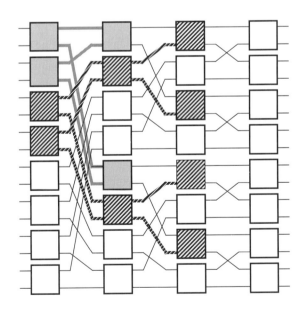

Figure 2.17. Buddy and constrained reachability property

Since the constrained reachability property holds in all the banyan topologies defined above, the four basic banyan networks are isomorphic to each other. In fact, a banyan network B can be obtained from another banyan network A by properly moving the SEs of A so that this latter network assumes the same interstage pattern as B. The isomorphism specification then requires also to state the inlet and outlet mapping between A and B, which is apparently given by Table 2.3 if network A is one of the four basic banyan topologies. For example, if A is the Baseline network and the isomorphic network B to be obtained is the Omega network with $N = 8$, Figure 2.18 shows the A-to-B mapping of SEs, inlets and outlets. In particular, the permutation ρ is first added at the inlets of the Baseline network, as specified in Table 2.3 (see Figure 2.18b) and then the SEs in stages 1 and 2 are moved so as to obtain the Omega topology. The mapping between SEs specifying the isomorphism between Φ and Ω can be obtained from Figure 2.18c and is given in Figure 2.18e. The inlet and outlet mappings are those shown in Figure 2.18d and they apparently consist in the permutations ρ and j.

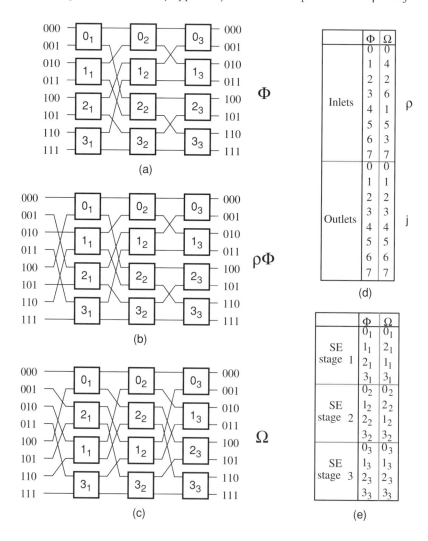

Figure 2.18. Example of network isomorphism

In some cases of isomorphic networks the inlet and outlet mapping is just the identity j if A and B are functionally equivalent, i.e. perform the same permutations. This occurs in the case of the $A = \Omega$, $B = \Gamma$ and $A = \Phi$, $B = \Phi^{-1}$. It is worth observing that the buddy and constrained reachability properties do not hold for all the banyan networks. In the example of Figure 2.14 the buddy property holds between stage 2 and 3, not between stage 1 and 2.

Other banyan networks have been defined in the technical literature, but their structures are either functionally equivalent to one of the three networks Ω, Σ and Γ, by applying, if necessary, external permutations analogously to the procedure followed in Table 2.3. Examples are the *Flip* network [Bat76] that is topologically identical to the reverse Omega network and the *Modified data manipulator* [Wu80a] that is topologically identical to a reverse SW-banyan.

Since each switching element can assume two states, the number of different states assumed by a banyan network is

$$2^{\frac{N}{2}\log_2 N} = \sqrt{N^N}$$

which also expresses the network of different permutations that the banyan network is able to set up. In fact, since there is only one path between any inlet and outlet, a specific permutation is set up by one and only one network state. The total number of permutations $N!$ allowed by a non-blocking network $N \times N$ can be expressed using the well-known Stirling's approximation of a factorial [Fel68]

$$N! \cong N^N e^{-N}\sqrt{2\pi N} \qquad (2.1)$$

which can be written as

$$\log_2 N! \cong N\log_2 N - 1.443N + 0.5\log_2 N \qquad (2.2)$$

For very large values of N, the last two terms of Equation 2.2 can be disregarded and therefore the factorial of N is given by

$$N! \cong 2^{N\log_2 N} = N^N$$

Thus the *combinatorial power* of the network [Ben65], defined as the fraction of network permutations that are set up by a banyan network out of the total number of permutations allowed by a non-blocking network, can be approximated by the value $N^{-N/2}$ for large N. It follows that the network blocking probability increases significantly with N.

In spite of such high blocking probability, the key property of banyan networks that suggests their adoption in high-speed packet switches based on the ATM standard is their packet self-routing capability: an ATM packet preceded by an address label, the *self-routing tag*, is given an I/O path through the network in a distributed fashion by the network itself. For a given topology this path is uniquely determined by the inlet address and by the routing tag, whose bits are used, one per stage, by the switching elements along the paths to route the cell to the requested outlet. For example, in an Omega network, the bit d_{n-h} of the self-routing tag $d_{n-1}d_{n-2}...d_1d_0$ indicates the outlet required by the packet at stage h ($d_h = 0$ means top outlet, $d_h = 1$ means bottom outlet)[1]. Note that the N paths leading from the different inlets to a given network outlet are traced by the same self-routing tag.

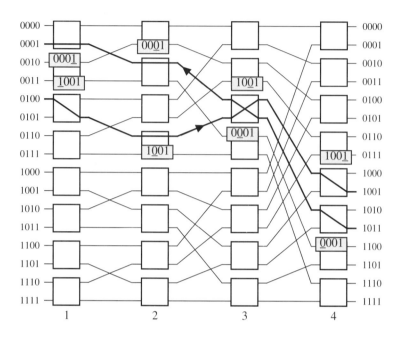

Figure 2.19. Reverse Baseline with example of self-routing

The self-routing rule for the examined topologies for a packet entering a generic network inlet and addressing a specific network outlet is shown in Table 2.2 ($I \rightarrow O$ connection). The table also shows the rule to self-route a packet from a generic network outlet to a specific network inlet ($O \rightarrow I$ connection). In this case the self-routing bit specifies the SE inlet to be selected stage by stage by the packet entering the SE on one of its outlets (bit 0 means now top inlet and bit 1 means bottom inlet). An example of self-routing in a reverse Baseline network is shown in Figure 2.19: the bold $I \rightarrow O$ path connects inlet 4 to outlet 9, whereas the bold $O \rightarrow I$ path connects outlet 11 to inlet 1.

As is clear from the above description, the operations of the SEs in the network are mutually independent, so that the processing capability of each stage in a switch $N \times N$ is $N/2$ times the processing capability of one SE. Thus, a very high parallelism is attained in packet processing within the interconnection network of an ATM switch by relying on space division techniques. Owing to the uniqueness of the I/O path and to the self-routing property, no centralized control is required here to perform the switching operation. However, some additional devices are needed to avoid the set-up of paths sharing one or more interstage links. This issue will be investigated while dealing with the specific switching architecture employing a banyan network.

---

1. If SEs have size $b \times b$ with $b = 2^x$ ($x = 2, 3, \ldots$) , then self-routing in each SE is operated based on $\log_2 b$ bits of the self-routing tag.

2.3.2. Sorting networks

Networks that are capable of sorting a set of elements play a key role in the field of interconnection networks for ATM switching, as they can be used as a basic building block in non-blocking self-routing networks.

Efficiency in sorting operations has always been a challenging research objective of computer scientists. There is no unique way of defining an optimum sorting algorithm, because the concept of optimality is itself subjective. A theoretical insight into this problem is given by looking at the algorithms which attempt to minimize the number of comparisons between elements. We simply assume that sorting is based on the comparison between two elements in a set of N elements and their conditional exchange. The information gathered during previous comparisons is maintained so as to avoid useless comparisons during the sorting operation. For example Figure 2.20 shows the process of sorting three elements 1, 2, 3, starting from an initial arbitrary relative ordering, say 1 2 3, and using pairwise comparison and exchange. A binary tree is then built since each comparison has two outcomes; let the left (right) subtree of node $A{:}B$ denote the condition $A < B$ $(B < A)$. If no useless comparisons are made, the number of tree leaves is exactly $N!$: in the example the leaves are exactly $3! = 6$ (note that the two external leaves are given by only two comparisons, whereas the others require three comparisons. An optimum algorithm is expected to minimizing the maximum number of comparisons

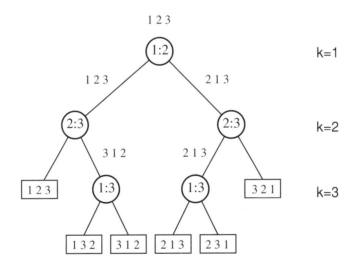

Figure 2.20. Sorting three elements by comparison exchange

required, which in the tree corresponds to minimize the number k of tree levels. By assuming the best case in which all the root-to-leaf paths have the same depth (they cross the same number of nodes), it follows that the minimum number of comparisons k required to sort N numbers is such that

$$2^k \geq N!$$

Based on Stirling's approximation of the factorial (Equation 2.2), the minimum number k of comparisons required to sort N numbers is on the order of $N\log_2 N$. A comprehensive survey of sorting algorithms is provided in [Knu73], in which several computer programs are described requiring a number of comparisons equal to $N\log_2 N$. Nevertheless, we are interested here in hardware sorting networks that cannot adapt the sequence of comparisons based on knowledge gathered from previous comparisons. For such "constrained" sorting the best algorithms known require a number of comparisons $N(\log_2 N)^2$ carried out in a total number $(\log_2 N)^2$ of comparison steps. These approaches, due to Batcher [Bat68], are based on the definition of parallel algorithms for sorting sequences of suitably ordered elements called merging algorithms. Repeated use of merging network enables to build full sorting networks.

2.3.2.1. Merging networks

A *merge network* of size N is a structure capable of sorting two ordered sequences of length $N/2$ into one ordered sequence of length N. The two basic algorithms to build merging networks are *odd–even merge* sorting and *bitonic merge* sorting [Bat68]. In the following, for the purpose of building sorting networks the sequences to be sorted will have the same size, even if the algorithms do not require such constraint.

The general scheme to sort two increasing sequences $\boldsymbol{a} = a_0, \dots, a_{N/2-1}$ and $\boldsymbol{b} = b_0, \dots, b_{N/2-1}$ with $a_0 \le a_1 \le \dots \le a_{N/2-1}$ and $b_0 \le b_1 \le \dots \le b_{N/2-1}$ by odd–even merging is shown in Figure 2.21. The scheme includes two mergers of size $N/2 \times N/2$, one

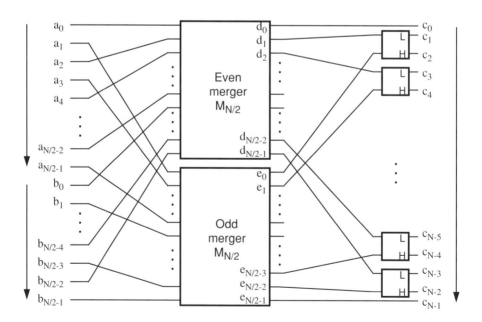

Figure 2.21. Odd–even merging

fed by the odd-indexed elements and the other by the even-indexed elements in the two sequences, followed by $N/2 - 1$ sorting (or comparison-exchange) elements 2×2, or *down-sorters*, routing the lower (higher) elements on the top (bottom) outlet. In Section 2.4.1 it is shown that the output sequence $c = c_0, ..., c_{N/2 - 1}$ is ordered and increasing, that is $c_0 \leq c_1 \leq ... \leq c_{N-1}$.

Since the odd–even merge sorter M_N of size $N \times N$ in Figure 2.21 uses two mergers $M_{N/2}$ of half size, it is possible to recursively build the overall structure that only includes 2×2 sorting elements, as shown in Figure 2.22 for $N = 16$.

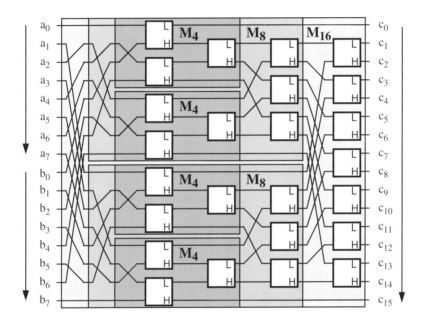

Figure 2.22. Odd–even merging network of size *N*=16

Based on the recursive construction shown in Figure 2.21, the number of stages $s(M_N)$ of the odd–even merge sorter is equal to

$$s[M_N] = \log_2 N$$

The total number $S[M_N]$ of sorting elements of this $N \times N$ merge sorter is computed recursively with the boundary condition $S[M_2] = 1$, that is

$$S[M_N] = 2S[M_{N/2}] + \frac{N}{2} - 1 = 2\left(2S[M_{N/4}] + \frac{N}{4} - 1\right) + \frac{N}{2} - 1 \qquad (2.3)$$

$$= \frac{N}{2} - 1 + 2\left(\frac{N}{4} - 1\right) + 4\left(\frac{N}{8} - 1\right) + \dots + \frac{N}{4}(2-1) + \frac{N}{2}$$

$$= \sum_{i=0}^{\log_2 N - 2} \left[2^i\left(\frac{N}{2^{i+1}} - 1\right)\right] + \frac{N}{2} = \frac{N}{2}(\log_2 N - 1) + 1$$

Note that the structure of the odd–even merge sorter is such that each element can be compared with the others a different number of times. In fact, the shortest I/O path through the network crosses only one element (i.e. only one comparison), whereas the longest path crosses $\log_2 N$ elements, one per stage.

Unlike the odd–even merge sorter, in the bitonic merge sorter each element is compared to other elements the same number of times (meaning that all stages contain the same number of elements), but this result is paid for by a higher number of sorting elements. A sequence of elements $a_0, a_1, \dots, a_{N-1}$ is said to be *bitonic* if an index j exists $(0 \le j \le N-1))$ such that the subsequences $a_0, \dots, a_j$ and $a_j, \dots, a_{N-1}$ are one monotonically increasing and the other monotonically decreasing. Examples of bitonic sequences are $(0,3,4,5,8,7,2,1)$ and $(8,6,5,4,3,1,0,2)$. A *circular bitonic* sequence is a sequence obtained shifting circularly the elements of a bitonic sequence by an arbitrary number of positions k $(k \in [0, N-1])$. For example the sequence $(3,5,8,7,4,0,1,2)$ is circular bitonic. In the following we will be interested in two specific *balanced* bitonic sequences, that is a sequence in which $a_0 \le a_1 \le \dots \le a_{N/2-1}$ and $a_{N/2} \ge a_{N/2+1} \ge \dots \ge a_{N-1}$ or $a_0 \ge a_1 \ge \dots \ge a_{N/2-1}$ and $a_{N/2} \le a_{N/2+1} \le \dots \le a_{N-1}$.

The $N \times N$ bitonic merger M_N shown in Figure 2.23 is able to sort increasingly a bitonic sequence of length N. It includes an initial shuffle permutation applied to the bitonic sequence, followed by $N/2$ sorting elements 2×2 (down-sorters) interconnected through a perfect unshuffle pattern to two bitonic mergers $M_{N/2}$ of half size $(N/2 \times N/2)$. Such a network performs the comparison between the elements a_i and $a_{i+N/2}$ $(0 \le i \le N/2-1)$ and generates two subsequences of $N/2$ elements each offered to a bitonic merger $M_{N/2}$. In Section 2.4.2 it is shown that both subsequences are bitonic and that all the elements in one of them are not greater than any elements in the other. Thus, after sorting the subsequence in each of the $M_{N/2}$ bitonic mergers, the resulting sequence $c_0, c_1, \dots, c_{N-1}$ is monotonically increasing.

The structure of the bitonic merger M_N in Figure 2.23 is recursive so that the bitonic mergers $M_{N/2}$ can be constructed using the same rule, as is shown in Figure 2.24 for $N = 16$. As in the odd–even merge sorter, the number of stages of a bitonic merge sorter is

$$s[M_N] = \log_2 N$$

but this last network requires a greater number $S[M_N]$ of sorting elements

$$S[M_N] = \frac{N}{2}\log_2 N$$

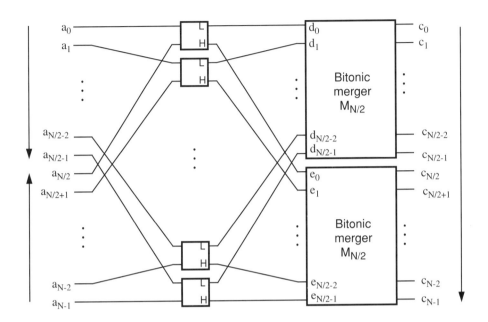

Figure 2.23. Bitonic merging

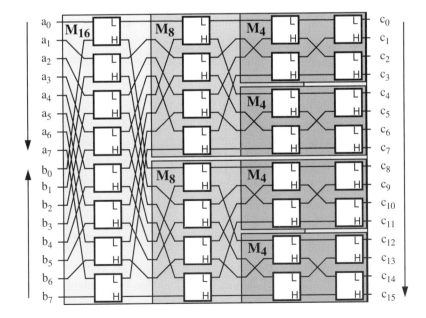

Figure 2.24. Bitonic merging network of size *N*=16

Interestingly enough, the bitonic merge sorter M_N has the same topology as the n–cube banyan network (shown in Figure 2.16 for $n = 4$), whose elements now perform the sorting function, that is the comparison-exchange, rather than the routing function.

Note that the odd–even merger and the bitonic merger of Figures 2.22 and 2.24, which generate an increasing sequence starting from two increasing sequences of half length and from a bitonic sequence respectively, includes only down-sorters. An analogous odd–even merger and bitonic merger generating a monotonically decreasing sequence $c_0, c_1, ..., c_{N-1}$ starting from two decreasing sequences of half length and from a bitonic sequence is again given by the structures of Figures 2.22 and 2.24 that include now only *up-sorters*, that is sorting elements that route the lower (higher) element on the bottom (top) outlet.

2.3.2.2. Sorting networks

We are now able to build *sorting networks* for arbitrary sequences using the well-known *sorting-by-merging* scheme [Knu73]. The elements to be sorted are initially taken two by two to form $N/2$ sequences of length 2 (step 1); these sequences are taken two by two and merged so as to generate $N/4$ sequences of length 4 (step 2). The procedure is iterated until the resulting two sequences of size $N/2$ are finally merged into a sequence of size N (step $n = \log_2 N$). Thus the overall sorting network includes $\log_2 N$ merging steps the i-th of which is accomplished by 2^{n-i} mergers M_{2^i}. The number of stages of sorting elements for such sorting network is then

$$s_N = \sum_{i=1}^{\log_2 N} i = \frac{\log_2 N (\log_2 N + 1)}{2} \tag{2.4}$$

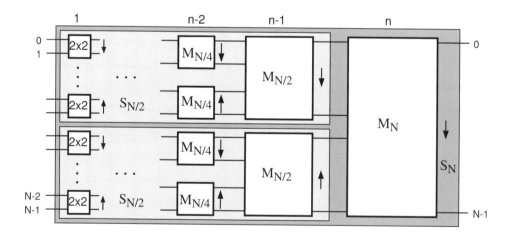

Figure 2.25. Sorting by merging

Such merging steps can be accomplished either with odd–even merge sorters, or with bitonic merge sorters. Figure 2.25 shows the first and the three last sorting steps of a sorting network based on bitonic mergers. Sorters with downward (upward) arrow accomplish increasing (decreasing) sorting of a bitonic sequence. Thus both down- and up-sorters are used in this network: the former in the mergers for increasing sorting, the latter in the mergers for decreasing sorting. On the other hand if the sorting network is built using an odd–even merge sorter, the network only includes down-sorters (up-sorters), if an increasing (decreasing) sorting sequence is needed. The same Figure 2.25 applies to this case with only downward (upward) arrows. The overall sorting networks with $N = 16$ are shown in Figure 2.26 for odd–even merging and in Figure 2.27 for bitonic merging. This latter network is also referred to as a Batcher network [Bat68].

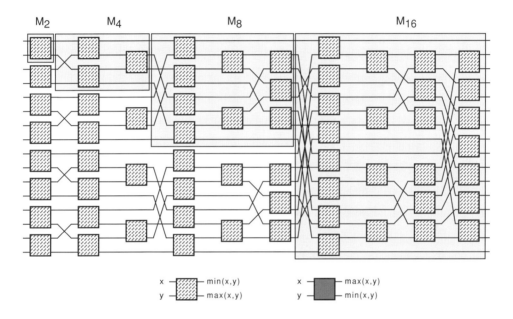

Figure 2.26. Odd–even sorting network for *N*=16

Given the structure of the bitonic merger, the total number of sorting elements of a bitonic sorting network is simply

$$S_N = \frac{N}{4}[\log_2^2 N + \log_2 N]$$

and all the I/O paths in a bitonic sorting network cross the same number of elements s_N given by Equation 2.4.

A more complex computation is required to obtain the sorting elements count for a sorting network based on odd–even merge sorters. In fact owing to the recursive construction of the sorting network, and using Equation 2.3 for the sorting elements count of an odd–even merger with size $(N/2^i) \times (N/2^i)$, we have

$$S_N = \sum_{i=0}^{\log_2 N - 1} 2^i S[M_{N/2^i}] = \sum_{i=0}^{\log_2 N - 1} 2^i \left[\frac{N}{2^{i+1}} \left(\log \frac{N}{2^i} - 1 \right) + 1 \right]$$

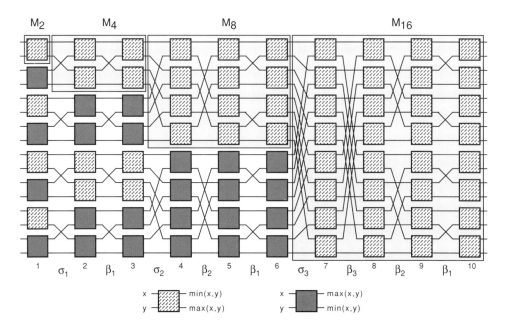

Figure 2.27. Bitonic sorting network for _N_=16

$$= \sum_{i=0}^{\log_2 N - 1} \frac{N}{2}(\log_2 N - 1 - i) + 2^i = \frac{N}{2}\left[\log_2 N(\log_2 N - 1) - \sum_{i=0}^{\log_2 N - 1} i\right] + \sum_{i=0}^{\log_2 N - 1} 2^i$$

$$= \frac{N}{2}\frac{\log_2 N(\log_2 N - 1)}{2} + N - 1 = \frac{N}{4}\left[\log_2^2 N - \log_2 N + 4\right] - 1$$

Given the structure of the odd–even merger, the number of elements crossed by an I/O path in the overall sorting network varies from $\log_2 N$ up to the value s_N given by Equation 2.4. In fact only one sorting element per merging step is crossed in the former case, whereas all the sorting stages are crossed in the latter case. Therefore, if t_D and τ denote the transmission time of each data unit and the sorting stage latency, respectively, the time T required to sort N fixed-size data unit ranges from T_{min} to T_{max} in the odd–even sorting with

$$T_{min} = t_D + \tau \log_2 N$$

$$T_{max} = t_D + \tau \frac{\log_2 N(\log_2 N + 1)}{2}$$

whereas in bitonic sorting is always given by

$$T = t_D + \tau \frac{\log_2 N(\log_2 N + 1)}{2}$$

Thus we have been able to build parallel sorting networks whose number of comparison–exchange steps grows as $(\log_2 N)^2$. Interestingly enough, the odd–even merge sorting network is the minimum-comparison network known for $N = 8$ and requires a number of comparisons very close to the theoretical lower bound for sorting networks [Knu73] (for example, the odd–even merge sorting network gives $S_{16} = 63$, whereas the theoretical bound is $S_{16} = 60$).

It is useful to describe the overall bitonic sorting network in terms of the interstage patterns. Let

$$s(j) = \sum_{i=1}^{j} i$$

denote the last stage of merge sorting step j, so that $s(j-1) + k$ is the stage index of the sorting stage k $(k = 1, ..., j)$ in merging step j $(j = 1, ..., n)$ (the boundary condition $s(0) = 0$ is assumed). If the interstage permutations are numbered according to the sorting stage they originate from (the interstage pattern i connects sorting stages i and $i + 1$), it is rather easy to see that the permutation $s(j)$ is the pattern σ_j and the permutation $s(j-1) + k$ is the pattern β_{j-k} $(1 \le k \le j)$. In other words, the interstage pattern between the last stage of merging step j and the first stage of merging step $j + 1$ $(1 \le j \le n - 1)$ is a shuffle pattern σ_j. Moreover the $j - 1$ interstage patterns at merging step j are butterfly $\beta_{j-1}, \beta_{j-2}, ..., \beta_1$. It follows that the sequence of permutation patterns of the 16×16 bitonic sorting network shown in Figure 2.27 is $\sigma_1, \beta_1, \sigma_2, \beta_2, \beta_1, \sigma_3, \beta_3, \beta_2, \beta_1$.

The concept of sorting networks based on bitonic sorting was further explored by Stone [Sto71] who proved that it is possible to build a parallel sorting network that only uses one stage of comparator-exchanges and a set of N registers interconnected by a shuffle pattern. The first step in this direction consists in observing that the sorting elements within each stage of the sorting network of Figure 2.27 can be rearranged so as to replace all the β_i patterns by perfect shuffle patterns. Let the rows be numbered 0 to $N/2 - 1$ top to bottom and the sorting element $x_{n-1}...x_1$ interface the network inlets $x_{n-1}...x_1 0$ and $x_{n-1}...x_1 1$. Let $r_i(x)$ denote the row index of the sorting element in stage i of the original network to be placed in row x of stage i in the new network and σ^0 indicate the identity permutation j. The rearrangement of sorting elements is accomplished by the following mapping:

$$r_{s(j-1)+k}(x_{n-1}...x_1) = \sigma^{j-k}(x_{n-1}...x_1)$$

For example, in stage $i = 4$, which is the first sorting stage of the third merging step $(j = 3, s(j-1) = 3, k = 1)$, the element in row 6 (110) is taken from the row of the original network whose index is given by $j - k = 2$ cyclic left rotations of the address 110, that gives 3 (011). The resulting network is shown in Figure 2.28 for $N = 16$, where the numbering of elements corresponds to their original position in the Batcher bitonic sorting network (it is worth observing that the first and last stage are kept unchanged). We see that the result of replacing the original β_i permutations by perfect shuffles is that the σ_i permutation $(1 \le i < n - 1)$ of the original sorting network has now become a permutation σ_{n-1}^{n-i}, that is the cascade of $n - i$ permutations σ_{n-1} (the perfect shuffle).

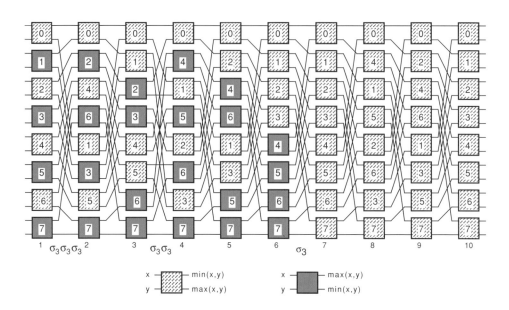

Figure 2.28. Modified bitonic sorting network for *N*=16

From such a modified Batcher sorting network, the Stone network is obtained in which all stages are preceded by a perfect shuffle [Sto71]. This network includes $\log_2 N$ subnetworks of $\log_2 N$ sorting stages and is shown in Figure 2.29. Each of these subnetworks performs one of

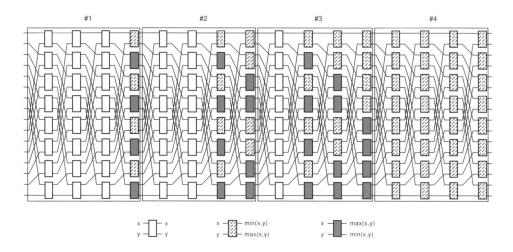

Figure 2.29. Stone sorting network for *N*=16

the sorting steps of the modified Batcher sorting network of Figure 2.28, the additional stages of elements in the straight state being only required to generate an all-shuffle sorting network. Each of the permutations σ_{n-1}^{n-i} of the modified Batcher sorting network is now replaced by a sequence of $n - i$ physical shuffles interleaved by $n - i - 1$ stages of sorting elements in the straight state. Note that the sequence of four shuffles preceding the first true sorting stage in the first subnetworks corresponds to an identity permutation ($\sigma_3^4 = j$ for $N = 16$). Therefore the number of stages and the number of sorting elements in a Stone sorting network are given by

$$s_N = \log_2^2 N$$

$$S_N = \frac{N}{2}\log_2^2 N$$

As above mentioned the interest in this structure lies in its implementation feasibility by means of the structure of Figure 2.30, comprising N registers and $N/2$ sorting elements interconnected by a shuffle permutation. This network is able to sort N data units by having the data units recirculate through the network $\log_2 N$ times and suitably setting the operation of each sorting element (straight, down-sorting, up-sorting) for each cycle of the data units. The sorting operation to be performed at cycle i is exactly that carried out at stage i of the full Stone sorting network. So a dynamic setting of each sorting element is required here, whereas each sorting element in the Batcher bitonic sorting network always performs the same type of sorting. The registers, whose size must be equal to the data unit length, are required here to enable the serial sorting of the data units $\log_2 N$ times independently of the latency amount of the sorting stage. So a full sorting requires $\log_2 N$ cycles of the data units through the single-stage network, at the end of which the data units are taken out from the network onto its N outlets. meaning that the sorting time T is given by

$$T = (t_D + \tau) \log_2^2 N$$

Note that the sorting time of the full Stone sorting network is

$$T = t_D + \tau \log_2^2 N$$

since the data units do not have to be stored before each sorting stage.

Analogously to the approach followed for multistage FC or PC networks, we assume that the cost of a sorting network is given by the cumulative cost of the sorting elements in the network, and that the basic sorting elements have a cost $C = 4$, due to the number of inlets and outlets. Therefore the sorting network cost index is given by

$$C = 4S_N$$

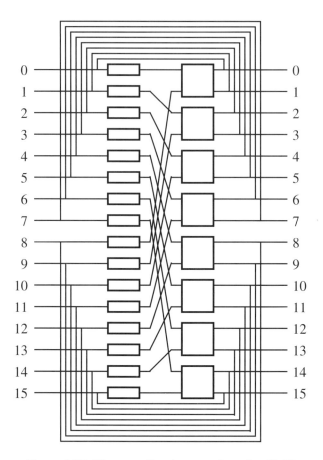

Figure 2.30. Stone sorting by one stage for N=16

2.4. Proof of Merging Schemes

In this section we prove that the networks implementing the odd–even merge sorting [Bat68] and bitonic sorting [Tur93] do indeed provide a sorted list starting from two sorted lists of half size.

2.4.1. Odd–even merge sorting

We now prove that the odd–even merge sorting scheme of Figure 2.21 actually generates an increasing sequence $c = (c_0, c_1, ..., c_{N-1})$ $(c_0 \leq c_1 \leq ... \leq c_{N-1})$ from two increasing sequences of half length $a = (a_0, a_1, ..., a_{N/2-1})$ and $b = (b_0, b_1, ..., b_{N/2-1})$ $(a_0 \leq a_1 \leq ... \leq a_{N/2-1}$ and $b_0 \leq b_1 \leq ... \leq b_{N/2-1})$. Let $d = (d_0, d_1, ..., d_{N/2-1})$ and

$e = (e_0, e_1, ..., e_{N/2-1})$ be the sorted sequences of the even-indexed and odd-indexed elements generated by the even merge and odd merge networks, respectively.

Let k of the $i+1$ elements $d_0, d_1, ..., d_i$ come from $a_0, a_2, a_4, ..., a_{N/2-2}$ and $i+1-k$ come from $b_0, b_2, b_4, ..., b_{N/2-2}$. The element d_i is not smaller than k elements in $a_0, a_2, a_4, ..., a_{N/2-2}$ and therefore is not smaller than $2k-1$ elements of the original sequence $\mathbf{a}$. Analogously, d_i is not smaller than $2i-1-2k$ elements of $\mathbf{b}$ and hence is not smaller than $2k-1+2i+1-2k = 2i$ elements of $\mathbf{c}$. Therefore

$$d_i \geq c_{2i-1} \tag{2.5}$$

An analogous reasoning on the i elements $e_0, e_1, ..., e_{i-1}$ gives the other condition

$$e_{i-1} \geq c_{2i-1} \tag{2.6}$$

Now, let k of the $2i+1$ elements $c_0, c_1, ..., c_{2i}$ come from $\mathbf{a}$ and $2i+1-k$ come from $\mathbf{b}$. If k is even, then c_{2i} is not smaller than

- k elements in $a_0, a_1, a_2, ..., a_{N/2-1}$
- $k/2$ elements in $a_0, a_2, a_4, ..., a_{N/2-2}$
- $2i+1-k$ elements in $b_0, b_1, b_2, ..., b_{N/2-1}$
- $i+1-k/2$ elements in $b_0, b_2, b_4, ..., b_{N/2-2}$
- $k/2 + i + 1 - k/2 = i+1$ elements in $d_0, d_1, d_2, ..., d_{N/2-1}$

so that

$$c_{2i} \geq d_i \tag{2.7}$$

An analogous reasoning on the odd-indexed elements of the two sequences $\mathbf{a}$ and $\mathbf{b}$ proves that c_{2i} is not smaller than i elements of $e_0, e_1, e_2, ..., e_{N/2-1}$ and hence

$$c_{2i} \geq e_{i-1} \tag{2.8}$$

If k is odd, the last two Equations 2.7-2.8 still hold.

Since all the elements of $\mathbf{d}$ and $\mathbf{e}$ must appear somewhere in the sequence $\mathbf{c}$ and $c_0 \leq c_1 \leq ... \leq c_{N-1}$, then Equations 2.5 and 2.6 imply

$$c_{2i-1} = min(d_i, e_{i-1})$$

whereas Equations 2.7 and 2.8 imply

$$c_{2i} = max(d_i, e_{i-1})$$

2.4.2. Bitonic merge sorting

Our objective here is to prove the correct sorting operation of the bitonic merging scheme of Figure 2.23, where $\mathbf{a} = a_0, a_1, ..., a_{N-1}$ is the input circular bitonic sequence and the generic down sorter i $(i \in [0, N/2-1])$ compares the elements a_i and $a_{i+N/2}$ of the input sequence. Let $d_i = min(a_i, a_{i+N/2})$ and $i_i = max(a_i, a_{i+N/2})$, so that the top merger receives the sequence $\mathbf{d} = d_0, d_1, ..., d_{N/2-1}$ and the bottom merger receives the sequence $e = e_0, e_1, ..., e_{N/2-1}$. Then we have to prove that the first-stage SEs send the $N/2$ smallest elements of **a** to the top bitonic merger and the $N/2$ largest elements of **a** to the bottom bitonic merger, that is $max(d_0, ..., d_{N/2-1}) \leq min(e_0, ..., e_{N/2-1})$, and that the two sequences **d** and **e** are both circular bitonic.

Let us consider without loss of generality a bitonic sequence **a** in which an index k ($k \in [0, N-1]$) exists so that $a_0 \leq a_1 \leq ... \leq a_k$ and $a_k \geq a_{k+1} \geq ... \geq a_{N-1}$. In fact a circular bitonic sequence obtained from a bitonic sequence **a** by a circular shift of j positions ($0 < j < N-1$) simply causes the same circular shift of the two sequences **d** and **e** without affecting the property $max(d_0, ..., d_{N/2-1}) \leq min(e_0, ..., e_{N/2-1})$. Two cases must be distinguished:

- $a_i \leq a_{i+N/2}$: the SE sends a_i to the top merger and $a_{i+N/2}$ to the bottom merger ($d_i = a_i$, $e_i = a_{i+N/2}$). This behavior is correct for all the occurrences of the index k, given that the input sequence contains at least $N/2$ elements no smaller than a_i and at least $N/2$ elements no larger than $a_{i+N/2}$. In fact:
  — $i \leq k \leq i + N/2$: in this case $a_i \leq x$ for $x \in \{a_{i+1}, ..., a_{i+N/2}\}$, so there are at least $N/2$ elements no smaller than a_i; moreover $a_{i+N/2} \geq x$ for $x \in \{a_0, ..., a_i\}$ and $a_{i+N/2} \geq x$ for $x \in \{a_{i+N/2+1}, ..., a_{N-1}\}$, so there are at least $N/2$ elements no larger than $a_{i+N/2}$.
  — $i \leq i + N/2 \leq k$: again $a_i \leq x$ for $x \in \{a_{i+1}, ..., a_{i+N/2}\}$, so there are at least $N/2$ elements no smaller than a_i; moreover $a_{i+N/2} \geq x$ for $x \in \{a_i, ..., a_{i+N/2-1}\}$, so there are at least $N/2$ elements no larger than $a_{i+N/2}$.

- $a_i \geq a_{i+N/2}$: the SE sends a_i to the bottom merger and $a_{i+N/2}$ to the top merger ($d_i = a_{i+N/2}$, $e_i = a_i$). This behavior is correct for all the occurrences of the index k, given that the input sequence contains at least $N/2$ elements no larger than a_i and at least $N/2$ elements no smaller than $a_{i+N/2}$. In fact:
  — $i \leq k \leq i + N/2$: in this case $a_i \geq x$ for $x \in \{a_0, ..., a_{i-1}\}$ and $a_i \geq x$ for $x \in \{a_{i+N/2}, ..., a_{N-1}\}$, so there are at least $N/2$ elements no larger than a_i; moreover $a_{i+N/2} \leq x$ for $x \in \{a_i, ..., a_{i+N/2-1}\}$, so there are at least $N/2$ elements no smaller than $a_{i+N/2}$.
  — $k \leq i \leq i + N/2$: again $a_i \geq x$ for $x \in \{a_{i+1}, ..., a_{i+N/2}\}$, so there are at least $N/2$ elements no larger than a_i; moreover $a_{i+N/2} \leq x$ for $x \in \{a_i, ..., a_{i+N/2-1}\}$, so there are at least $N/2$ elements no smaller than $a_{i+N/2}$.

We now show that each of the two mergers receives a bitonic sequence. Let i be the largest index for which $a_i \leq a_{i+N/2}$. Then, the top merger receives the sequence $a_0, ..., a_i, a_{i+N/2+1}, ..., a_{N-1}$ and the bottom merger receives $a_{i+1}, ..., a_{i+N/2}$, that is two subsequences of the original bitonic sequence. Since each subsequence of a bitonic sequence is still bitonic, it follows that each merger receives a bitonic sequence.

2.5. References

[Bat68]   K.E. Batcher, "Sorting networks and their applications", *AFIPS Proc. of Spring Joint Computer Conference*, 1968, pp. 307-314.

[Bat76]   K. E. Batcher, "The flip network in STARAN", *Proc. of Int. Conf. on Parallel Processing*, Aug. 1976, pp. 65-71.

[Ben65]   V.E. Benes, *Mathematical Theory of Connecting Networks and Telephone Traffic*, Academic Press, New York, 1965.

[Dia81]   D.M. Dias, J.R. Jump, "Analysis and simulation of buffered delta networks", *IEEE Trans. on Comput.*, Vol. C-30, No. 4, Apr. 1981, pp. 273-282.

[Fel68]   W. Feller, *An Introduction to Probability Theory and Its Applications*, John Wiley & Sons, New York, 3rd ed., 1968.

[Gok73]   L.R. Goke, G.J. Lipovski, "Banyan networks for partitioning multiprocessor systems", *Proc. of First Symp. on Computer Architecture*, Dec. 1973, pp. 21-30.

[Knu73]   D.E. Knuth, *The Art of Computer Programming, Vol. 3: Sorting and Searching*, Addison-Wesley, Reading, MA, 1973.

[Kru86]   C.P. Kruskal, M. Snir, "A unified theory of interconnection networks", *Theoretical Computer Science*, Vol. 48, No. 1, pp. 75-94.

[Law75]   D.H. Lawrie, "Access and alignment of data in an array processor", *IEEE Trans. on Comput.*, Vol. C-24, No. 12, Dec. 1975, pp. 1145-1155.

[Pat81]   J.H. Patel, "Performance of processor-memory interconnections for multiprocessors", *IEEE Trans. on Comput.*, Vol C-30, Oct. 1981, No. 10, pp. 771-780.

[Pea77]   M.C. Pease, "The indirect binary n-cube microprocessor array", *IEEE Trans. on Computers*, Vol. C-26, No. 5, May 1977, pp. 458-473.

[Ric93]   G.W. Richards, "Theoretical aspects of multi-stage networks for broadband networks", Tutorial presentation at INFOCOM 93, San Francisco, Apr.-May 1993.

[Sie81]   H.J. Siegel, R.J. McMillen, "The multistage cube: a versatile interconnection network", *IEEE Comput.*, Vol. 14, No. 12, Dec. 1981, pp. 65-76.

[Sto71]   H.S. Stone, "Parallel processing with the perfect shuffle", *IEEE Trans on Computers*, Vol. C-20, No. 2, Feb. 1971, pp.153-161.

[Tur93]   J. Turner, "Design of local ATM networks", Tutorial presentation at INFOCOM 93, San Francisco, Apr.-May 1993.

[Wu80a]   C-L. Wu, T-Y. Feng, "On a class of multistage interconnection networks", *IEEE Trans. on Comput.*, Vol. C-29, No. 8, August 1980, pp. 694-702.

[Wu80b]   C-L. Wu, T-Y. Feng, "The reverse exchange interconnection network", *IEEE Trans. on Comput.*, Vol. C-29, No. 9, Sep. 1980, pp. 801-811.

2.6. Problems

2.1   Build a table analogous to Table 2.3 that provides the functional equivalence to generate the four basic and four reverse banyan networks starting now from the reverse of the four basic banyan networks, that is from reverse Omega, reverse SW-banyan, reverse n-cube and reverse Baseline.

2.2   Draw the network defined by $P(0) = P(n) = j, P(h) = \sigma^{n-h}$ $(1 \le h \le n-1)$ for a network with size $N = 8$; determine (a) if this network satisfies the construction rule of a banyan network (b) if the buddy property is satisfied at all stages (c) if it is a delta network, by determining the self-routing rule stage by stage.

2.3   Repeat Problem 2.2 for $N = 16$.

2.4   Repeat Problem 2.2 for a network that is defined by $P(0) = P(n) = j, P(1) = \beta_1$, $P(h) = \beta_{n-h+1}$ $(2 \le h \le n-1)$ for $N = 8$.

2.5   Repeat Problem 2.4 for $N = 16$.

2.6   Find the permutations $P(0)$ and $P(n)$ that enable an SW-banyan network to be obtained with $N = 16$ starting from the network $P(1) = \beta_1, P(h) = \beta_{n-h+1}$ $(2 \le h \le n-1)$.

2.7   Determine how many bitonic sorting networks of size $N = 16$ can be built (one is given in Figure 2.27) that generate an increasing output sequence considering that one network differs from the other if at least one sorting element in a given position is of different type (down-sorter, up-sorter) in the two networks.

2.8   Find the value of the stage latency τ in the Stone sorting network implemented by a single sorting stage such that the registers storing the packets cycle after cycle would no more be needed.

2.9   Determine the asymptotic ratio, that is for $N \to \infty$, between the cost of an odd–even sorting network and a Stone sorting network.

Chapter 3   *Rearrangeable Networks*

The class of rearrangeable networks is here described, that is those networks in which it is always possible to set up a new connection between an idle inlet and an idle outlet by adopting, if necessary, a rearrangement of the connections already set up. The class of rearrangeable networks will be presented starting from the basic properties discovered more than thirty years ago (consider the Slepian–Duguid network) and going through all the most recent findings on network rearrangeability mainly referred to banyan-based interconnection networks.

Section 3.1 describes three-stage rearrangeable networks with full-connection (FC) interstage pattern by providing also bounds on the number of connections to be rearranged. Networks with interstage partial-connection (PC) having the property of rearrangeability are investigated in Section 3.2. In particular two classes of rearrangeable networks are described in which the self-routing property is applied only in some stages or in all the network stages. Bounds on the network cost function are finally discussed in Section 3.3.

3.1. Full-connection Multistage Networks

In a two-stage FC network it makes no sense talking about rearrangeability, since each I/O connection between a network inlet and a network outlet can be set up in only one way (by engaging one of the links between the two matrices in the first and second stage terminating the involved network inlet and outlet). Therefore the rearrangeability condition in this kind of network is the same as for non-blocking networks.

Let us consider now a three-stage network, whose structure is shown in Figure 3.1. A very useful synthetic representation of the paths set up through the network is enabled by the matrix notation devised by M.C. Paull [Pau62]. A *Paull matrix* has r_1 rows and r_3 columns, as many as the number of matrices in the first and last stage, respectively (see Figure 3.2). The matrix entries are the symbols in the set $\{1, 2, ..., r_2\}$, each element of which represents one

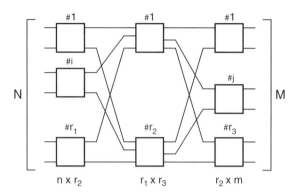

Figure 3.1. Three-stage FC network

of the middle-stage matrices. The symbol a in the matrix entry (i, j) means that an inlet of the first-stage matrix i is connected to an outlet of the last-stage matrix j through the middle-stage matrix a. The generic matrices i and j are also shown in Figure 3.1. Each matrix entry can contain from 0 up to r_2 distinct symbols; in the representation of Figure 3.2 a connection between i and j crosses matrix a and three connections between k and j are set up through matrices b, c and d.

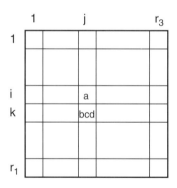

Figure 3.2. Paull matrix

Based on its definition, a Paull matrix always satisfies these conditions:

- each row contains at most $min(n, r_2)$ distinct symbols;
- each column contains at most $min(r_2, m)$ distinct symbols.

In fact, the number of connections through a first-stage (last-stage) matrix cannot exceed either the number of the matrix inlets (outlets), or the number of paths (equal to the number of middle-stage matrices) available to reach the network outlets (inlets). Furthermore, each symbol cannot appear more than once in a row or in a column, since only one link connects matrices of adjacent stages.

The most important theoretical result about three-stage rearrangeable networks is due to D. Slepian [Sle52] and A.M. Duguid [Dug59].

Slepian–Duguid theorem. A three-stage network is rearrangeable if and only if

$$r_2 \geq \max(n, m)$$

Proof. The original proof is quite lengthy and can be found in [Ben65]. Here we will follow a simpler approach based on the use of the Paull matrix [Ben65, Hui90]. We assume without loss of generality that the connection to be established is between an inlet of the first-stage matrix i and an outlet of the last-stage matrix j. At the call set-up time at most $n-1$ and $m-1$ connections are already supported by the matrices i and j, respectively. Therefore, if $r_2 > \max(n-1, m-1)$ at least one of the r_2 symbols is missing in row i and column j. Then at least one of the following two conditions of the Paull matrix holds:

1. There is a symbol, say a, that is not found in any entry of row i or column j.

2. There is a symbol in row i, say a, that is not found in column j and there is a symbol in column j, say b, that is not found in row i.

If Condition 1 holds, the new connection is set up through the middle-stage matrix a. Therefore a is written in the entry (i, j) of the Paull matrix and the established connections need not be rearranged. If only Condition 2 holds, the new connection $i-j$ can be set up only after rearranging some of the existing connections. This is accomplished by choosing arbitrarily one of the two symbols a and b, say a, and building a chain of symbols in this way (Figure 3.3a): the symbol b is searched in the same column, say j_2, in which the symbol a of row i appears. If this symbol b is found in row, say, i_3, then a symbol a is searched in this row. If such a symbol a is found in column, say j_4, a new symbol b is searched in this column. This chain construction continues as long as a symbol a or b is not found in the last column or row visited. At this point we can rearrange the connections identified by the chain $i, j_2, i_3, j_4, i_5, \ldots$ replacing symbol a with b in rows $i, i_3, i_5, \ldots$ and symbol b with symbol a in columns $j_2, j_4, \ldots$. By this approach symbols a and b still appear at most once in any row or column and symbol a no longer appears in row i. So, the new connection $i-j$ can be routed through the middle-stage matrix a (see Figure 3.3b).

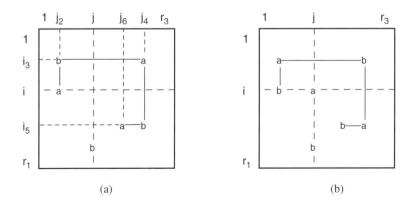

(a)   (b)

Figure 3.3. Connections rearrangement by the Paull matrix

This rearrangement algorithm works only if we can prove that the chain does not end on an entry of the Paull matrix belonging either to row i or to column j, which would make the rearrangement impossible. Let us represent the chain of symbols in the Paull matrix as a graph in which nodes represent first- and third-stage matrices, whereas edges represent second-stage matrices. The graphs associated with the two chains starting with symbols a and b are represented in Figure 3.4, where c and k denote the last matrix crossed by the chain in the second and first/third stage, respectively. Let "open (closed) chain" denote a chain in which the first and last node belong to a different (the same) stage. It is rather easy to verify that an open chain crosses the second stage matrices an odd number of times, whereas a closed chain makes it an even number of times. Hence, an open (closed) chain includes an odd (even) number of edges. We can prove now that in both chains of Figure 3.4 $k \neq i, j$. In fact if $c = a$, $k \neq j$ by assumption of Condition 2, and $k \neq i$ since $k \neq i$ would result in a closed chain C_1 with an odd number of edges or in an open chain C_2 with an even number of edges, which is impossible. Analogously, if $c = b$, $k \neq i$ by assumption of Condition 2 and $k \neq j$, since $k \neq j$ would result in an open chain C_1 with an even number of edges or in a closed chain C_2 with an odd number of edges, which is impossible.   ❏

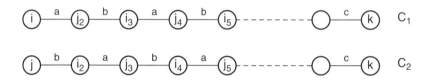

Figure 3.4. Chains of connections through matrices a and b

It is worth noting that in a squared three-stage network the Slepian–Duguid rule for a rearrangeable network becomes $r_2 = n$. The cost index C for a squared rearrangeable network ($N = M$, $n = m$, $r_1 = r_3$) is

$$C = 2nr_2r_1 + r_1^2r_2 = 2n^2r_1 + nr_1^2 = 2Nn + \frac{N^2}{n}$$

The network cost for a given N depends on the number n. By taking the first derivative of C with respects to n and setting it to 0, we find the condition providing the minimum cost network, that is

$$n = \sqrt{\frac{N}{2}} \tag{3.1}$$

Interestingly enough, Equation 3.1 that minimizes the cost of a three-stage rearrangeable network is numerically the same as Equation 4.2, representing the approximate condition for the cost minimization of a three-stage strict-sense non-blocking network. Applying Equation 3.1 to partition the N network inlets into r_1 groups gives the minimum cost of a three-stage RNB network:

$$C = 2\sqrt{2}N^{\frac{3}{2}} \tag{3.2}$$

Thus a Slepian–Duguid rearrangeable network has a cost index roughly half that of a Clos non-blocking network, but the former has the drawback of requiring in certain network states the rearrangement of some connections already set up.

From the above proof of rearrangeability of a Slepian–Duguid network, there follows this theorem:

Theorem. The number of rearrangements at each new connection set-up ranges up to

$$\varphi_M = 2\min(r_1, r_3) - 2.$$

Proof. Let (i_a, j_a) and (i_b, j_b) denote the two entries of symbols a and b in rows i and j, respectively, and, without loss of generality, let the rearrangement start with a. The chain will not contain any symbol in column j_b, since a new column is visited if it contains a, absent in j_b by assumption of Condition 2. Furthermore, the chain does not contain any symbol in row i_b since a new row is visited if it contains b but a second symbol b cannot appear in row i_b. Hence the chain visits at most $r_1 - 1$ rows and $r_3 - 1$ columns, with a maximum number of rearrangements equal to $r_1 + r_3 - 2$. Actually φ_M is only determined by the minimum between r_1 and r_3, since rows and columns are visited alternatively, thus providing $\varphi_M = 2\min(r_1, r_3) - 2$. ❏

Paull [Pau62] has shown that φ_M can be reduced in a squared network with $n_1 = r_1$ by applying a suitable rearrangement scheme and this result was later extended to networks with arbitrary values of r_1.

Paull theorem. The maximum number of connections to be rearranged in a Slepian–Duguid network is

$$\varphi_M = \min(r_1 r_3) - 1$$

Proof. Following the approach in [Hui90], let us assume first that $r_1 \geq r_3$, that is columns are less than rows in the Paull matrix. We build now two chains of symbols, one starting from symbol a in row i $(a, b, a, b, \ldots)$ and another starting from symbol b in column j $(b, a, b, a, \ldots)$. In the former case the chain $i, j_2, i_3, j_4, i_5, \ldots$ is obtained, whereas in the other case the chain is $j, i_2, j_3, i_4, j_5, \ldots$. These two chains are built by having them grow alternatively, so that the lengths of the two chains differ for at most one unit. When either of the two chains cannot grow further, that chain is selected to operate rearrangement. The number of growth steps is at most $r_3 - 2$, since at each step one column is visited by either of the two chains and the starting columns including the initial symbols a and b are not visited. Thus $\varphi_M = r_3 - 1$, as also the initial symbol of the chain needs to be exchanged. If we now assume that $r_1 \leq r_3$, the same argument is used to show that $\varphi_M = r_1 - 1$. Thus, in general no more than $\min(r_1 r_3) - 1$ rearrangements are required to set up any new connection request between an idle network inlet and an idle network outlet. ❏

The example of Figure 3.5 shows the Paull matrix for a three-stage network 24×25 with $r_1 = 4$ and $r_3 = 5$. The rearrangeability condition for the network requires $r_2 = 6$; let these matrices be denoted by the symbols $\{a, b, c, d, e, f\}$. In the network state represented by Figure 3.5a a new connection between the matrices 1 and 1 of the first and last stage is requested. The middle-stage matrices c and d are selected to operate the rearrangement according to Condition 2 of the Slepian–Duguid theorem (Condition 1 does not apply here). If the rearrangement procedure is based on only one chain and the starting symbol is c in row 1, the final state represented in Figure 3.5b is obtained (new connections are in italicized bold), with a total of 5 rearrangements. However, by applying the Paull theorem and thus generating two alternatively growing chains of symbols, we realize that the chain starting from symbol d in column 1 stops after the first step. So the corresponding total number of rearrangements is 2 (see Figure 3.5c). Note that we could have chosen the symbol e rather than d since both of them are missing in column 1. In this case only one connection would have been rearranged rather than two as previously required. Therefore minimizing the number of rearrangements in practical operations would also require to optimal selection of the pair of symbols in the Paull matrix, if more than one choice is possible, on which the connection rearrangement procedure will be performed.

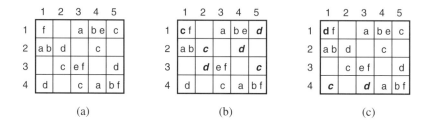

Figure 3.5. Example of application of the Paull matrix

In the following for the sake of simplicity we will assume a squared network, that is $N = M$, $n_1 = m_s = n$, unless specified otherwise.

3.2. Partial-connection Multistage Networks

We have shown that banyan networks, in spite of their blocking due to the availability of only one path per I/O pair, have the attractive feature of packet self-routing. Furthermore, it is possible to build rearrangeable PC networks by using banyan networks as basic building block. Thus RNB networks can be further classified as

- *partially self-routing*, if packet self-routing takes place only in a portion of this network;
- *fully self-routing*, if packet self-routing is applied at all network stages.

These two network classes will be examined separately in the next sections.

3.2.1. Partially self-routing PC networks

In a PC network with partial self-routing some stages apply the self-routing property, some others do not. This means that the processing required to set up the required network permutation is partially distributed (it takes place directly in the self-routing stages) and partially centralized (to determine the switching element state in all the other network stages).

Two basic techniques have been proposed [Lea91] to build a rearrangeable PC network with partial self-routing, both providing multiple paths between any couple of network inlet and outlet:

- *horizontal extension* (HE), when at least one stage is added to the basic banyan network.
- *vertical replication* (VR), when the whole banyan network is replicated several times;

Separate and joined application of these two techniques to build a rearrangeable network is now discussed.

3.2.1.1. Horizontal extension

A network built using the HE technique, referred to as *extended banyan network* (EBN), is obtained by means of the *mirror imaging* procedure [Lea91]. An EBN network of size $N \times N$ with $n + m$ stages ($m \leq n - 1$) is obtained by attaching to the first network stage of a banyan network m switching stages whose connection pattern is obtained as the mirror image of the permutations in the last m stage of the original banyan network. Figure 3.6 shows a 16×16 EBN SW-banyan network with $m = 1 - 3$ additional stages. Note that adding m stages means making available 2^m paths between any network inlet and outlet. Packet self-routing takes place in the last $n = \log_2 N$ stages, whereas a more complex centralized routing control is required in the first m stages. It is possible to show that by adding $m = n - 1$ stages to the original banyan network the EBN becomes rearrangeable if this latter network can be built recursively as a three-stage network.

A simple proof is reported here that applies to the $(2\log_2 N - 1)$-stage EBN network built starting from the recursive banyan topology SW-banyan. Such a proof relies on a property of permutations pointed out in [Ofm67]:

Ofman theorem. It is always possible to split an arbitrary permutation of size N into two subpermutations of size $N/2$ such that, if the permutation is to be set up by the $N \times N$ network of Figure 3.7, then the two subpermutations are set up by the two non-blocking $N/2 \times N/2$ central subnetworks and no conflicts occur at the first and last switching stage of the overall network.

This property can be clearly iterated to split each permutation of size $N/2$ into two subpermutations of size $N/4$ each set up by the non-blocking $N/4 \times N/4$ subnetworks of Figure 3.7 without conflicts at the SEs interfacing these subnetworks. Based on this property it becomes clear that the EBN becomes rearrangeable if we iterate the process until the "central" subnetworks have size 2×2 (our basic non-blocking building block). This result is obtained after $n - 1$ serial steps of decompositions of the original permutation that generate $N/2$ permutations of size $N/2^{n-1} = 2$. Thus the total number of stages of 2×2 switching elements becomes $2(n-1) + 1 = 2n - 1$, where the last unity represents the "central" 2×2 subnetworks (the resulting network is shown in Figure 3.6c). Note that the first and last stage of SEs are connected to the two central subnetworks of half size by the butterfly β_{n-1} pattern.

If the reverse Baseline topology is adopted as the starting banyan network to build the $(2\log_2 N - 1)$-stage EBN, the resulting network is referred to as a *Benes network* [Ben65]. It is interesting to note that a Benes network can be built recursively from a three-stage full-connection network: the initial structure of an $N \times N$ Benes network is a Slepian–Duguid network with $n_1 = m_3 = 2$. So we have $N/2$ matrices of size 2×2 in the first and third

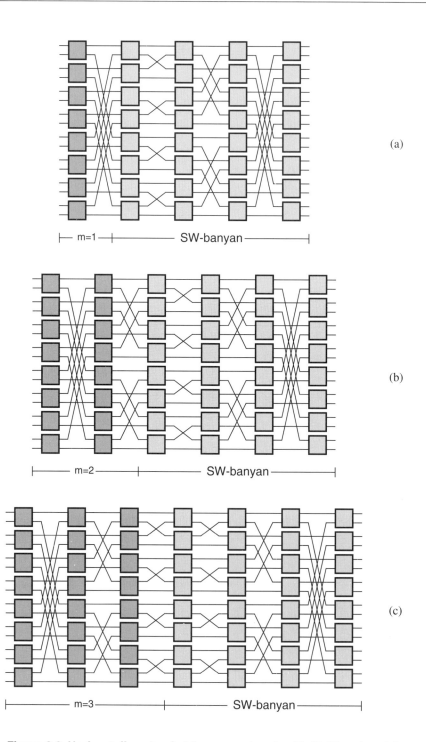

Figure 3.6. Horizontally extended banyan network with N=16 and m=1-3

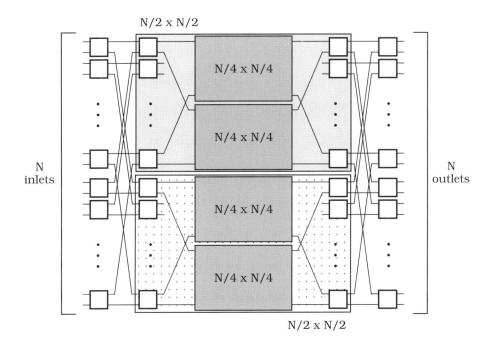

Figure 3.7. Recursive network construction for the Ofman theorem

stage and two $N/2 \times N/2$ matrices in the second stage interconnected by an EGS pattern that provides full connectivity between matrices in adjacent stages. Then each of the two $N/2 \times N/2$ matrices is again built as a three-stage structure of $N/4$ matrices of size 2×2 in the first and third stage and two $N/4 \times N/4$ matrices in the second stage. The procedure is iterated until the second stage matrices have size 2×2. The recursive construction of a 16×16 Benes network is shown in Figure 3.8, by shadowing the $B_{N/n}$ subnetworks ($n = 2, 4, 8$) recursively built.

The above proof of rearrangeability can be applied to the Benes network too. In fact, the recursive network used with the Ofman theorem would be now the same as in Figure 3.7 with the interstage pattern β_{n-1} replaced by σ_{n-1}^{-1} at the first stage and σ_{n-1} at the last stage. This variation would imply that the permutation of size N performed in the network of Figure 3.6c would give in the recursive construction of the Benes network the same setting of the first and last stage SEs but two different permutations of size $N/2$. The Benes network is thus rearrangeable since according to the Ofman theorem the recursive construction of Figure 3.7 performs any arbitrary permutation.

Thus an $N \times N$ Benes network has $s = 2\log_2 N - 1$ stages of 2×2 SEs, each stage including $N/2$ SEs. Therefore the number of its SEs is

$$S = N\log_2 N - \frac{N}{2}$$

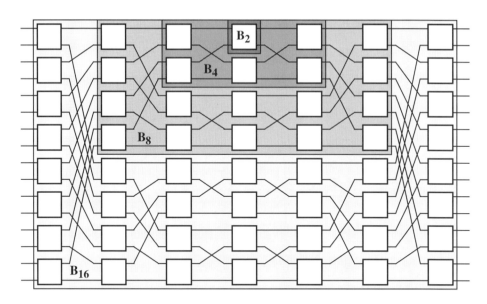

Figure 3.8. Benes network

with a cost index (each 2×2 SE accounts for 4 crosspoints)

$$C = 4N\log_2 N - 2N \qquad (3.3)$$

If the number of I/O connections required to be set up in an $N \times N$ network is N, the connection set is said to be *complete*, whereas an *incomplete* connection set denotes the case of less than N required connections (apparently, since each SE always assumes either the straight or the cross state, N I/O physical connections are always set up). The number of required connections is said to be the *size* of the connection set. The set-up of an incomplete/complete connection set through a $N \times N$ Benes network requires the identification of the states of all the switching elements crossed by the connections. This task is accomplished in a Benes network by the recursive application of a serial algorithm, known as a *looping algorithm* [Opf71], to the three-stage recursive Benes network structure, until the states of all the SEs crossed by at least one connection have been identified. The algorithm starts with a three-stage $N \times N$ network with first and last stage each including $N/2$ elements and two middle $N/2 \times N/2$ networks, called upper (U) and lower (L) subnetworks. By denoting with *busy* (*idle*) a network termination, either inlet or outlet, for which a connection has (has not) been requested, the looping algorithm consists of the following steps:

1. **Loop start**. In the first stage, select the unconnected busy inlet of an already connected element, otherwise select a busy inlet of an unconnected element; if no such inlet is found the algorithm ends.

2. **Forward connection**. Connect the selected network inlet to the requested network outlet through the only accessible subnetwork if the element is already connected to the other
   subnetwork, or through a randomly selected subnetwork if the element is not yet connected; if the other outlet of the element just reached is busy, select it and go to step 3; otherwise go to step 1.

3. **Backward connection**. Connect the selected outlet to the requested network inlet
   through the subnetwork not used in the forward connection; if the other inlet of the element just reached is busy and not yet connected, select it and go to step 2; otherwise go to
   step 1.

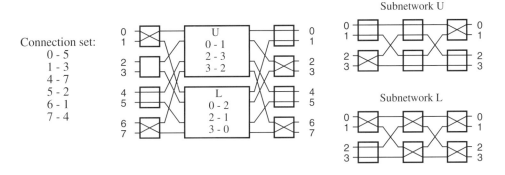

Figure 3.9. Example of application of the looping algorithm

Depending on the connection set that has been requested, several loops of forward and
backward connections can be started. Notice that each loop always starts with an unconnected
element in the case of a complete connection set. Once the algorithm ends, the result is the
identification of the SE state in the first and last stage and two connection sets of maximum
size $N/2$ to be set up in the two subnetworks $N/2 \times N/2$ by means of the looping algorithm. An example of application of the looping algorithm in an 8×8 network is represented
in Figure 3.9 for a connection set of size 6 (an SE with both idle terminations is drawn as
empty). The first application of the algorithm determines the setting of the SEs in the first and
last stage and two sets of three connections to be set up in each of the two central subnetworks
4×4. The looping algorithm is then applied in each of these subnetworks and the resulting
connections are also shown in Figure 3.9. By putting together the two steps of the looping
algorithm, the overall network state of Figure 3.10 is finally obtained. Parallel implementations
of the looping algorithm are also possible (see, e.g., [Hui90]), by allocating a processing capability to each switching element and thus requiring their complete interconnection for the
mutual cooperation in the application of the algorithm. We could say that the looping algorithm is a constructive proof of the Ofman theorem.

A further refining of the Benes structure is represented by the *Waksman network* [Wak68]
which consists in predefining the state of a SE (*preset* SE) per step of the recursive network
construction (see Figure 3.10 for $N = 16$). This network is still rearrangeable since, compared
to a Benes network, it just removes the freedom of choosing the upper or lower middle network for the first connection crossing the preset SE. Thus the above looping algorithm is now

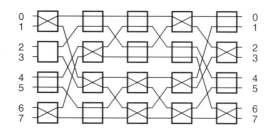

Figure 3.10. Overall Benes network resulting from the looping algorithm

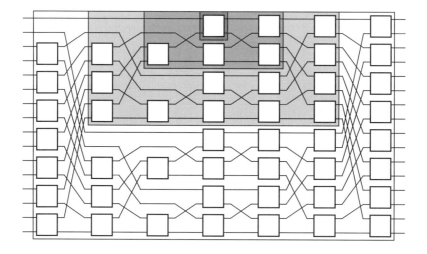

Figure 3.11. Waksman network

modified in the loop start, so that the busy inlets of the preset element must have already been connected before starting another loop from non-preset elements. Apparently, in the forward connection there is now no freedom of selecting the middle subnetwork when the loop starts from a preset element. Figure 3.12 gives the Waksman network state establishing the same connection set of size 6 as used in the Benes network of Figure 3.10.

It is rather easy to verify that the number of SEs in a Waksman network is

$$S = \left[\frac{N}{2}(2\log_2 N - 1) - \sum_{i=0}^{\log_2 N - 2} 2^i \right] = N\log_2 N - N + 1 \qquad (3.4)$$

with a cost index

$$C = 4N\log_2 N - 4N + 4 \qquad (3.5)$$

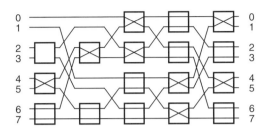

Figure 3.12. Overall Waksman network resulting from the looping algorithm

3.2.1.2. Vertical replication

By applying the VR technique the general scheme of a $N \times N$ *replicated banyan network* (RBN) is obtained (Figure 3.13). It includes N *splitters* $1 \times K$, K banyan networks $N \times N$ and N *combiners* $K \times 1$ connected through EGS patterns. In this network arrangement the packet self-routing takes place within each banyan network, whereas a more complex centralized control of the routing in the splitters has to take place so as to guarantee the rearrangeability condition.

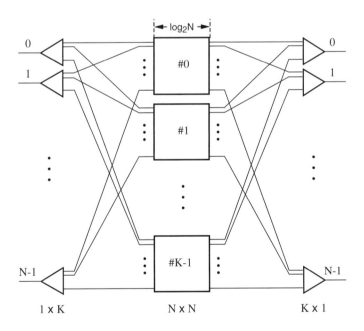

Figure 3.13. Vertically replicated banyan network

A rather simple reasoning to identify the number K of banyans that guarantees the network rearrangeability with the VR technique relies on the definition of the *utilization factor* (UF)

[Agr83] of the links in the banyan networks. The UF u_k of a generic link in stage k is defined as the maximum number of I/O paths that share the link. Then, it follows that the UF is given by the minimum between the number of network inlets that reach the link and the number of network outlets that can be reached from the link. Given the banyan topology, it is trivial to show that all the links in the same interstage pattern have the same UF factor u, e.g. in a 32×32 banyan network $u_k = 2, 4, 4, 2$ for the links of stage $k = 1, 2, 3, 4$, respectively (switching stages, as well as the interstage connection patterns following the stage, are always numbered 1 through $n = \log_2 N$ from the inlet side to the outlet side of the network).

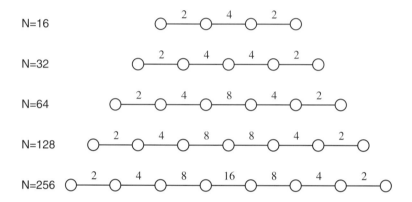

Figure 3.14. Utilization factor of a banyan network

Figure 3.14 represents the utilization factor of the banyan networks with $n = 4 - 8$, in which each node represents the generic SE of a stage and each branch models the generic link of an interstage connection pattern (the nodes terminating only one branch represent the SEs in the first and last stage). Thus, the maximum UF value of a network with size N is

$$u_{max} = 2^{\left\lfloor \frac{n}{2} \right\rfloor}$$

meaning that up to u_{max} I/O connections can be requiring a link at the "center" of the network. Therefore the following theorem holds.

Theorem. A replicated banyan network with size $N \times N$ including K planes is rearrangeable if and only if

$$K \geq u_{max} = 2^{\left\lfloor \frac{n}{2} \right\rfloor} \tag{3.6}$$

Proof. The necessity of Condition 3.6 is immediately explained considering that the u_{max} connections that share the same "central" network link must be routed onto u_{max} distinct banyan networks, as each banyan network provides a single path per I/O network pair.

The sufficiency of Condition 3.6 can be proven as follows, by relying on the proof given in [Lea90], which is based on graph coloring techniques. The n-stage banyan network can be seen as including a "first shell" of switching stages 1 and n, a "second shell" of switching stages 2 and $n-1$, and so on; the innermost shell is the $\lfloor n/2 \rfloor$-th. Let us represent in a graph the SEs of the first shell, shown by nodes, and the I/O connections by edges. Then, an arbitrary permutation can be shown in this graph by drawing an edge per connection between the left-side nodes and right-side nodes. In order to draw this graph, we define an algorithm that is a slight modification of the looping algorithm described in Section 3.2.1.1. Now a busy inlet can be connected to a busy outlet by drawing an edge between the two corresponding terminating nodes. Since we may need more than one edge between nodes, we say that edges of two colors can be used, say red and blue. Then the looping algorithm is modified saying that in the loop forward the connection is done by a red edge if the node is still unconnected, by a blue edge if the node is already connected; a red (blue) edge is selected in the backward connection if the right-side node has been reached by a blue (red) edge.

The application of this algorithm implies that only two colors are sufficient to draw the permutation so that no one node has two or more edges of the same color. In fact, the edges terminating at each node are at most two, since each SE interfaces two inlets or two outlets. Furthermore, on the right side the departing edge (if the node supports two connections) has always a color different from the arriving edge by construction. On the left side two cases must be distinguished: a red departing edge and a blue departing edge. In the former case the arriving edge, if any, must be blue since colors are alternated in the loop and the arriving edge is either the second, the fourth, the sixth, and so on in the chain. In the latter case the departing edge is by definition different from the already drawn edge, which is red since the node was initially unconnected. Since two colors are enough to build the graph, two parallel banyan planes, including the stages 1 and n of the first shell, are enough to build a network where no two inlets (outlets) share the same link outgoing from (terminating on) stage 1 (n). Each of these banyan networks is requested to set up at most $n/2$ connections whose specification depends on the topology of the selected banyan network.

This procedure can be repeated for the second shell, which includes stages 2 and $n-1$, thus proving the same property for the links outgoing from and terminating on these two stages. The procedure is iterated $\lfloor n/2 \rfloor$ times until no two I/O connections share any interstage link. Note that if n is even, at the last iteration step outgoing links from stage $n/2$ are the ingoing links of stage $n/2+1$. An example of this algorithm is shown in Figure 3.15 for a reverse Baseline network with $N = 16$; red edges are represented by thin line, blue edges by thick lines. For each I/O connection at the first shell the selected plane is also shown (plane I for red edges, plane II for blue connections). The overall network built with the looping algorithm is given in Figure 3.16. From this network a configuration identical to that of RBN given in Figure 3.13 can be obtained by "moving" the most central 1×2 splitters and 2×1 combiners to the network edges, "merging" them with the other splitters/combiners and replicating correspondingly the network stages being crossed. Therefore $1 \times K$ splitters and $K \times 1$ combiners are obtained with K replicated planes (as many as the number of "central" stages at the end of the looping algorithm), with K given by Equation 3.6. Therefore the sufficiency condition has been proven too.   ❑

The number of planes making a $N \times N$ RBN rearrangeable is shown in Table 3.1.

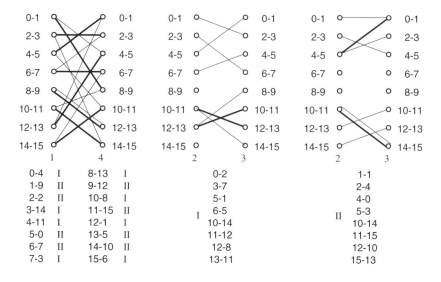

Figure 3.15. Example of algorithm to set up connections in a reverse Baseline network

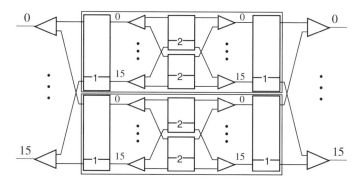

Figure 3.16. Overall RNB network resulting from *t* looping algorithm with *N*=16

By means of Equation 3.6 (each 2×2 SE accounts for 4 crosspoints), the cost index for a rearrangeable network is obtained:

$$C = 4 \cdot 2^{\left\lfloor \frac{n}{2} \right\rfloor} \frac{N}{2} \log_2 N + 2N \cdot 2^{\left\lfloor \frac{n}{2} \right\rfloor} = 2N 2^{\left\lfloor \frac{n}{2} \right\rfloor} (\log_2 N + 1) \qquad (3.7)$$

where the last term in the sum accounts for the cost of splitters and combiners.

Thus we can easily conclude that the HE technique is much more cost–effective than the VR technique in building a rearrangeable network $N \times N$: in the former case the cost grows with $N \log_2 N$ (Equations 3.3 and 3.5), but in the latter case with $N^{3/2} \log_2 N$ (Equation 3.7).

Table 3.1. Replication factor in a rearrangeable VR banyan network

N	K
8	2
16	4
32	4
64	8
128	8
256	16
512	16
1024	32

3.2.1.3. Vertical replication with horizontal extension

The VR and HE techniques can be jointly adapted to build a rearrangeable network [Lea91]: a non-rearrangeable EBN is first selected with $m < n - 1$ additional stages and then some parts of this network are vertically replicated so as to accomplish rearrangeability. As in the case of an EBN network, the initial banyan topology is either the SW-banyan or the reverse Baseline network, so that the horizontally extended network has a recursive construction and the looping algorithm for network rearrangeability can be applied. In particular, an EBN network $N \times N$ with m additional stages determines 2 "central" subnetworks of size $N/2 \times N/2$ of $n - 1$ stages, 4 "central" subnetworks of size $N/4 \times N/4$ of $n - 2$ stages, ..., 2^m "central" subnetworks of size $N/2^m \times N/2^m$ of $n - m$ stages.

Theorem. An extended banyan network with size $N \times N$ and m additional stages is rearrangeable if and only if each of the 2^m "central" subnetworks of size $2^{n-m} \times 2^{n-m}$ is replicated by a factor

$$K \geq 2^{\left\lfloor \frac{n-m}{2} \right\rfloor}$$

and N splitters $1 \times 2^{\lfloor (n-m)/2 \rfloor}$ and N combiners $2^{\lfloor (n-m)/2 \rfloor} \times 1$ are provided to provide full accessibility to the replicated networks. An example of such rearrangeable VR/HE banyan network is given in Figure 3.17 for $N = 16$, $m = 2$.

Proof. In a banyan network with m additional stages, applying recursively the Ofman theorem m times means that the original permutation of size N can always be split into 2^m subpermutations of size 2^{n-m} each to be set up by a $2^{n-m} \times 2^{n-m}$ network (in the example of Figure 3.17 we can easily identify these four 4×4 subnetworks). Then by applying the rearrangeability condition of a pure VR banyan network, each of these subnetworks has to be replicated

$$2^{\lfloor (n-m)/2 \rfloor}$$

times in order to obtain an overall rearrangeable network. ❏

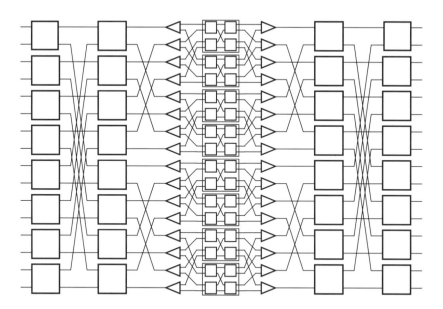

Figure 3.17. Vertical replication of an extended banyan network with *N*=16 and *m*=2

Table 3.2 gives the replication factor of a rearrangeable VR/HE banyan network with $n = 3 - 10$ and $m = 0 - 9$. Note that the diagonal with $m = n - 1$ gives the Benes network and the row $m = 0$ gives a pure VR rearrangeable network.

Table 3.2. Replication factor in a rearrangeable VR/HE banyan network

	N=8	16	32	64	128	256	512	1024
m=0	2	4	4	8	8	16	16	32
1	2	2	4	4	8	8	16	16
2	1	2	2	4	4	8	8	16
3		1	2	2	4	4	8	8
4			1	2	2	4	4	8
5				1	2	2	4	4
6					1	2	2	4
7						1	2	2
8							1	2
9								1

The cost function for a replicated-extended banyan network having m additional stages $(0 \leq m \leq n-1)$ is

$$C = 4 \left[\frac{N}{2} 2m + 2^m 2^{\left\lfloor \frac{n-m}{2} \right\rfloor} \frac{2^{n-m}}{2} (n-m) \right] + 2N 2^{\left\lfloor \frac{n-m}{2} \right\rfloor} = 4Nm + 2N(n-m+1) 2^{\left\lfloor \frac{n-m}{2} \right\rfloor}$$

In fact we have $2m$ "external" stages, interfacing through N splitters/combiners 2^m subnetworks with size $2^{n-m} \times 2^{n-m}$ and $n-m$ stages, each replicated $2^{\lfloor (n-m)/2 \rfloor}$ times. Note that such cost for $m = 0$ reduces to the cost of a rearrangeable RBN, whereas for the case of a rearrangeable EBN ($m = n-1$) the term representing the splitters/combiners cost has to be removed to obtain the cost of a Benes network.

Thus the joint adoption of VR and HE techniques gives a network whose cost is intermediate between the least expensive pure horizontally extended network ($m = n-1$) and the most expensive pure vertically replicated network ($m = 0$). Note however that, compared to the pure VR arrangement, the least expensive HE banyan network is less fault tolerant and has a higher frequency of rearrangements due to the blocking of a new call to be set up. It has been evaluated through computer simulation in [Lea91] that in a pure VR banyan network with $N = 16$ ($m = 0, K = 4$) the probability of rearrangement of existing calls at the time of a new call set-up is about two orders of magnitude greater than in a pure HE banyan network ($m = 3, K = 1$).

It is worth noting that an alternative network jointly adopting VR and HE technique can also be built by simply replicating $2^{\lfloor (n-m)/2 \rfloor}$ times the whole EBN including $n + m$ stages, by thus moving splitters and combiners to the network inlets and outlets, respectively. Nevertheless, such a structure, whose cost function is

$$C = 4 \left[\frac{N}{2} 2^{\left\lfloor \frac{n-m}{2} \right\rfloor} (n+m) \right] + 2N 2^{\left\lfloor \frac{n-m}{2} \right\rfloor} = 2N(n+m+1) 2^{\left\lfloor \frac{n-m}{2} \right\rfloor}$$

is more expensive than the preceding one.

3.2.1.4. Bounds on PC rearrangeable networks

Some results are now given to better understand the properties of rearrangeable networks. In particular attention is first paid to the necessary conditions that guarantee the rearrangeability of a network with arbitrary topology and afterwards to a rearrangeable network including only shuffle patterns.

A general condition on the number of stages in an arbitrary multistage rearrangeable network can be derived from an extension of the utilization factor concept introduced in Section 3.2.1.2 for the vertically replicated banyan networks. The channel graph is first drawn for the multistage networks for a generic I/O couple: all the paths joining the inlet to the outlet are drawn, where SEs (as well as splitters/combiners) and interstage links are mapped onto nodes and branches, respectively. A multistage squared network with regular topology is always assumed here, so that a unique channel graph is associated with the network, which hence does not depend on the specific inlet/outlet pair selected. Figure 3.18 shows the channel graph associated with an RBN with $K = 4$ (a), to an EBN with $m = 2$ (b), to a Benes network (c), and to a Shuffle-exchange network with $s = 7$ stages (d), all with size 16×16.

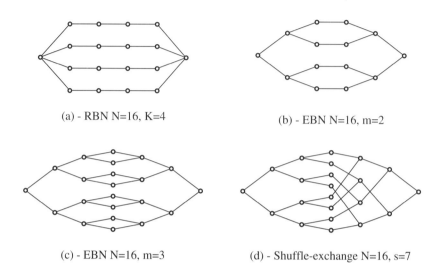

(a) - RBN N=16, K=4   (b) - EBN N=16, m=2

(c) - EBN N=16, m=3   (d) - Shuffle-exchange N=16, s=7

Figure 3.18. Channel graphs for networks with N=16

The *normalized utilization factor* (NUF) $U_k = u_k/r_k$ of the generic link in stage k of a multistage network is now introduced [Dec92] where r_k is the number of branches of stage k of the network channel graph and u_k indicates, as before, the maximum number of I/O paths that can share the link of stage k. Again, the network is assumed to have a regular topology so that all the links in a stage have the same UF and NUF. All the I/O paths in the channel graph include in general $s + c$ nodes, where s is the number of switching stages in the multistage network and c is the number of splitting/combining stages (if any). It follows that the number of links (branches) of an I/O path in the network (graph) is $s + c - 1$, numbered starting from 1.

Theorem. A multistage network with regular topology, that is with only one channel graph for all the I/O pairs, is rearrangeable only if

$$U_k \leq 1 \qquad (1 \leq k \leq s + c - 1) \tag{3.8}$$

Proof. The necessity of Condition 3.8 is immediately explained considering that if u_k inlets (outlets) share the same link at stage k, at least r_k distinct paths must be available at stage k to set up the u_k connections terminating the u_k inlets (outlets). Therefore the channel graph, which represents all the paths available to each I/O pair, must contain $r_k \geq u_k$ edges at stage k so that the u_k conflicting inlets can be routed on different interstage links. ❑

By looking again at the example of Figure 3.18a, representing the channel graph of an RBN with 4 banyan planes, it follows that $U_k \leq 1$ for each stage since $u_k = 1, 2, 4, 4, 2, 1$ and $r_k = 4, 4, 4, 4, 4, 4$, thus confirming rearrangeability of the network. In the case of Figure 3.18b, an EBN network with $m = n - 2$, we have $u_k = 2, 4, 8, 4, 2$ and $r_k = 2, 4, 4, 4, 2$, thus restating that the network is not rearrangeable since at stage 3 the NUF is $U_3 = 2$. It is easy to verify that the two networks whose channel graphs are shown in

Figure 3.18c and 3.18d satisfy the necessary condition for rearrangeability, since $U_k = 1$ for each stage k. Nevertheless, observe that the former network is rearrangeable as the looping algorithm guarantees that any arbitrary permutation can always be set up, whereas for the latter network nothing can be said as an analogous constructive algorithm is not known.

A lower bound on the number of SEs in a rearrangeable network is obtained using the theoretical argument by Shannon [Sha50]. In order for an $N \times N$ network with S two-state SEs to be rearrangeable non-blocking, the number of its states must be at least equal to the number of permutations of N addresses onto N positions, thus

$$2^{S_{min}} = N! \tag{3.9}$$

Note that an EBN supports multiple paths per I/O pair, so that two different network states can determine the same network permutation. Therefore Equation 3.9 only provides a lower bound on the number of SEs needed in a rearrangeable network. This number S_{min} of SEs is obtained, by means of Stirling's approximation of Equation 2.1, from

$$S_{min} = \log_2 N! \cong N\log_2 N - 1.443N + 0.5\log_2 N \tag{3.10}$$

Interestingly enough, the SE count of a Waksman network, given by Equation 3.4, is very close to this theoretical minimum.

Let us pay attention to EBN networks with all stages having the same number of SEs. If the network has s stages with $N/2$ SEs 2×2 per stage $(S = sN/2)$, Stirling's approximation gives a theoretical bound on the minimum number s_{min} of stages for a rearrangeable network:

$$s_{min} = \left\lceil \frac{S_{min}}{N/2} \right\rceil = 2\log_2 N - 2$$

which holds for $N \geq 4$ [Sov83].

This lower bound can be improved using the normalized utilization factor. In fact the maximum utilization factor of an EBN with m additional $(1 \leq m \leq n)$ stages is $2^{\lfloor (n+m)/2 \rfloor}$, whereas the number of I/O paths is 2^m, so that the necessary condition on the NUF requires that

$$2^{\left\lfloor \frac{n+m}{2} \right\rfloor} \leq 2^m$$

which is satisfied for $m \geq n - 1$. Therefore the lower bound on the number of network stages in a rearrangeable EBN becomes

$$s_{min} = 2\log_2 N - 1$$

Thus, a Benes network reaches the lower bound in network stages.

We have seen that each stage added to a banyan network implies doubling the paths available per I/O pair. The use of the mirror imaging procedure for the extension of a banyan network such that the resulting network has a recursive structure has enabled us to identify rearrangeable networks with number of elements very close to the theoretical minimum. If the overall target network is arranged to include only shuffle interstage patterns, which is called a

Shuffle-exchange network, the looping algorithm cannot be applied any more. Therefore in general more stages than in a Benes rearrangeable network are required in a shuffle-exchange network to provide rearrangeability.

The first rearrangeability condition about Shuffle-exchange networks can be derived directly from the study of sorting networks carried out in Section 2.3.2. We have shown that one of the possible implementations of a sorting network is the Stone structure [Sto71], which is the cascade of $\log_2^2 N$ shuffle stages interleaved by shuffle patterns. Therefore we can state that $\log_2 N$ is the number of shuffle stages to be crossed that is sufficient to guarantee network rearrangeability; the required permutation is set up directly by the sorting operations of each single stage. Later it was proven [Par80] that a Shuffle-exchange network with $3\log_2 N$ stages is rearrangeable, although no algorithm has been shown to set up the connection pattern through the network. A better result was later provided [Wu81] giving a sufficient condition for network rearrangeability through the following theorem.

Theorem. A Shuffle-exchange network with size $N \times N$ is rearrangeable if the number of its stages is $3\log_2 N - 1$.

Proof. Let us consider the network of Figure 3.19, obtained from a Benes network preceded by a permutation ρ. Since the Benes network is rearrangeable, so is such a modified Benes network since the added permutation simply changes the mapping between network state and performed permutation. This network is now modified by properly exchanging the position

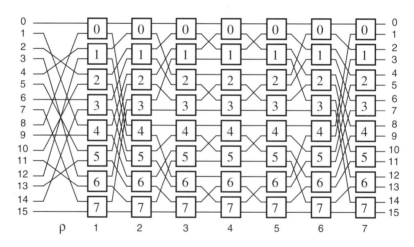

Figure 3.19. Modified Benes network

of SEs in all the stages other than the last one so as to obtain a new network that performs the same permutations and is therefore rearrangeable. We use the same notations introduced in Section 2.3.2.2 for the mapping of a bitonic sorting network into a Stone network, so that $r_i(x)$ denotes the row index of the SE in stage i in the old topology to be moved onto row x of the same stage in the new topology. The new rearrangeable topology is then given by the following mapping in the first n stages and in the last $n-1$ stages:

$$r_1(x_{n-1}...x_1) = x_1...x_{n-1}$$

$$r_{i+1}(x_{n-1}...x_1) = x_i...x_1 x_{i+1}...x_{n-1} \quad (1 \leq i \leq n-2)$$

$$r_{i+n}(x_{n-1}...x_1) = x_{i+1}...x_{n-1} x_i...x_1 \quad (1 \leq i \leq n-2)$$

$$r_{2n-1}(x_{n-1}...x_1) = r_n(x_{n-1}...x_1) = x_{n-1}...x_1$$

Therefore, the last two stages ($2n-2$ and $2n-1$) are kept unchanged, whereas the SEs in the first two stages are exchanged in the same way. The network of Figure 3.20 has thus been obtained, where each of the $2n-1$ stages is preceded by a shuffle permutation and a δ permutation separates the first n stages (an Omega network) from the last $n-1$ stages. It is

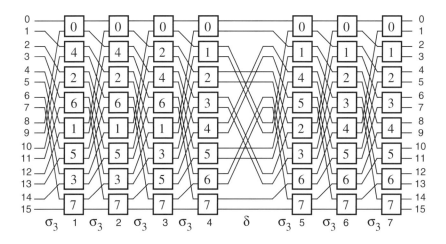

Figure 3.20. Rearrangeable network implementable by only shuffle patterns

proven in [Wu81] that the δ permutation can be realized by a series of two permutations W_1 and W_2 and that an Omega network performs not only the permutations realized by a sequence of an Omega network and a W_1 permutation but also a W_2 permutation. So the permutations realized by the cascade of an Omega network and a δ permutation are set up by the series of two Omega networks. Therefore, since an Omega network includes n stages, each preceded by a shuffle pattern, a total number of $n + n + (n-1) = 3n-1$ stages interleaved by a perfect shuffle gives a rearrangeable network. ❏

Such a Shuffle-exchange network is shown in Figure 3.21 for $N = 16$ and thus includes 11 stages (the shuffle pattern preceding the first switching stage has been omitted since it does not affect the network rearrangeability). A better result about Shuffle-exchange networks was obtained in [Hua86] showing that $3\log_2 N - 3$ stages are enough to provide network rearrangeability. The specific case of $N = 8$ was studied in [Rag87] for which five stages are shown to be sufficient for network rearrangeability. The corresponding algorithm to set up the permutation is also given. The previous bound on network stages for a network with arbitrary size was further enhanced in [Var88] by proving that $3\log_2 N - 4$ is the more stringent sufficient condition. All these bounds have been found based on approaches that relate these Shuffle-exchange networks to the recursive Benes network for which the looping algorithm provides the permutation set-up.

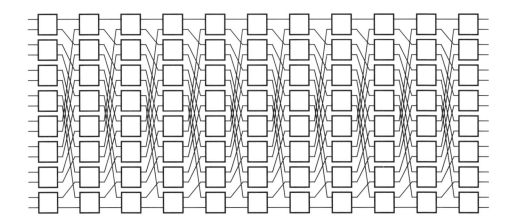

Figure 3.21. Rearrangeable shuffle-exchange network

3.2.2. Fully self-routing PC networks

A new class of partial-connection networks is now described in which the self-routing property is accomplished in all the network stages. Therefore, unlike the networks described in the previous section, here no centralized processing is required to set up a given network permutation, that is to route all the packets offered to the networks to each of the requested (different) destinations. The processing is distributed over all the network stages.

The basic concept to guarantee that a self-routing banyan network, although inherently blocking, never enters any blocking states is to offer to the network a configuration of packets whose respective addresses are such that blocking never occurs. We are going to prove that if the packets enter the banyan network as a cyclically compact and monotone sequence, to be defined next, then the banyan network is non-blocking[1]. The usefulness of the sorting networks described in Section 2.3.2 will become clear from the contiguity feature of the elements in the above sequences.

Let $x = x_0, ..., x_{N-1}$ represent a *sequence* of N elements in which $x_i = e$, i.e. element i is empty, or $x_i \in [0, N-1]$ and the number k of non-empty elements is said to be the *size* of the sequence. Each non-empty element of $\mathbf{x}$ appears at most once in the sequence. A sequence $\mathbf{x}$ of size k $(k \geq 2)$ is said to be *cyclically compact and monotone* (CCM) if and only if a *base m* exists such that $x_{(m+j) \bmod N}$ is non-empty for each $0 \leq j \leq k-1$ and

$$x_{(m+i) \bmod N} < x_{(m+j) \bmod N}$$

---

1. We are assuming that all packets are offered at the same time and take the same time to be transferred inside the network.

(*increasing CCM sequence*) or

$$x_{(m+i)\,\mathrm{mod}\,N} > x_{(m+j)\,\mathrm{mod}\,N}$$

(*decreasing CCM sequence*) for $i, j \in [0, k-1]$ and $i < j$. A subset of CCM sequences is given by the *compact and monotone sequences* (CM) of size k in which the base m is such that $m \le (m+k-1)\,\mathrm{mod}\,N$. For example, the 8-element sequences $X_1 = (5, 6, e, e, e, 0, 1, 3)$, $X_2 = (6, 5, 4, 3, 2, 1, 0, 7)$ and $X_3 = (e, e, e, e, e, 5, 4, 1)$ are an increasing CCM sequence of size 5 and base 5, a decreasing CCM sequence of size 8 and base 7, a decreasing CM sequence of size 3 and base 5, respectively.

We now look at the non-blocking conditions of a banyan network when it receives a set of packets represented by a sequence **x** in which $x_i = j$ and $x_i = e$ mean that inlet i receives a packet addressing outlet j ($j \in [0, N-1]$) and no packet, respectively. Since all the non-empty elements in the sequence are different by definition, then each outlet is addressed by at most one packet. The following theorem and its proof are reported in [Hui90], [Lee88] and [Kim92]. Other approaches to prove the same theorem can be found in [Nar88], [Fer93].

Theorem. An Omega network is non-blocking[1] for each set of input packets represented by a CCM sequence of arbitrary size $m \le N$.

Proof. The Omega network topology enables identification of the sequence of interstage links engaged by an inlet/outlet connection. Each of these links will be identified by the address of the SE outlet it originates from. Let us consider the inlet/outlet pair x–y with $x = x_{n-1}\ldots x_0$ and $y = y_{n-1}\ldots y_0$. Owing to the shuffle pattern preceding each stage, including the first one, the SE engaged at stage 1 by the network path is $x_{n-2}\ldots x_0$. At stage 1 the self-routing rule is accomplished using the output address bit y_{n-1}, so that the outlet address of the link engaged at stage 1 is $x_{n-2}\ldots x_0 y_{n-1}$. The same rule is applied for the second stage and the outlet address engaged has address $x_{n-3}\ldots x_0 y_{n-1} y_{n-2}$. The link engaged at stage k is therefore identified by the address $x_{n-k-1}\ldots x_0 y_{n-1}\ldots y_{n-k}$. Iteration of this rule n times provides the SE address at stage n $y = y_{n-1}\ldots y_0$, which is what we would expect. In general we could say that the link engaged at the generic stage k is obtained by taking a "window" of size n of the string $x_{n-1}\ldots x_0 y_{n-1}\ldots y_{n-1}$ starting from its $(k+1)$-th bit (the first n bit give the inlet address). An example of this rule is given in Figure 3.22 for the I/O pair 0101–0111.

The proof is first given for a CM sequence and later extended to a CCM one. Let a–b and a'–b' be two inlet/outlet pairs with $x = x_{n-1}\ldots x_0$ ($x = a, a', b, b'$) taken from a CM sequence; therefore $a' > a$. Without loss of generality, we set also $b' > b$. Proving the theorem for $b' < b$ simply require us to exchange b with b'. Let us suppose that these two connections share the output link at stage k, that is

$$a_{n-k-1}\ldots a_0 b_{n-1}\ldots b_{n-k} = a'_{n-k-1}\ldots a'_0 b'_{n-1}\ldots b'_{n-k} \tag{3.11}$$

Owing to the monotone and compact features of the CM sequence, it follows that

$$a' - a \le b' - b \tag{3.12}$$

---

1. This theorem states only the RNB condition since the connections are set up all together.

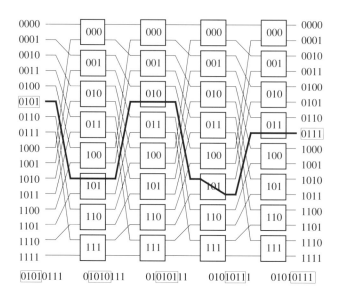

Figure 3.22. Interstage link labels in an Omega network

Equation 3.11 enables us to write the following inequalities:

$$a' - a = a'_{n-1} \ldots a'_0 - a_{n-1} \ldots a_0 = 2^{n-k}(a'_{n-1} \ldots a'_{n-k} - a_{n-1} \ldots a_{n-k}) \geq 2^{n-k} \qquad (3.13)$$

$$b' - b = b'_{n-1} \ldots b'_0 - b_{n-1} \ldots b_0 = b'_{n-k-1} \ldots b'_0 - b_{n-k-1} \ldots b_0 \leq 2^{n-k} - 1 \qquad (3.14)$$

which applied into Equation 3.12 give the result $0 \leq -1$. Therefore the assumption that a–b and a'–b' share the interstage link at stage k can never be verified and the two I/O paths are link independent, so proving the non-blocking condition of the Omega network for a CM sequence.

Let us consider now the case of a CCM sequence, in which $a' < a$. Now due to the cyclic compactness (modulo N) of the non-empty elements in the sequence, Equation 3.12 becomes

$$(a' - a) \bmod N \leq b' - b \qquad (3.15)$$

The inequality to be used now for the first member is

$$(a' - a) \bmod N = N + a' - a = 1a'_{n-1} \ldots a'_0 - 0a_{n-1} \ldots a_0$$

$$= 2^{n-k}(1a'_{n-1} \ldots a'_{n-k} - 0a_{n-1} \ldots a_{n-k}) \geq 2^{n-k} \qquad (3.16)$$

Equations 3.14 and 3.16 used in Equation 3.15 lead to the same inequality $0 \leq -1$ so that the non-blocking condition of the Omega network is finally proven for a CCM sequence. ❑

It is worth observing that this non-blocking condition applies also to the n-cube network which performs the same permutations as the Omega network.

Now we are able to construct a fully self-routing rearrangeable network, using the concept introduced so far. A sorting network $N \times N$ is able to sort up to N packets received in an arbitrary configuration and its output turns out to be a CM sequence, either increasing or decreasing, depending of the sorting order of the network. Thus, based on the preceding theorem, the cascade of a sorting network and an Omega network is a non-blocking structure, that based on the operation of the sorting elements and routing elements in the two networks is self-routing at all stages.

As far as the sorting network is concerned we will always refer to the Batcher bitonic sorting operations described in Section 2.3.2. In fact having the same "length" for all the I/O paths through the network is such an important feature for easing the packet alignment for comparison purposes in each stage that the non-minimum number of sorting elements of the structure becomes irrelevant. Figure 3.23 shows the block scheme of a non-blocking network, in which sorting and routing functions are implemented by a Batcher network and an n-cube (or Omega) network. This non-blocking network includes an overall number s of sorting/routing stages:

$$s = \frac{1}{2}\log_2 N (\log_2 N + 1) + \log_2 N = \frac{1}{2}\log_2^2 N + \frac{3}{2}\log_2 N$$

with a total number of sorting/switching elements equal to $s(N/2)$ and a cost index

$$C = N[\log_2^2 N + 3\log_2 N]$$

Thus a 32×32 Batcher–banyan network includes 20 stages with a total number of 320 elements, whereas 33,280 elements compose a $1,024 \times 1,024$ network distributed in 55 stages.

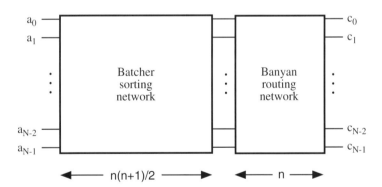

Figure 3.23. Depth of interconnection network based on sorting-routing

We point out that such a Batcher-banyan network guarantees the absence of _internal conflicts_ for interstage links between different I/O paths. Thus, in a packet environment we have to guarantee that the packet configuration offered to the rearrangeable network is free from _external conflicts_, that is each network outlet is addressed by at most one of the packets received at the network inputs. An example of switching operated by a Batcher-banyan network for $N = 8$ without external conflicts is given in Figure 3.24.

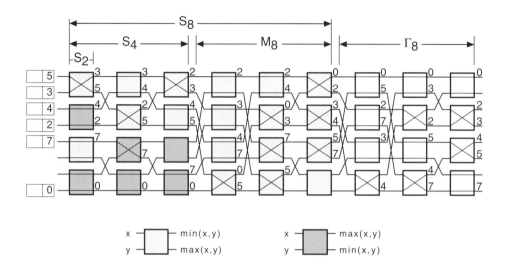

Figure 3.24. Switching example in a sorting-routing network

3.2.3. Fully self-routing PC networks with output multiplexing

An important role in ATM switching is played by those interconnection networks capable of transferring more than one packet in a slot from different switch inlets to the same switch outlet. From a network theory standpoint, we say that such networks are capable of setting up several connections from different network inlets to the same network outlet. Therefore such a network is said to be provided with an *output multiplexing* or *output speed-up* capability. A network outlet is said to have a capacity, or a speed-up factor, equal to K, if it can support up to K connections at the same time. A network outlet that supports 0 or K connections at a give time is said to be *idle* or *full*, whereas the outlet is said to be *busy* if the number of supported connections is less than K.

An interconnection network is defined as *rearrangeable K-non-blocking*, or *K-rearrangeable*, if it can always set-up a new connection between an idle network inlet and a non-full network outlet by eventually rearranging the existing connections. This property is clearly an extension of the traditional definition of a rearrangeable network, which is obtained from a K-rearrangeable network assuming $K = 1$. The general structure of a multistage K-rearrangeable $N \times N$ network[1] suitable to application in ATM networks is represented in Figure 3.25. It is an extension of the fully-self-routing rearrangeable network described in Section 3.2.2. It includes an $N \times N$ sorting network, followed by an $N \times NK$ routing network that interfaces N output multiplexers by means of N sets of K connections each. From a physical standpoint, a circuit-switching application of a K-rearrangeable network implies that the rate on the network outlets must be K times higher than the rate on the network inlets. In a packet switching

---

1. Unless stated otherwise, $N = 2^n$ $(n = 2, 3, ...)$, that is the network size is assumed to be a power of two.

environment, in which we are more interested in view of ATM applications, the rate on the network inlets and outlets is usually the same, given that we provide a queueing capability in the output multiplexers to store K packets that can be received at the same time.

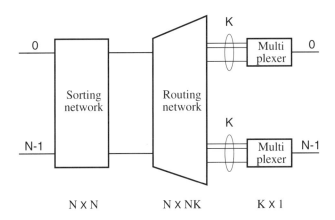

Figure 3.25. General structure of *K*-rearrangeable network

As in the case of rearrangeable networks without output speed-up, the sorting network is usually implemented as a Batcher bitonic sorting network. Unless specified otherwise, each packet is given a self-routing tag that is the juxtaposition of the network destination address (*DA*) and of a routing index (*RI*). Both fields *RI* and *DA* carry an integer, the former in the interval $[0, K-1]$ and coded by $k' = \log_2 K$ bits, the latter in the interval $[0, N-1]$ and coded by $n = \log_2 N$ bits. Packets with the same *DA* are given different RIs.

The most simple implementation (*implementation a*) of the routing network is an $NK' \times NK'$ banyan structure where $K' = 2^{\lceil \log_2 K \rceil}$ (Figure 3.26). So, when K is a power of 2, $K' = K$. In such a banyan network, only the first N out of NK' inlets are used. Owing to the EGS connection pattern between the banyan network outlets and the N multiplexers each with K inlets, only the first NK outlets of the routing network are used. Packet self-routing in this routing network can still be accomplished given that each packet is preceded by a self-routing tag (*RI,DA*), that is the juxtaposition of the routing index *RI* (the most significant bits of the routing tag) and destination address (or network outlet) *DA* (the least significant bits of the routing tag). The packets with the same *DA* are at most K and all carry a different *RI*. It is rather easy to verify that such cell tagging provides the desired conflict-free self-routing in the banyan network, given that it receives a cyclically compact and monotone (CCM) sequence of packets (see Section 3.2.2) that is guaranteed by the sorting network. Owing to the EGS pattern up to K packets can be received by the generic multiplexer i $i = 0, ..., N-1$, each packet carrying a different *RI* coupled with $DA = i$, and therefore the overall network is K-rearrangeable. The cost function of such a K-rearrangeable network is

$$C = 4\frac{Nn(n+1)}{2} + 4\frac{NK'}{2}(n+k') = N[\log_2^2 N + (2K'+1)\log_2 N + 2K'\log_2 K']$$

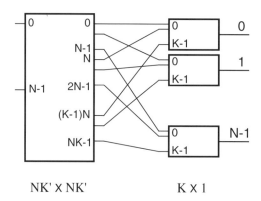

$$NK' \times NK' \qquad\qquad K \times 1$$

Figure 3.26. Implementation a of *K*-rearrangeable network

A different implementation (*implementation b*) of the routing network is shown in Figure 3.27, which basically differs from the previous solution in the pattern interconnecting the $NK' \times NK'$ banyan network to the N output multiplexers. Now the K adjacent outlets $K'i, \ldots K'i + K - 1$ of the banyan network are connected to the multiplexer i $(i = 0, \ldots, N-1)$. Thus now the $NK' - NK$ idle outlets of the banyan network are not adjacent. The packet self-routing is accomplished by associating a self-routing tag (DA, RI) to each packet, so that field DA now carries the most significant bits of the tag. Also in this case the self-routing property of this structure can be immediately verified as well as in K-rearrangeability. The importance of this solution lies in the network simplification enabled in the case $K' = K$, that is when the speed-up factor is a power of two. If the n–cube topology is selected for the banyan network, then it can be easily seen that the setting of the switching elements in the last k' stages is not important for a correct packet routing to the addressed multiplexers. Therefore, a new *truncated banyan network* is obtained by removing the last k' stages of SEs of the original network, by still maintaining the sequence of its interstage patterns. In the truncated banyan network the self-routing tag is simply the destination address DA (a formal proof can be found in [Lie89]). Adopting a truncated banyan network rather than a full banyan network simply means that the $n \leq K$ packets addressing an output multiplexer can be received in a different order on the K inlets of the multiplexer. The cost function of this K-rearrangeable network with truncated banyan is

$$C = 4\frac{Nn(n+1)}{2}\frac{}{2} + 4\frac{NK'}{2}n = N[\log_2^2 N + (2K' + 1)\log_2 N]$$

Adopting the truncated n-cube network is still possible also if $K < K'$. However in this case we always need K' connections to each output multiplexer, since the K packets addressing multiplexer i can emerge on any combination of K' outlets $K'i, \ldots, K'(i+1) - 1$ of the truncated banyan network.

Another implementation of the routing network (*implementation c*) is derived observing that the $N \times NK$ banyan network can be split into a set of smaller networks, for example with size

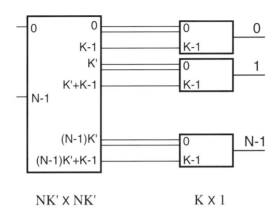

Figure 3.27. Implementation b of K-rearrangeable network

$N \times N$ each. The conceptual structure of an $N \times N$ interconnection network of Figure 3.28 is thus obtained. It includes a Batcher bitonic sorting network $N \times N$, K n-cube banyan networks $N \times N$, each feeding the N multiplexers, and N expanders $1 \times K$, the j-th of which interfaces the outlet j of the sorting network with the inlet j of the K banyan networks. It can be shown that if the expanders are implemented as binary trees with k' stages of 1×2 switching elements (analogously to the crossbar binary tree described in Section 2.1.2), then the interconnection network is K-rearrangeable and fully self-routing by means of a self-routing tag (RI,DA) preceding each packet. Field RI would be used in the k'-stage binary trees and field DA in the banyan networks. In fact the packets addressing the same banyan network emerge on adjacent sorting network outlets (they have the same RI) and their addresses are all different (packets with the same DA have been given different RIs). Therefore each banyan network will receive a set of packets on adjacent lines with increasing DAs, that is a CM sequence, by virtue of the operation performed by the upstream sorting network. The cost function of this implementation of a K-rearrangeable network is

$$C = 4\frac{Nn\,(n+1)}{2}\frac{}{2} + 2\,(K'-1)\,N + 4K\frac{N}{2}n = N\,[\log_2^2 N + (2K+1)\log_2 N + 2K' - 2]$$

The network of Figure 3.28 can be simplified for an arbitrary integer value of K into the structure of Figure 3.29 (*implementation c*), by replacing the expanders with an EGS pattern to interconnect the sorting network with the K banyan networks. In this example, K is assumed to be a power of two, so that the last $N - N/K$ inlets are idle in each banyan network. If $K \neq 2^i$ (i integer) then the idle inlets of the banyan networks are still the last ones, but their number is not the same in all the networks. In this implementation the self-routing tag of each packet only includes the field DA, so that at most K adjacent packets with the same DA can emerge from the sorting network.

Theorem. The multistage interconnection network of Figure 3.29 is K-rearrangeable.

Proof. For the proof we will refer to an ATM packet-switching environment in which the information units are the packets, even if the proof holds for circuit switching as well. In gen-

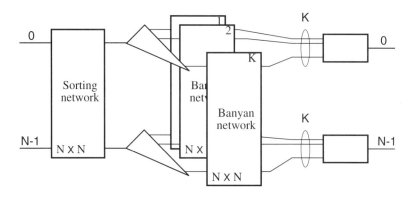

Figure 3.28. Implementation c of *K*-rearrangeable network

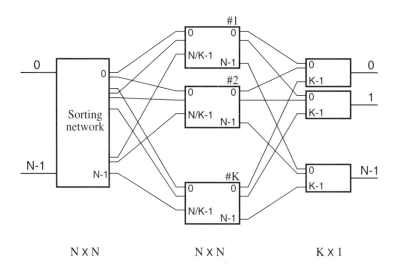

Figure 3.29. Implementation c' of *K*-rearrangeable network

eral the interconnection network is requested to switch $m \leq N$ packets, of which at most K address the same network outlet. Then, a packet sequence with monotonic non-decreasing addresses appear at the first m outlets of the sorting network. Owing to the topological property of the EGS pattern, these m packets are distributed cyclically to the K banyan networks, so that no two packets with the same address can enter the same banyan network and each banyan network receives on adjacent inlets a subsequence of the original m-packet sequence. Therefore each banyan network receives a CM (compact and monotone) sequence of packets. As already proved in Section 3.2.2, each of the K banyan networks is then non-blocking, that is free from internal conflicts, and thus the overall network is K-rearrangeable since it can switch up to K packets to the same network. ❑

For example, if $N = 8$, $K = 3$, the sequence of packets $(3,0,4,3,4,6,e,3)$ offered to the network (e means empty packet) is sorted as $(0,3,3,3,4,4,6,e)$ and the banyan networks #1, #2 and #3 receive the CM sequences $(0,3,6)$, $(3,4,e)$ and $(3,4)$, respectively. It is therefore clear that such implementation c' does not require any additional routing index other than the destination address DA to be fully self-routing. The cost function of this K-rearrangeable network is

$$C = 4\frac{Nn(n+1)}{2} \frac{1}{2} + 4K\frac{N}{2}n = N[\log_2^2 N + (2K+1)\log_2 N]$$

which is thus the smallest among the three different solutions presented.

3.3. Bounds on the Network Cost Function

The existence of upper bounds on the cost function of rearrangeable multistage networks is now investigated, where the network cost is again only determined by the number of crosspoints required in the $N \times N$ network (for simplicity only squared networks are considered with $N = 2^n$). We have already shown that the cost function is of the type $\alpha N(\log_2 N)^\beta$ with $\beta = 1$: in fact the cost function of both the Benes and Waksman networks is

$$C(N, N) \leq 4N\log_2 N + O(N)$$

Pippenger [Pip78] proved that using 3×3 basic matrices, rather than 2×2 matrices as in the Benes and Waksman network, gives a a slight reduction of the coefficient α with a cost function equal to

$$C(N, N) \leq 6N\log_3 N + O(N(\log_2 N)^{1/2}) = 3.79N\log_2 N + O(N(\log_2 N)^{1/2}) \quad (3.17)$$

where the equality $\log_b N = (\log_2 N)/(\log_2 b)$ has been used. The same asymptotic bound was earlier reported in [Ben65].

The cost function of various rearrangeable networks for a wide range of network sizes is shown in Figure 3.30. The cost of the Benes and Waksman networks is basically the smallest one for any network size and their value is practically the same as the bound provided by Pippenger (Equation 3.17). The three-stage Slepian–Duguid network is characterized by a cost very close to the minimum for small network sizes, say up to $N = 128$, whereas it becomes more expensive than the previous ones for larger network sizes. The Batcher-banyan network, which is even more expensive than the crossbar network for $N \leq 32$, has a lower cost than the Slepian–Duguid network for $N \geq 4096$.

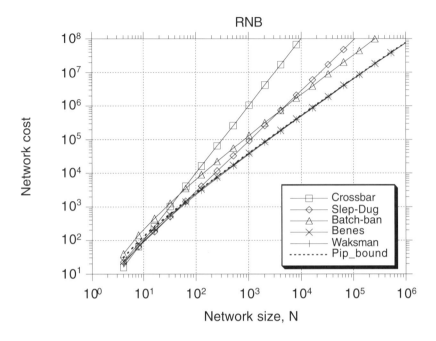

Figure 3.30. Cost function of rearrangeable networks

3.4. References

[Agr83]   D.P. Agrawal, "Graph theoretical analysis and design of multistage interconnection networks", *IEEE Trans. on Comput.*, Vol. C-32, No. 7, July 1983, pp. 637-648.

[Ben65]   V.E. Benes, *Mathematical Theory of Connecting Networks and Telephone Traffic*, Academic Press, New York, 1965.

[Dec92]   M. Decina, A. Pattavina, "High speed switching for ATM networks", *Tutorial presentation at INFOCOM 92*, Florence, Italy, May 1992.

[Dug59]   A.M. Duguid, "Structural properties of switching networks", *Brown Univ. Progr. Rept.*, BTL-7, 1959.

[Fer93]   G. Ferrari, M. Lenti, A. Pattavina, "Non-blocking conditions of self-routing multistage interconnection networks", *Int. J. of Digital and Analog Commun. Systems*, Vol. 6, No. 3, July-Sep. 1993, pp. 109-113.

[Hua86]   S.-T. Huang, S.K. Tripathi, "Finite state model and compatibility theory: new analysis tool for permutation networks", *IEEE Trans. on Comput.*, Vol. C-24, No. 12, Dec. 1975, pp. 1145-1155.

[Hui90]   J.Y. Hui, *Switching and Traffic Theory for Integrated Broadband Networks*, Kluwer Academic Press, Norwell, MA, 1990.

[Kim92]   H.S. Kim, A. Leon-Garcia, "Nonblocking property of reverse banyan networks", *IEEE Trans. on Commun.*, Vol. 40, No. 3, Mar. 1992, pp. 472-476.

[Law75]   D.H. Lawrie, "Access and alignment of data in an array processor", *IEEE Trans. on Comput.*, Vol. C-24, No. 12, Dec. 1975, pp. 1145-1155.

[Lea90]   C.-T. Lea, "Multi-log$_2$N networks and their applications in high-speed electronic and photonic switching systems, *IEEE Trans. on Commun.*, Vol. 38, No. 10, Oct. 1990, pp. 1740-1749.

[Lea91]   C.-T. Lea, D.-J. Shyy, "Tradeoff of horizontal decomposition versus vertical stacking in rearrangeable nonblocking networks", *IEEE Trans. on Commun.*, Vol. 39, No. 6, June 1991, pp. 899-904.

[Lee88]   T.T. Lee, "Nonblocking copy networks for multicast packet switching", *IEEE Trans. on Commun.*, Vol. 6, No. 9, Dec. 1988, pp. 1455-1467.

[Lie89]   S.C. Liew, K.W. Lu, "A three-stage architecture for very large packet switches", *Int. J. of Digital and Analog Cabled Systems*, Vol. 2, No. 4, Oct.-Dec. 1989, pp. 303-316.

[Nar88]   M.J. Narasimha, "The Batcher-banyan self-routing network: universality and simplification", *IEEE Trans. on Commun.*, Vol. 36, No. 10, Oct. 1988, pp. 1175-1178.

[Ofm67]   J.P. Ofman, "A universal automaton", *Trans. Moscow Mathematical Society*, Vol. 14, 1965; translation published by American Mathematical Society, Providence, RI, 1967, pp. 200-215.

[Opf71]   D.C. Opferman, N.T. Tsao-Wu, "On a class of rearrangeable switching networks - Part I: control algorithms", *Bell System Tech. J.*, Vol. 50, No 5, May-June 1971, pp. 1579-1600.

[Par80]   D.S. Parker, "Notes on shuffle/exchange-type switching networks", *IEEE Trans. on Comput.*, Vol C-29, Mar. 1980, No. 3, pp. 213-222.

[Pau62]   M.C. Paull, "Reswitching of connection networks", *Bell System Tech. J.*, Vol. 41, No. 3, May 1962, pp. 833-855.

[Pip78]   N. Pippenger, "On rearrangeable and non-blocking switching networks", *J. of Comput. and System Science*, Vol. 17, No. 4, Sept. 1978, pp.145-162.

[Rag87]   C.S. Raghavendra, A. Varma, "Rearrangeability of the five-stage shuffle/exchange network for N=8", *IEEE Trans. on Commun.*, Vol. 35, No. 8, Aug. 198, pp. 808-812.

[Sha50]   C.E. Shannon, "Memory requirements in a telephone exchange", *Bell System Tech. J.*, Vol. 29, July 1950, pp. 343-349.

[Sle52]   D. Slepian, "Two theorems on a particular crossbar switching network", unpublished manuscript, 1952.

[Sov83]   F. Sovis, "Uniform theory of the shuffle-exchange type permutation networks", *Proc. of 10-th Annual Symp. on Comput. Architecture*, 1983, pp.185-191.

[Sto71]   H.S. Stone, "Parallel processing with the perfect shuffle", *IEEE Trans on Comput.*, Vol. C-20, No. 2, Feb. 1971, pp.153-161.

[Var88]   A. Varma, C.S. Raghavendra, "Rearrangeability of multistage shuffle/exchange networks", *IEEE Trans. on Commun.*, Vol. 36, No. 10, Oct. 1988, pp. 1138-1147.

[Wak68]   A. Waksman, "A permutation network", *J. of ACM*, Vol. 15, No. 1, Jan, 1968, pp. 159-163.

[Wu81]   C-L. Wu, T-Y. Feng, "The universality of the shuffle-exchange network", *IEEE Trans. on Comput.*, Vol. C-30, No. 5, May 1981, pp. 324-332.

3.5. Problems

3.1   Without relying on the formal arguments shown in the proof of the Slepian–Duguid theorem, prove by simply using Figure 3.3 that the chain of symbols in the Paull matrix never ends on row i and column j.

3.2   By relying on the banyan network properties, prove that a Baseline EBN with $n-1$ additional stages is rearrangeable.

3.3   Find a network state that sets up the permutation 0-4, 1-12, 2-5, 3-8, 4-13, 5-0, 6-6, 7-15, 8-1, 9-7, 10-10, 11-2, 12-14, 13-3, 14-9, 15-11 in a Benes network with $N = 16$; determine the number of different network states that realize the requested permutation.

3.4   Repeat Problem 3.3 for a Waksman network.

3.5   Count the number of different network states that enable to set up the incomplete connection set $i - (i + 8)$ for $i = 0, 1, \ldots, 7$ in a Benes network with $N = 16$.

3.6   Repeat Problem 3.5 for a Waksman network.

3.7   Compute the cost of a rearrangeable VR/HE banyan network with m additional stages $(m \leq n - 1)$ in which the replication factor $2^{\lfloor (n-m)/2 \rfloor}$ is applied to the central subnetworks of size $N/2^{m-h} \times N/2^{m-h}$ with $0 \leq h \leq m$.

3.8   Draw the channel graph for an RBN with $N = 32$ and $K = 6$ as well as for a VR/HE banyan network with $N = 32$, $m = 1$ and $K = 4$ by determining whether the necessary condition for network rearrangeability based on the NUF parameter is satisfied.

3.9   Provide an intuitive explanation based on the channel graph and associated NUF values for the minor asymptotic cost of a rearrangeable network based on the HE technique rather than on the VR technique.

3.10   Prove the non-blocking condition of an n-cube network for a CCM sequence of size N using the bitonic merging principle. Extend the proof to CCM sequences of arbitrary size $m \leq N$.

3.11   Draw a four-stage interconnection network with interstage FC pattern in which the first stage includes 25 splitters 1×4 and the last stage 16 combiners 3×1 by determining if such network is rearrangeable.

Chapter 4   *Non-blocking Networks*

The class of strict-sense non-blocking networks is here investigated, that is those networks in which it is always possible to set up a new connection between an idle inlet and an idle outlet independently of the network permutation at the set-up time. As with rearrangeable networks described in Chapter 3, the class of non-blocking networks will be described starting from the basic properties discovered more than thirty years ago (consider the Clos network) and going through all the most recent findings on network non-blocking mainly referred to banyan-based interconnection networks.

Section 4.1 describes three-stage non-blocking networks with interstage full connection (FC) and the recursive application of this principle to building non-blocking networks with an odd number of stages. Networks with partial connection (PC) having the property of non-blocking are investigated in Section 4.2, whereas Section 4.3 provides a comparison of the different structures of partially connected non-blocking networks. Bounds on the network cost function are finally discussed in Section 4.4.

4.1. Full-connection Multistage Networks

We investigate here how the basic FC network including two or three stages of small crossbar matrices can be made non-blocking. The study is then extended to networks built by recursive construction and thus including more than three stages.

4.1.1. Two-stage network

The model of two-stage FC network, represented in Figure 2.11, includes r_1 matrices $n \times r_2$ at the first stage and r_2 matrices $r_1 \times m$ at the second stage. This network clearly has full accessibility, but is blocking at the same time. In fact, if we select a couple of arbitrary matrices at the first and second stage, say A_i and B_j, no more than one connection between the n inlets of A_i and the m outlets of B_j can be set up at a given time. Since this limit is due to the single link between matrices, a non-blocking two-stage full-connection network is then easily obtained by properly "dilating" the interstage connection pattern, that is by providing d links between any couple of matrices in the two stages (Figure 4.1). Also such an FC network is a subcase of an EGS network with $m_i = dr_{i+1}$ $(i = 1, ..., s-1)$. The minimum link dilation factor required in a non-blocking network is simply given by

$$d = \min(n, m)$$

since no more than $\min(n, m)$ connections can be set up between A_i and B_j at the same time. The network cost for a two-stage non-blocking network is apparently d times the cost of a non-dilated two-stage network. In the case of a squared network $(N = M, \ n = m, \ r_1 = r_2 = r)$ using the relation $N = rn$, we obtain a cost index

$$C = ndr_2 r_1 + dr_1 mr_2 = 2n^2 r^2 = 2N^2$$

that is the two-stage non-blocking network doubles the crossbar network cost.

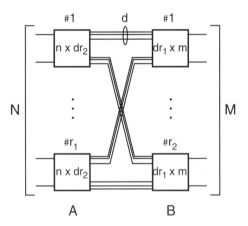

Figure 4.1. FC two-stage dilated network

Thus, the feature of smaller matrices in a two-stage non-blocking FC network compared to a single crossbar network is paid by doubling the cost index, independent of the value selected for the parameter n.

4.1.2. Three-stage network

The general scheme of a three-stage network is given in Figure 4.2, in which, as usual, n and m denote the number of inlets and outlets of the first- (A) and third- (C) stage matrices, respectively. Adopting three stages in a multistage network, compared to a two-stage arrangement, introduces a very important feature: different I/O paths are available between any couple of matrices A_i and C_j each engaging a different matrix in the second stage (B). Two I/

O paths can share interstage links, i.e. when the two inlets (outlets) belong to the same A (C) matrix. So, a suitable control algorithm for the network is required in order to set up the I/O path for the new connections, so as not to affect the I/O connections already established.

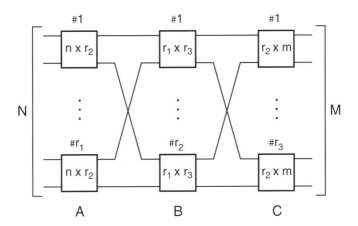

Figure 4.2. FC three-stage network

Full accessibility is implicitly guaranteed by the full connection of the two interstage patterns. Thus, $r_2 = 1$ formally describes the full accessibility condition.

The most general result about three-stage non-blocking FC networks with arbitrary values for $n_1 = n$ and $m_3 = m$ is due to C. Clos [Clo53].

Clos theorem. A three-stage network (Figure 4.2) is strict-sense non-blocking if and only if

$$r_2 \geq n + m - 1 \tag{4.1}$$

Proof. Let us consider two tagged matrices in the first (A) and last (C) stage with maximum occupancy but still allowing the set up of a new connection. So, $n - 1$ and $m - 1$ connections are already set up in the matrices A and C, respectively, and one additional connection has to be set up between the last idle input and output in the two tagged matrices (see Figure 4.3). The worst network loading condition corresponds to assuming an engagement pattern of the second stage matrices for these $(n - 1) + (m - 1)$ paths such that the $n - 1$ second-stage matrices supporting the $n - 1$ connections through matrix A are different from the $m - 1$ second-stage matrices supporting the $m - 1$ connections through matrix C. This also means that the no connection is set up between matrices A and C. Since one additional second-stage matrix is needed to set up the required new connection, $(n - 1) + (m - 1) + 1 = n + m - 1$ matrices in the second stage are necessary to make the three-stage network strictly non-blocking. ❑

The cost index C a squared non-blocking network ($N = M$, $n = m$, $r_1 = r_3$) is

$$C = 2nr_1r_2 + r_1^2 r_2 = 2nr_1(2n - 1) + r_1^2(2n - 1) = (2n - 1)\left(2N + \frac{N^2}{n^2}\right)$$

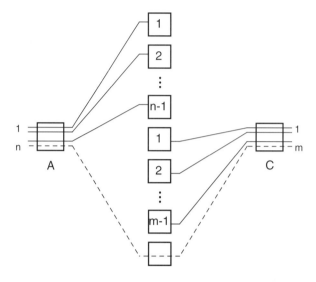

Figure 4.3. Worst case occupancy in a three-stage network

The network cost for a given N thus depends on the number r_1 of first-stage matrices, that is on the number of inlets per first-stage matrix since $N = nr_1$. By taking the first derivative of C with respect to n and setting it to 0, we easily find the solution

$$n \cong \sqrt{\frac{N}{2}} \tag{4.2}$$

which thus provides the minimum cost of the three-stage SNB network, i.e.

$$C = 4\sqrt{2}N^{\frac{3}{2}} - 4N \tag{4.3}$$

Unlike a two-stage network, a three-stage SNB network can become cheaper than a crossbar (one-stage) network. This event occurs for a minimum cost three-stage network when the number of network inlets N satisfies the condition $\sqrt{N} > 2 + 2\sqrt{2}$ (as is easily obtained by equating the cost of the two networks). Interestingly enough, even only $N = 24$ inlets are enough to have a three-stage network cheaper than the crossbar one. By comparing Equations 4.3 and 3.2, giving the cost of an SNB and RNB three-stage network respectively, it is noted that the cost of a non-blocking network is about twice that of a rearrangeable network.

4.1.3. Recursive network construction

Networks with more than three stages can be built by iterating the basic three stage construction. Clos showed [Clo53] that a five-stage strict-sense non-blocking network can be recursively built starting from the basic three-stage non-blocking network by designing each matrix of the second-stage as a non-blocking three-stage network. The recursion, which can be repeated an arbitrary number of times to generate networks with an odd number of stages s, enables the construction of networks that become less expensive when N grows beyond certain thresholds (see [Clo53]). Nevertheless, note that such new networks with an odd number of stages $s \geq 5$ are no longer connected multistage networks. In general a squared network (that is specular across the central stage) with an odd number of stages s $(s \geq 3)$ requires $(s-1)/2$ parameters to be specified that is

$$n_1 = m_s, \ n_2 = m_{s-1}, \ ..., \ n_{(s-1)/2} = m_{(s+3)/2}$$

(recall that according to the basic Clos rule $m_i = 2n_i - 1$ $(i = 1, ..., (s-1)/2)$. For a five-stage network $(s = 5)$ the optimum choice of the two parameters n_1, n_2 can be determined again by computing the total network cost and by taking its first derivative with respect to n_1 and n_2 and setting it to 0. Thus the two conditions

$$N = \frac{2n_1 n_2^3}{n_2 - 1}$$

$$N = \frac{n_1 n_2^2 (2n_1^2 + 2n_2 - 1)}{(2n_2 - 1)(n_1 - 1)}$$

(4.4)

are obtained from which n_1 and n_2 are computed for a given N.

Since such a procedure is hardly expandable to larger values of s, Clos also suggested a recursive general dimensioning procedure that starts from a three-stage structure and then according to the Clos rule (Equation 4.1) expands each middle-stage matrix into a three-stage structure and so on. This structure does not minimize the network cost but requires just one condition to be specified, that is the parameter n_1, which is set to

$$n_1 = N^{\frac{2}{s+1}}$$

(4.5)

The cost index of the basic three-stage network built using Equation 4.5 is

$$C_3 = (2\sqrt{N} - 1)3N = 6N^{\frac{3}{2}} - 3N$$

(4.6)

The cost index of a five-stage network (see Figure 4.4) is readily obtained considering that $n_1 = N^{1/3}$, so that each of the $2n_1 - 1$ three-stage central blocks has a size $N^{2/3} \times N^{2/3}$ and thus a cost given by Equation 4.6 with N replaced by $N^{2/3}$. So, considering the additional cost of the first and last stage the total network cost is

$$C_5 = \left(2N^{\frac{1}{3}} - 1\right)^2 3N^{\frac{2}{3}} + \left(2N^{\frac{1}{3}} - 1\right)2N = 16N^{\frac{4}{3}} - 14N + 3N^{\frac{2}{3}}$$

(4.7)

Again, a seven-stage network is obtained considering that $n_1 = N^{1/4}$ so that each of the $2n_1 - 1$ five-stage central blocks has a size $N^{3/4} \times N^{3/4}$ and thus a cost index given by

Equation 4.7 with N replaced by $N^{3/4}$. So, considering the additional cost of the first and last stage the total network cost is

$$C_7 = \left(2N^{\frac{1}{4}} - 1\right)^3 3N^{\frac{1}{2}} + \left(2N^{\frac{1}{4}} - 1\right)^2 2N^{\frac{3}{4}} + \left(2N^{\frac{1}{4}} - 1\right)2N \qquad (4.8)$$

$$= 36N^{\frac{5}{4}} - 46N + 20N^{\frac{3}{4}} - 3N^{\frac{1}{2}}$$

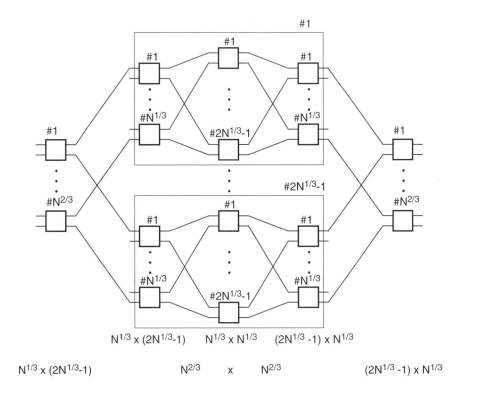

Figure 4.4. Recursive five-stage Clos network

This procedure can be iterated to build an s-stage recursive Clos network (s odd) whose cost index can be shown to be

$$C_s = 2 \sum_{k=2}^{\frac{s+1}{2}} \left(2N^{\frac{2}{s+1}} - 1\right)^{\frac{s+3}{2} - k} N^{\frac{2k}{s+1}} + \left(2N^{\frac{2}{s+1}} - 1\right)^{\frac{s-1}{2}} N^{\frac{4}{s+1}} \qquad (4.9)$$

which reduces to [Clo53]

$$C_{2t+1} = \frac{n^2(2n-1)}{n}[(5n-3)(2n-1)^{t-1} - 2n^t]$$

with $s = 2t+1$ and $N = n^{t+1}$. An example of application of this procedure for some values of network size N with a number of stages ranging from 1 to 9 gives the network costs of Table 4.1. It is observed that it becomes more convenient to have more stages as the network size increases.

Table 4.1. Cost of the recursive Clos s-stage network

N	$s = 1$	$s = 3$	$s = 5$	$s = 7$	$s = 9$
100	10,000	5,700	6,092	7,386	9,121
200	40,000	16,370	16,017	18,898	23,219
500	250,000	65,582	56,685	64,165	78,058
1,000	1,000,000	186,737	146,300	159,904	192,571
2,000	4,000,000	530,656	375,651	395,340	470,292
5,000	25,000,000	2,106,320	1,298,858	1,295,294	1,511,331
10,000	100,000,000	5,970,000	3,308,487	3,159,700	3,625,165

As already mentioned, there is no known analytical solution to obtain the minimum cost network for arbitrary values of N; moreover, even with small networks for which three or five stages give the optimal configuration, some approximations must be introduced to have integer values of n_i. By means of exhaustive searching techniques the minimum cost network can be found, whose results for some values of N are given in Table 4.2 [Mar77]. The minimum cost network specified in this table has the same number of stages as the minimum-cost network with (almost) equal size built with the recursive Clos rule (see Table 4.1). However the former network has a lower cost since it optimizes the choice of the parameters n_i. For example, the five-stage recursive Clos network with $N = 1000$ has $n_1 = n_2 = 10$ (see Figure 4.4), whereas the minimum-cost network with $N = 1001$ has $n_1 = 11$, $n_2 = 7$.

Table 4.2. Minimum cost network by exhaustive search

N	s	n_1	n_2	n_3	C_s
100	3	5			5,400
500	5	10	5		53,200
1001	5	11	7		137,865
5,005	7	13	7	5	1,176,175
10,000	7	20	10	5	2,854,800

4.2. Partial-connection Multistage Networks

A partial-connection multistage network can be built starting from four basic techniques:

- *vertical replication* (VR) of banyan networks, in which several copies of a banyan network are used;

- *vertical replication coupled with horizontal extension* (VR/HE), in which the single planes to be replicated include more stages than in a basic banyan network;

- *link dilation* (LD) of a banyan network, in which the interstage links are replicated a certain number of times;

- *EGS network*, in which the network is simply built as a cascade of EGS permutations.

In general, the control of strict-sense non-blocking networks requires a centralized control based on a storage structure that keeps a map of all the established I/O paths and makes it possible to find a cascade of idle links through the network for each new connection request between an idle inlet and an idle outlet.

4.2.1. Vertical replication

Let us first consider the adoption of the pure VR technique that results in the overall replicated banyan network (RBN) already described in the preceding section (see also Figure 3.13). The procedure to find out the number K of networks that makes the RBN strict-sense non-blocking must take into account the worst case of link occupation considering that now calls cannot be rearranged once set up [Lea90].

Theorem. A replicated banyan network of size $N \times N$ with K planes is strict-sense non-blocking if and only if

$$K \geq \begin{cases} \dfrac{3}{2} \, 2^{\frac{n}{2}} - 1 & n \text{ even} \\[2ex] 2^{\frac{n+1}{2}} - 1 & n \text{ odd} \end{cases} \tag{4.10}$$

Proof. A *tagged I/O path*, say the path 0-0, is selected, which includes $n-1$ interstage *tagged links*. All the other *conflicting I/O paths* that share at least one interstage link with the tagged I/O path are easily identified. Such link sharing for the tagged path is shown by the double tree of Figure 4.5 for $N = 64, 128$, thus representing the cases of n even ($n = 6$) and n odd ($n = 7$) (the stage numbering applies also to the links outgoing from each switching stage). Each node (branch) in the tree represents an SE (a link) of the original banyan network and the branches terminated on one node only represent the network inlets and outlets. Four subtrees can be identified in the double tree with n even (Figure 4.5a), two on the inlet side and two on the outlet side, each including $2^{n/2-1}$ "open branches": the subtree terminating the inlets (outlets) 0-3 and 4-7 are referred to as *upper subtree* and *lower subtree*, respectively. It is quite simple to see that the worst case of link occupancy is given when the inlets (outlets) of the upper inlet (outlet) subtree are connected to outlets (inlets) other than those in the outlet

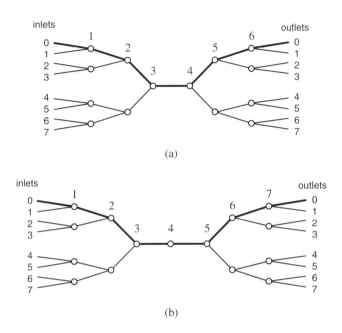

Figure 4.5. Double tree for the proof of non-blocking condition

(inlet) subtrees by engaging at least one tagged link. Moreover, since an even value of n implies that we have a central branch not belonging to any subtree, the worst loading condition for the tagged link in the central stage (stage 3 in the figure) is given when the inlets of lower inlet subtree are connected to the outlets of the lower outlet subtree. In the upper inlet subtree the tagged link of stage 1 is shared by one conflicting I/O path originating from the other SE inlet (the inlet 1), the tagged link of stage 2 is shared by two other conflicting paths originated from inlets not accounted for (the inlets 2 and 3), and the tagged link of stage $(n-2)/2$ (the last tagged link of the upper inlet subtree) is shared by $2^{(n-2)/2-1}$ conflicting paths originated from inlets which have not already been accounted for. We have two of these upper subtrees (on inlet and outlet side); furthermore the "central" tagged link at stage $n/2$ is shared by $2^{(n-2)/2}$ conflicting I/O paths (those terminated on the lower subtrees). Then the total number of conflicting I/O paths is

$$n_c = 2\left(2^0 + 2^1 + \ldots + 2^{\frac{n-2}{2}-1}\right) + 2^{\frac{n-2}{2}} = \frac{3}{2}2^{\frac{n}{2}} - 2 \tag{4.11}$$

The number of planes sufficient for an RBN with n even to be strictly non-blocking is then given by $n_c + 1$ as stated in Equation 4.10, since in the worst case each conflicting I/O path is routed onto a different plane, and the unity represents the additional plane needed by the tagged path to satisfy the non-blocking condition. An analogous proof applies to the case of n odd (see Figure 4.5b for $N = 128$), which is even simpler since the double tree does not have a central link reaching the same number of inlets and outlets. Thus the double tree includes only two subtrees, each including $2^{(n-1)/2}$ "open branches". Then the total number of conflicting I/O paths is

$$n_c = 2\left(2^0 + 2^1 + \ldots + 2^{\frac{n-1}{2}-1}\right) = 2^{\frac{n+1}{2}} - 2 \tag{4.12}$$

and the number of planes sufficient for an RBN with n odd to be strictly non-blocking is given by $n_c + 1$, thus proving Equation 4.10. The proof of necessity of the number of planes stated in the theorem immediately follows from the above reasoning on the worst case. In fact it is rather easy to identify a network state in which $K - 1$ connections are set up, each sharing one link with the tagged path and each routed on a different plane. ❏

So, the cost function of a strictly non-blocking network based on pure VR is

$$C = 4K\frac{N}{2}\log_2 N + 2NK = 2NK(\log_2 N + 1)$$

The comparison between the vertical replication factor required in a rearrangeable network and in a non-blocking network, shown in Table 4.3 for $N = 2^n$ ($n = 3 - 10$), shows that the strict-sense non blocking condition implies a network cost that is about 50% higher than in a rearrangeable network of the same size.

Table 4.3. Replication factor in rearrangeable and strictly non-blocking VR banyan networks

	RNB	SNB
$N = 8$	2	3
16	4	5
32	4	7
64	8	11
128	8	15
256	16	23
512	16	31
1024	32	47

4.2.2. Vertical replication with horizontal extension

The HE technique can be used jointly with the VR technique to build a non-blocking network by thus allowing a smaller replication factor. The first known result is due to Cantor [Can70, Can72] who assumed that each plane of the $N \times N$ overall (Cantor) network is an $N \times N$ Benes network. Therefore the same vertical replication scheme of Figure 3.13 applies here where the "depth" of each network is now $2\log_2 N - 1$ stages.

Theorem. A VR/HE banyan network of size $N \times N$ built by replicating K times a Benes network is strict-sense non-blocking if and only if

$$K \geq \log_2 N \qquad (4.13)$$

(an example of a Cantor network is shown in Figure 4.6).

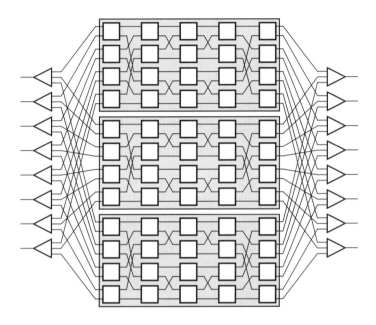

Figure 4.6. Cantor network for *N*=8

Proof. Also in this case the paths conflicting with the tagged I/O path are counted so that the required number of planes to provide the non-blocking condition is computed. Unlike a banyan network, a Benes network makes available more than one path from each inlet to each outlet. Since each added stage doubles the I-O paths, a Benes network provides $2^{\log_2 N - 1} = N/2$ paths per I/O pair. Figure 4.7 shows for $N = 16$ these 8 *tagged paths* each including 6 *tagged links* for the *tagged I/O pair* 0-0, together with the corresponding channel graph (each node of the channel graph is representative of a *tagged SE*, i.e. an SE along a tagged path). The two tagged links outgoing from stage 1 are shared by the tagged inlet and by only one other inlet (inlet 1 in our example), which, upon becoming busy, makes unavailable 2^{m-1} paths for the I/O pair 0-0 (see also the channel graph of the example). The four tagged links of the tagged paths outgoing from stage 2 are shared by four network inlets in total, owing to the buddy property. In fact the two SEs originating the links are reached by the same first-stage SEs. Out of these four inlets, one is the tagged inlet and another has already been accounted for as engaging the first-stage link. Therefore only two other inlets can engage one of the four tagged links at stage 2, and each of these makes unavailable 2^{m-2} tagged paths. In general, there are 2^i tagged links outgoing from SEs of stage i ($1 \leq i \leq m$), which are accessed by only 2^i network inlets owing to the constrained reachability property of a banyan

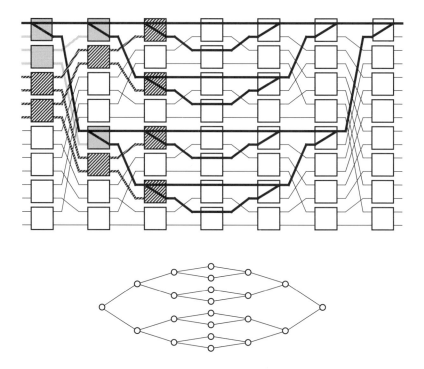

Figure 4.7. Extended banyan network with *m*=3

network. This property also implies that the network inlets reaching the tagged SEs at stage $i-1$ are a subset of the inlets reaching the tagged SEs at stage i. Therefore, the 2^i tagged links outgoing from stage i can be engaged by 2^{i-1} inlets not accounted for in the previous stages. Each of these inlets, upon becoming busy, makes unavailable 2^{m-i} tagged paths. Therefore, the total number of paths that can be blocked by non-tagged inlets is

$$n_{bi} = 1 \cdot 2^{m-1} + 2 \cdot 2^{m-2} + \dots + 2^{m-1} \cdot 1 = \sum_{i=1}^{m} 2^{i-1} \cdot 2^{m-i} = m2^{m-1}$$

where $m = \log_2 N - 1$ is the number of additional stages in each Benes network compared to a banyan network. Based on the network symmetry across the central stage of the Benes networks, the same number of paths $n_{bo} = n_{bi}$ is made unavailable by non-tagged outlets. The worst case of tagged path occupancy is given when the tagged paths made unavailable by the non-tagged inlets and by the non-tagged outlets are disjoint sets. It is rather easy to show that this situation occurs in a Cantor network owing to the recursive construction of the Benes network, which results in a series–parallel channel graph of the Cantor network (the channel graph of the Cantor network of Figure 4.6 is shown in Figure 4.8). In order for the Cantor network to be non-blocking, the number of its tagged I/O paths must not be smaller than the total number of blocked paths plus one (the path needed to connect the tagged I/O pair). The total number of tagged I/O paths is clearly given by K times the number of tagged I/O paths

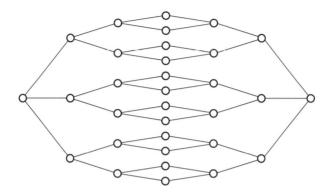

Figure 4.8. Channel graph of the Cantor network with *N*=8

per Benes plane, that is $N/2$. The total number n_b of blocked I/O paths in the Cantor network is still $n_b = n_{bi} + n_{bo}$ in spite of the K planes, since the generic network inlet i is connected to the inlet i of the K Benes planes and can make only one of them busy. Therefore

$$K \frac{N}{2} \geq n_b + 1 = 2 (\log_2 N - 1) 2^{\log_2 N - 2} + 1 = (\log_2 N - 1) \frac{N}{2} + 1$$

which, owing to the integer values assumed by K, gives the minimum number of planes sufficient to guarantee the strictly non-blocking condition

$$K \geq \log_2 N$$

thus completing the proof of sufficiency. The proof of necessity of at least $\log_2 N$ planes for the network non-blocking stated in the theorem immediately follows from the above reasoning on the worst case. In fact it is rather easy to identify a network state in which all the tagged paths in $K-1$ planes are inaccessible to a new connection request. ❏

The cost index for the $N \times N$ Cantor network is

$$C = 4\log_2 N \frac{N}{2} (2\log_2 N - 1) + 2N\log_2 N = 4N\log_2^2 N$$

where the last term in the sum accounts for the crosspoint count in the expansion and concentration stages.

So, the asymptotic growth of the cost index in a Cantor network is $O(N(\log_2 N)^2)$, whereas in a rearrangeable Benes network it is $O(N\log_2 N)$. Notice however that the higher cost of strict-sense non-blocking networks over rearrangeable networks is accompanied by the extreme simplicity of control algorithms for the former networks. In fact choosing an I/O path for a new connection only requires the knowledge of the idle–busy condition of the interstage links available to support that path.

We further observe that the pure VR non-blocking network has an asymptotic cost $O(N^{3/2}\log_2 N)$ whereas the cost of the Cantor network is $O(N(\log_2 N)^2)$. Therefore, the

Cantor network is asymptotically cheaper than a pure VR strictly non-blocking network. However, owing to the different coefficient α of the term with the highest exponent ($\alpha_{Cantor} = 4$, $\alpha_{VR} \cong 3$), the pure VR banyan network is cheaper than the Cantor network for $\log_2 N < 6$.

A more general approach on how to combine the VR and HE techniques has been described in [Shy91]. Analogously to the approach described for rearrangeable networks, now we build a replicated-extended banyan network by vertically replicating K times an EBN with m additional stages.

Theorem. A VR/HE banyan network with K planes each configured as an EBN where the m additional stages are added by means of the mirror imaging technique is strict-sense non-blocking if and only if

$$K \geq \begin{cases} \dfrac{3}{2} 2^{\frac{n-m}{2}} + m - 1 & n+m \; even \\[2em] 2^{\frac{n-m+1}{2}} + m - 1 & n+m \; odd \end{cases} \tag{4.14}$$

Proof. The proof developed for the Cantor network is easily extended to a VR/HE banyan network. Consider for example the extended banyan network of Figure 4.9 in which $N = 16$, $m = 2$, in which the tagged paths for the I/O pair 0-0 are drawn in bold (the corresponding channel graph is also shown in the figure). Now the number of tagged I/O paths available in an EBN plane is 2^m, which becomes $N/2$ for $m = n - 1$ (the Benes network). The number of the tagged I/O paths made unavailable by the non-tagged inlets is computed analogously to the Cantor network by taking into account the different total number of paths in the EBN plane. Therefore the non-tagged inlets, upon becoming busy, engage a tagged link outgoing from stage $1, 2, ..., m$ and thus make unavailable 2^{m-1}, 2^{m-2}, ..., 2^0 tagged paths. Unlike the Cantor network we have now other tagged links to be taken into account that originate from stage $m + 1$, $m + 2$, ..., until the centre of the EBN is reached. If $n + m$ is even, the engagement of a tagged link outgoing from stage $m + 1$, ..., $(n + m)/2 - 1$ makes unavailable only one tagged path. Analogously to the pure VR network, we have to consider that an even value of $n + m$ implies that the "central" tagged links (i.e., those of stage $(n + m)/2$) reach the same number of inlets and outlets, so that the paths made unavailable by these links must not be counted twice. An example is again represented by Figure 4.9, where the central links are those of stage 3. Therefore the number of blocked paths in an EBN plane with $n + m$ even is now

$$n_b = 2 \left[\sum_{i=1}^{m} 2^{i-1} \cdot 2^{m-i} + \sum_{i=m+1}^{\frac{n+m}{2}-1} 2^{i-1} \right] + 2^{\frac{n+m}{2}-1} = m2^m + \sum_{i=m+1}^{\frac{n+m}{2}-1} 2^i + 2^{\frac{n+m}{2}-1} \tag{4.15}$$

Note that it is legitimate to double the number of blocked paths from each side of the plane (the first term in Equation 4.15). In fact in the worst case the blocked paths from one side of the plane are disjoint from the blocked paths originated from the other side of the

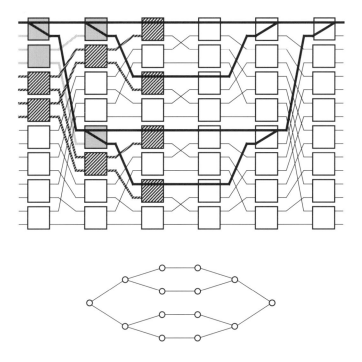

Figure 4.9. Extended banyan network with *m*=2

plane. This fact can be easily verified on the channel graph of Figure 4.9. In order for the over-all network to be non-blocking, the number of I/O paths in the VR/HE banyan network must not be smaller than the total number of blocked paths plus one (the path needed to connect the tagged I/O pair). As in the previous proof of the Cantor network, the total number of blocked paths in the VR/HE banyan network is still n_b, since the generic network inlet i is connected to the inlet i of the K EBN planes and can make busy only one of them. Therefore

$$K\,2^m \geq 2^m \left[m + \sum_{i=1}^{\frac{n+m}{2}-1-m} 2^i + 2^{\frac{n+m}{2}-1-m} \right] + 1$$

which reduces to

$$K \geq m + \sum_{i=1}^{\frac{n-m}{2}-1} 2^i + 2^{\frac{n-m}{2}-1} + 2^{-m} = \frac{3}{2}\,2^{\frac{n-m}{2}} + m - 2 + 2^{-m}$$

thus proving the assertion (Equation 4.14) for $n + m$ even.

In the case of $n + m$ odd we do not have the "central" tagged links, so that the total number of blocked paths is now

$$
n_b = 2 \left[\sum_{i=1}^{m} 2^{i-1} \cdot 2^{m-i} + \sum_{i=m+1}^{\frac{n+m-1}{2}} 2^{i-1} \right] = m2^m + \sum_{i=m+1}^{\frac{n+m-1}{2}} 2^i
\tag{4.16}
$$

The number of planes sufficient to make the VR/HE banyan network non-blocking is given by

$$
K 2^m \geq 2^m \left[m + \sum_{i=1}^{\frac{n+m-1}{2} - m} 2^i \right] + 1
$$

which reduces to

$$
K \geq m + \sum_{i=1}^{\frac{n-m-1}{2}} 2^i + 2^{-m} = 2^{\frac{n-m+1}{2}} + m - 2 + 2^{-m}
$$

thus proving the assertion (Equation 4.14) for $n + m$ odd.

The proof of necessity of the number of planes K for the network non-blocking stated in the theorem immediately follows from the above reasoning on the worst case. In fact it is rather easy to identify a network state in which all the tagged paths in $K - 1$ planes are inaccessible to a new connection request. ❏

Table 4.4 gives the required replication factor K for networks with $n = 3 - 10$, $m = 0 - 9$. Note that the value K for $m = n - 1$ corresponds to a Cantor network, whereas the row $m = 0$ gives a pure VR non-blocking network. The cost function of such a VR/HE banyan network is

$$
C = 4K \frac{N(\log_2 N + m)}{2} + 2NK = 2NK(\log_2 N + m + 1)
$$

It is very interesting to note that there are some network configurations cheaper than the Cantor network (those given in bold in Table 4.4). In fact the cheapest non-blocking network is given by a structure with the same vertical replication factor as the Cantor network, but each plane is two stages "shorter" (it includes $m = n - 3$ additional stages).

4.2.3. Link dilation

An SNB banyan network can also be obtained through link dilation (LD), that is by replicating the interstage links by a factor K_d, meaning that the network includes SEs of size $2K_d \times 2K_d$. Figure 4.10 shows a *dilated banyan network* (DBN) 16×16 with Baseline topology and dilation factor $K_d = 2$. It is rather simple to find the dilation factor of a banyan network that makes it

Table 4.4. Replication factor in a non-blocking VR/HE banyan network

	$N = 8$	16	32	64	128	256	512	1024
$m = 0$	**3**	5	7	11	15	23	31	47
1	3	**4**	6	8	12	16	24	32
2	3	4	**5**	7	9	13	17	25
3		4	5	**6**	8	10	14	18
4			5	6	**7**	9	11	15
5				6	7	**8**	10	12
6					7	8	**9**	11
7						8	9	**10**
8							9	10
9								10

non-blocking. In fact this factor is simply given by the maximum utilization factor u_{max} of the network defined in Section 3.2.1.2, that is the maximum number of I/O paths that cross a generic interstage link (that is the "central" ones). Thus a dilated banyan network is non-blocking when its interstage links are dilated by a factor

$$K_d \geq u_{max} = 2^{\left\lfloor \frac{n}{2} \right\rfloor}$$

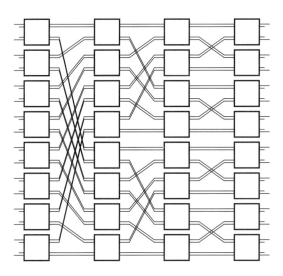

Figure 4.10. Dilated banyan network with *N*=16 and *K*=2

For example a 64×64 DBN is non-blocking if all links are dilated by a factor $K_d = 8$. The cost index for a DBN, expressing as usual the crosspoint count, is

$$C = \frac{N}{2} \log_2 N \left[2 \cdot 2^{\left\lfloor \frac{n}{2} \right\rfloor} \right]^2$$

Hence the cost of a DBN grows with $N^2 \log_2 N$, that is it is always more expensive than a crossbar network.

A cheaper dilated banyan network is obtained by observing that the utilization factor UF of the links is lower in stages close to the inlets or outlets and grows as we approach the center of the network, as is represented in Figure 3.14. Thus if we dilate each interstage link by a factor

$$K_d = u$$

equal to the utilization factor of the link in that stage, a SNB network is obtained, which is referred to as *progressively dilated banyan network* (PDBN). A representation of the PDBN structure for $N = 16 - 256$ is given in Figure 4.11 showing the link dilation factor of each stage.

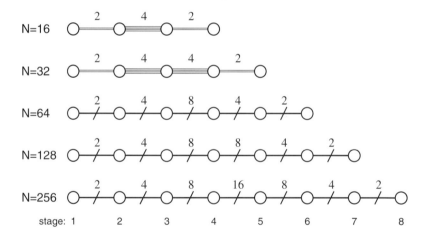

Figure 4.11. Dilation factor in non-blocking PDBN

4.2.4. EGS networks

A different approach to obtain a strict-sense non-blocking network that still relies on the use of very simple 2×2 SEs has been found recently by Richards relying on the use of the extended generalized shuffle (EGS) pattern [Ric93]. The general structure of an EGS network $N \times N$ includes N splitters $1 \times F$, s stages of $NF/2$ SEs of size 2×2 and N combiners $F \times 1$, mutually interconnected by EGS patterns (see Figure 4.12). We say that such a network has $s + 2$ stages where switching takes place, numbered 0 through $s + 1$, so that splitters and combiners accomplish the switching in stage 0 and $s + 1$, respectively, whereas the traditional switching

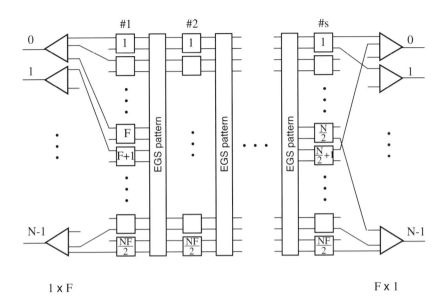

Figure 4.12. EGS network

takes place in the other s stages. Accordingly the EGS pattern following stage i $(0 \le i \le s)$ is referred to as interstage pattern of stage i.

Theorem. An EGS network with fanout F and s stages $(s \ge 1)$ is strict-sense non-blocking if

$$F \ge \begin{cases} 2^{n-s}\left(\dfrac{3}{2} 2^{\frac{s}{2}} - 1\right) & s \text{ even} \\[4mm] 2^{n-s}\left(2^{\frac{s+1}{2}} - 1\right) & s \text{ odd} \end{cases} \qquad (s \le n) \qquad (4.17)$$

$$F \ge \begin{cases} \dfrac{3}{2} 2^{\frac{2n-s}{2}} + s - n - 1 & s \text{ even} \\[4mm] 2^{\frac{2n-s+1}{2}} + s - n - 1 & s \text{ odd} \end{cases} \qquad (s > n) \qquad (4.18)$$

Proof. The proof of the condition for $s > n$ is rather simple, if we observe the analogy between this EGS network and a VR/HE banyan network of the same size $N \times N$ with $m = s - n$ additional stages and a number of planes $K = F$. The channel graphs of the two networks are shown in Figure 4.13 for $N = 8$, $F = 3$, $s = 5$. It can be verified that the suf-

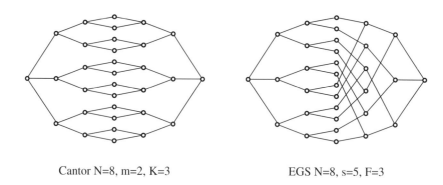

Cantor N=8, m=2, K=3   EGS N=8, s=5, F=3

Figure 4.13. Channel graphs for Cantor and EGS networks with *N*=8

ficiency proof of the non-blocking condition of the VR/HE banyan network applies as well to the EGS network. In fact the total number of available I/O paths is the same in the two networks. Moreover, owing to the buddy property[1] relating splitters and SEs of the first stage, as well as SEs of the last stage and combiners, the number of conflicting I/O paths in the EGS network must be counted starting from only one SE of the first stage (network inlets side) and from only one SE of the last stage (network outlets side), exactly as in the VR/HE banyan network. If we sum the number of paths blocked by connections originating from the network inlets and from the network outlets which are in conflict with the tagged I/O connection, the same number n_b as in the VR/HE banyan network (Equations 4.15 and 4.16) is obtained. Therefore, the equation expressing the minimum number of planes K in a non-blocking VR/HE banyan network (Equation 4.14) is the same given now for the fanout F of an EGS network, since $n + m = s$. Nevertheless, owing to the non series–parallel channel graph of EGS networks, it is no more true that the paths blocked by the two sides of the network are disjoint in the worst case. Therefore the above non-blocking condition is only sufficient and not necessary.

In order to prove the non-blocking condition for $s \leq n$, let us refer to Figure 4.14, showing an EGS network with $N = 8$, $F = 4$, $s = 2$. Owing to the EGS interstage patterns, the generic SE of the first stage reaches 2^{s-1} adjacent SEs in the last stage and thus 2^s adjacent combiners, i.e. adjacent network outlets. Therefore, the first stage can be said to include $F2^{s-1}$ adjacent groups of 2^{n-s} adjacent SEs each reaching all the $N = 2^n$ network outlets. Owing to the EGS pattern between splitters and first stage, each network inlet has access to F adjacent SEs of the stage, so that the total number of paths available for each specific I/O pair is $F/2^{n-s}$. Also in this case the buddy property enables us to say that the I/O paths conflicting with the tagged I/O path (0-0 in Figure 4.14) can be counted with reference to only one generic tagged path originating at stage 1 and terminating at stage s. In this case the number of

---

1. Here the buddy property, originally defined for a network including only 2×2 SEs, must be applied by straightforward extension considering that the two adjacent stages (0 and 1, s and $s + 1$) interface to each other having upstream elements with a number of outlets different from the number of inlets in the downstream SEs.

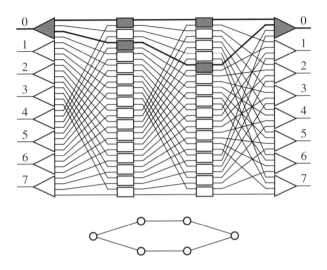

Figure 4.14. EGS network with N=8, F=4, s=2

conflicting paths is computed using the same approach adopted for the non-blocking VR banyan network (Section 4.2.1) now referred to a truncated banyan network with s stages rather
than n stages $(s \leq n)$. Therefore, using Equations 4.11 and 4.12, the fanout value guaranteeing the non-blocking condition is provided by relating the available and the conflicting I/O
paths, that is

$$\frac{F}{2^{n-s}} \geq n_c + 1 = \begin{cases} \frac{3}{2} \, 2^{\frac{s}{2}} - 2 + 1 & s \text{ even} \\ \\ 2^{\frac{s+1}{2}} - 2 + 1 & s \text{ odd} \end{cases}$$

thus proving the assertion (Equation 4.17) for the sufficiency condition. In this case of $s \leq n$ the
minimum fanout required for the sufficiency condition is also necessary, since the count of the
conflicting paths has been done with reference to a real worst case of network occupancy.   ❏

The fanout value F of the splitters so obtained are given in Table 4.5 for a $2^n \times 2^n$ network
with $n = 3 - 10$, $s = 1 - 19$. We compare first this table with Table 4.4 giving the number
of planes K necessary in a network jointly using VR and HE techniques to be strictly non-
blocking for a given number m of additional stages in each banyan plane, and we discover a
very interesting result. If $s \geq n$, an EGS network with s stages and a VR/HE banyan network
with $n + m = s$ stages require the same fanout of the splitters to be non-blocking (recall that
the splitter fanout equals the number of planes in a VR/HE banyan network). Note that the
two networks have the same cost, since they include the same number and type of components, the only difference lying in the interstage connection pattern.

Table 4.5. Splitter fanout of a non-blocking EGS network

	N = 8	16	32	64	128	256	512	1024
s = 1	4	8	16	32	64	128	256	512
2	4	8	16	32	64	128	256	512
3	**3**	6	12	24	48	96	192	384
4	3	5	10	20	40	80	160	320
5	3	**4**	7	14	28	56	112	224
6		4	6	11	22	44	88	176
7		4	**5**	8	15	30	60	120
8			5	7	12	23	46	92
9			5	**6**	9	16	31	62
10				6	8	13	24	47
11				6	**7**	10	17	32
12					7	9	14	25
13					7	**8**	11	18
14						8	10	15
15						8	**9**	12
16							9	11
17							9	**10**
18								10
19								10

In fact the EGS network has a cost function

$$C = 4s\frac{NF}{2} + 2NF = 2NF(s+1)$$

The minimum cost network is given in this case by $s = 2n - 3$ (bold elements in Table 4.5), which corresponds to the analogous condition $m = n - 3$ found in a VR/HE banyan network. The EGS network represents however a more general approach to the definition of multistage non-blocking networks, since it allows the definition of networks with an arbitrary number of stages $s \geq 1$ whereas the minimum number of stages in a VR/HE banyan network is always n.

Interestingly enough, smaller fanouts can be obtained for some EGS network configurations with $s > n$ [Ric93] (recall that if the number of stages exceeds n, Equation 4.18 expresses only a sufficient condition for the network non-blocking). For example for $N = 512$ a fanout $F = 8$ provides a non-blocking EGS network, whereas the better result given by Table 4.5 is $F = 9$.

A useful insight in the equivalence between EGS networks and other networks can be obtained by Figures 4.15 and 4.16, showing an EGS and a VR/HE banyan network 8×8

with $s = 3$ and $s = 5$ stages, respectively. The former figure shows that the two networks are isomorphic, since the mapping of inlets and outlets of the networks is the identity, whereas the mapping between SEs of the same stage is shown in the figure itself. Note that the channel graph of the two networks is the same, that is of the series–parallel type, and the two networks are also functionally equivalent as they are non-blocking. The same network isomorphism does not hold in the case of the latter figure, since no mapping can be found between inlets, outlets and SEs that makes the two networks identical. An explanation of this fact lies in the difference between the channel graphs of the two networks, shown in Figure 4.13: the VR/HE banyan network has a series–parallel channel graph, the EGS network does not. Nevertheless the two networks are functionally equivalent, since both of them are non-blocking.

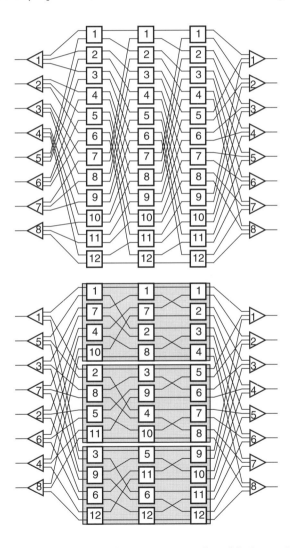

Figure 4.15. Equivalence between networks with three stages

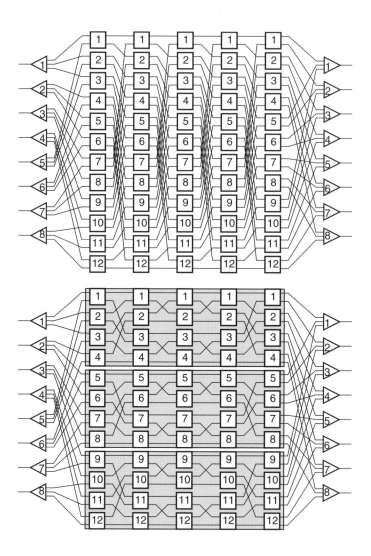

Figure 4.16. Equivalence between networks with five stages

4.3. Comparison of Non-blocking Networks

Figure 4.17 [Dec92] summarizes graphically the main results obtained for strictly non-blocking networks referred to a network size 512×512: a triangle indicates the internal expansion factor, that is the operation of splitters and combiners, whereas a rectangle represents the internal switching. The rectangle height, equal to the expansion factor, indicates the number of different inlets of the switching part accessed by each of the 512 inlets, whereas the rectangle

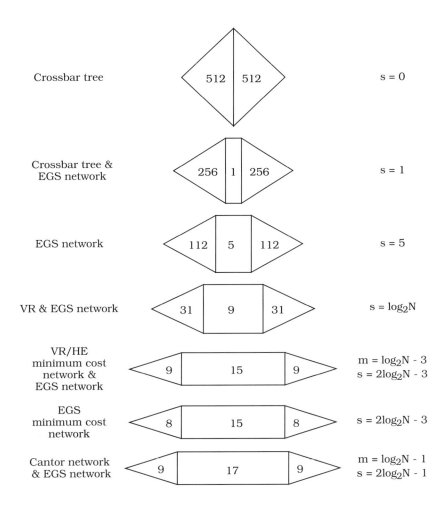

Figure 4.17. Alternative solutions for a non-blocking network with *N*=512

width indicates the number of switching stages. The expansion factor has been denoted with K in a replicated banyan network and with F in an EGS network.

Adopting the crossbar tree network means making available an expansion factor equal to the network size with no switching stages. The crossbar tree with only one switching stage reduces the expansion factor to 256. The expansion factor can be further reduced by making available more paths through the network: an EGS network requires an expansion factor of 112 (see Table 4.5) with $s = 5$ switching stages. If the value of s is further increased up to 9, the EGS network requires a fanout $F = 31$. This expansion factor is the same characterizing the pure vertically replicated banyan network, that is $K = 31$ (see Table 4.3). Increasing further the stage number means adopting horizontal expansion coupled with vertical replication: the "largest" network is the Cantor network whose number of switching stages is $s = 17$ with an expansion factor $K = 9$ (see Table 4.4). We have shown how the cheapest VR/HE banyan network has two stages less than the Cantor network, that is $s = 15$, which requires the same expansion factor $K = 9$ (see Table 4.4). An EGS network with minimum cost according to Table 4.5 is characterized by the same network height and width as the previous VR/HE banyan network, that is $F = 9$ and $s = 15$. Only the EGS networks enables the building of an even cheaper network: the fanout can be reduced to $F = 8$ while keeping the same number of stages $s = 15$ [Ric93].

4.4. Bounds on the Network Cost Function

The existence of upper bounds on the cost function of non-blocking multistage networks is now investigated, where the network cost is again given by the number of crosspoints required in the $N \times N$ network (for simplicity, unless stated otherwise, only squared networks with $N = 2^n$ are considered). We have seen that the full-connection three-stage network described by the Clos rule $(r_2 = 2n - 1)$ has minimum cost when $n \cong \sqrt{N/2}$. When N exceeds a certain threshold, a recursive decomposition of the middle-stage matrices based on the same condition provides again a cheaper network. For such a network there is no closed-form solution for the network cost, although Cantor [Can72] showed that this cost C is less than

$$Ne^{2\sqrt{\log N}}$$

A deeper insight in the problem of finding network cost bounds is provided [Can72] when we assume a three-stage squared network very close to the recursive structure described in Section 4.1.2 having $r_2 = 2n$ middle-stage matrices and a partition of the N inlets into $\sqrt{N}$ first-stage matrices with $\sqrt{N}$ inlets each (thus now $n = \sqrt{N}$). Let this network be denoted as $[L_{a, 2a}, L_{a, a}, L_{2a, a}]$ with $a = \sqrt{N}$, where $A_{x, y}, B_{u, v}, C_{z, w}$ denote the sizes $x \times y$, $u \times v$ and $z \times w$ of the matrices at the first (A), second (B) and third (C) stage, respectively. Note that such notation completely describes a three-stage full-connection network, since obviously $r_1 = u$, $N = xu$, and $r_3 = v$ and $M = wv$ ($x = w$ and $u = v$ in a squared network). If $C(x, y)$ denotes the cost of an $x \times y$ network, then the three-stage network has a cost

$$C(a^2, a^2) = aC(a, 2a) + 2aC(a, a) + aC(a, 2a) \leq 6aC(a, a)$$

since by obvious consideration $C(a, 2a) \leq 2C(a, a)$. Now we iterate the three-stage construction to each $a \times a$ matrix and obtain for an $N \times N$ network

$$C(N, N) = C(2^n, 2^n) \leq 6 \cdot 2^{n/2} C(2^{n/2}, 2^{n/2}) = 6^2 \cdot 2^{n/2 + n/4} C(2^{n/4}, 2^{n/4})$$

$$= 6^{\log_2 n} \cdot 2^{n-1} C(2, 2) = n^{\log_2 6} \frac{N}{2} C(2, 2) = \alpha N (\log_2 N)^{\log_2 6}$$

Thus the cost index has an asymptotic growth described by a function of the type $aN(\log_2 N)^\beta$ with α and β constant. In the case just examined $\beta = \log_2 6 = 2.58$.

Cantor also showed that lower exponents β of the $\log_2 N$ factor can be obtained, even if his proof does not always show how to build such network. He considered a rectangular network $[L_{a, 2a}, L_{b, 2b}, L_{2a, a}]$ having ab inlets and $2ab$ outlets. The cost for such network is

$$C(ab, 2ab) = bC(a, 2a) + 2aC(b, 2b) + 2bC(a, 2a)$$

which for $a = b$ and iterative decomposition of each matrix gives

$$C(N, N) \leq C(N, 2N) \leq \alpha N (\log_2 N)^{\log_2 5}$$

So, this network is less expensive than the previous one since we have decreased the exponent of $\log_2 N$ to $\beta = \log_2 5 = 2.32$. By further theoretical considerations, Cantor proved also that such an exponent can be decreased to $\beta = 2.27$.

We have already seen that by means of a recursive design of a partial-connection network (the Cantor network) the cost function becomes

$$C(N, N) = 4N(\log_2 N)^2$$

thus with an exponent $\beta = 2$, which is smaller than all the β values previously found, and a multiplying constant $\alpha = 4$. Given the same exponent $\beta = 2$, a further improvement was found by Pippenger [Pip78], who was able to reduce the multiplying constant α. He showed that a network based on 5×5 basic matrices, rather than 2×2 in Cantor's construction, has a cost that asymptotically (for very large N) becomes

$$C(N, N) \leq 16N(\log_5 N)^2 + O(N(\log_2 N)^{3/2}) = 2.97N(\log_2 N)^2 + O(N(\log_2 N)^{3/2})$$

where the equality $\log_b N = (\log_2 N) / (\log_2 b)$ has been applied.

Nevertheless, such a bound can be further improved by theoretical non-constructive approaches, that is without providing the physical architecture that guarantees the bound. Bassalygo and Pinsker [Bas73] proved that the asymptotic bound for a non-blocking network is

$$C(N, N) \leq 67N\log_2 N + O(N)$$

which was further improved by Pippenger [Pip78] to

$$C(N, N) \leq 90N\log_3 N + O(N(\log_2 N)^{1/2}) = 57N\log_2 N + O(N(\log_2 N)^{1/2}) \quad (4.19)$$

A graphical comparison of the network cost for the different non-blocking structures examined in this chapter is given in Figure 4.18. The crossbar network is the least expensive only for very small network sizes $(N \leq 16)$, as already discussed in Section 4.1.2. For larger values of N the recursive Clos network built using Equation 4.5 becomes less expensive; in particular a network with three stages is the most economical for $N \leq 128$, with five stages for sizes up to $N = 4096$, with seven stages for larger sizes. The Cantor network becomes the most convenient solution only for very large networks. Further we observe that the Pippenger bound given by Equation 4.19 approaches the cost of the best solutions only for very large values of N. This is not surprising if we recall that the bound is asymptotic, that is it holds as N approaches infinity.

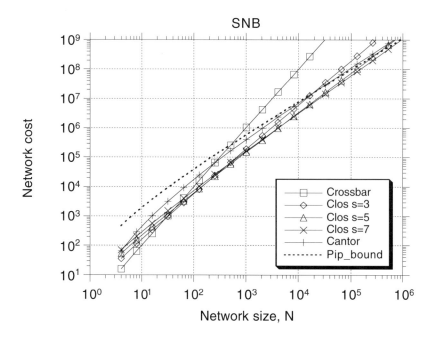

Figure 4.18. Cost of non-blocking networks

4.5. References

[Bas73]   L.A. Bassalygo, M.S. Pinsker, "Complexity of an optimum nonblocking switching network without reconnections", *Problemy Peredachi Informatsii*, Vol. 9, No. 1, Jan.-Mar. 1973, pp. 84–87; translated into English in *Problems of Information Transmission*, Vol. 9, No. 1, Nov. 1974, pp. 64-66.

[Can70]   D. Cantor, "On construction of non-blocking switching networks", *Proc. of Symp. on Comput.-Commun. Networks and Teletraffic*, Polytechnic Institute, Brooklyn, NY, 1972, pp. 253-255.

[Can72]   D. Cantor, "On non-blocking switching networks", *Networks*, Vol. 1, 1972, pp. 367-377.

[Clo53]   C. Clos, "A study of non-blocking switching networks", *Bell System Tech. J.*, Vol 32, Mar. 1953, pp. 406-424.

[Dec92]   M. Decina, A. Pattavina, "High speed switching for ATM networks", Tutorial presentation at INFOCOM 92, Florence, Italy, May 1992.

[Lea90]   C.-T. Lea, "Multi-log$_2$N networks and their applications in high-speed electronic and photonic switching systems, *IEEE Trans. on Commun.*, Vol. 38, No. 10, Oct. 1990, pp. 1740-1749.

[Mar77]   M.J. Marcus, "The theory of connecting networks and their complexity: a review", *Proc. of the IEEE*, Vol. 65, No. 9, Sept. 1977, pp. 1263-1271.

[Pip78]   N. Pippenger, "On rearrangeable and non-blocking switching networks", *J. of Computer and System Science*, Vol. 17, No. 4, Sep. 1978, pp.145-162.

[Ric93]   G.W. Richards, "Theoretical aspects of multi-stage networks for broadband networks", Tutorial presentation at INFOCOM 93, San Francisco, Apr.-May 1993.

[Shy91]   D.-J. Shyy, C.-T. Lea, "Log₂(N,m,p) strictly nonblocking networks", *IEEE Trans. on Commun.*, Vol. 39, No. 10, Oct. 1991, pp. 1502-1510.

4.6. Problems

4.1   Derive the cost function of a two-stage fully connected squared network $N \times N$ without link dilation, explaining why it gives a cost smaller than the crossbar network.

4.2   Draw a nine-stage non-blocking network using the recursive Clos construction. Determine the network cost function C_9 and verify that it agrees with Equation 4.9.

4.3   Provide an intuitive explanation based on the expressions of the cost function C_{2t+1} given in Equations 4.6, 4.7, 4.8 and found in Problem 4.2 for the fact that the minimum cost of the recursive Clos network is given by a number of stages that increases with the network size N.

4.4   Compute the cost of a five-stage non-blocking minimum-cost Clos network and derive Equation 4.4 providing the values for n_1 and n_2 in a minimum-cost network.

4.5   Derive the expression of the cost function for a PDBN with size $N \times N$.

4.6   Draw the channel graph of a non-blocking EGS network with $N = 32$ and $s = 8$ or $s = 5$.

4.7   Prove that an EGS network with $N = 16$, $s = 4$, $F = 5$ is isomorphic to an RBN of the same size with replication factor $K = 5$, by specifying the mapping between the two networks.

4.8   Show how Equation 4.9 can be derived.

4.9   Draw an EGS network with $N = 8$, $s = 3$, $F = 2$ saying if it satisfies the conditions to be non-blocking or rearrangeable.

Chapter 5   *The ATM Switch Model*

The B-ISDN envisioned by ITU–T is expected to support a heterogeneous set of narrowband and broadband services by sharing as much as possible the functionalities provided by a unique underlying transport layer based on the ATM characteristics. As already discussed in Section 1.2.1, two distinctive features characterize an ATM network: (i) the user information is transferred through the network in small fixed-size packets, called *cells*[1], each 53 bytes long, divided into a *payload* (48 bytes) for the user information and a *header* (5 bytes) for control data; (ii) the transfer mode of user information is *connection-oriented*, that is cells are transferred onto virtual links previously set up and identified by a label carried in the cell header. Therefore from the standpoint of the switching functions performed by a network node, two different sets of actions can be identified: operations accomplished at virtual call set up time and functions performed at cell transmission time.

At call set-up time a network node receives from its upstream node or user–network interface (UNI) a request to set up a virtual call to a given end-user with certain traffic characteristics. The node performs a connection acceptance control procedure, not investigated here, and if the call is accepted the call request is forwarded to the downstream node or UNI of the destination end-user. What is important here is to focus on the actions executed within the node in preparation of the next transfer of ATM cells on the virtual connection just set up. The identifier of the virtual connection entering the switching node carried by the call request packet is used as a new entry in the routing table to be used during the data phase for the new virtual connection. The node updates the table by associating to that entry identifier a new exit identifier for the virtual connection as well as the address of the physical output link where the outgoing connection is being set up.

At cell transmission time the node receives on each input link a flow of ATM cells each carrying its own virtual connection identifier. A table look-up is performed so as to replace in

---

1. The terms cell and packet will be used interchangeably in this section and in the following ones to indicate the fixed-size ATM packet.

the cell header the old identifier with the new identifier and to switch the cell to the switch output link whose address is also given by the table.

Both *virtual channels* (VC) and *virtual paths* (VP) are defined as virtual connections between adjacent routing entities in an ATM network. A logical connection between two end-users consists of a series of $n+1$ virtual connections, if n switching nodes are crossed; a virtual path is a bundle of virtual channels. Since a virtual connection is labelled by means of a hierarchical key VPI/VCI (*virtual path identified/virtual channel identifier*) in the ATM cell header (see Section 1.5.3), a switching fabric can operate either a full VC switching or just a VP switching. The former case corresponds to a full ATM switch, while the latter case refers to a simplified switching node with reduced processing where the minimum entity to be switched is a virtual path. Therefore a VP/VC switch reassigns a new VPI/VCI to each virtual cell to be switched, whereas only the VPI is reassigned in a VP switch, as shown in the example of Figure 5.1.

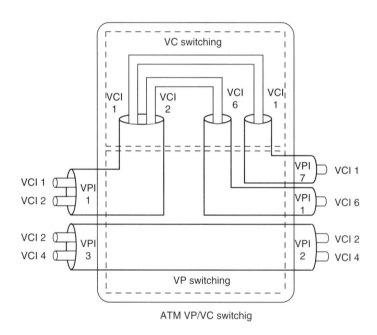

Figure 5.1. VP and VC switching

A general model of an ATM switch is defined in Section 5.1 on which the specific architectures described in the following sections will be mapped. A taxonomy of ATM switches is then outlined in Section 5.2 based on the identification of the key parameters and properties of an ATM switch.

5.1. The Switch Model

Research in ATM switching has been developed worldwide for several years showing the feasibility of ATM switching fabrics both for small-to-medium size nodes with, say, up to a few hundreds of inlets and for large size nodes with thousands of inlets. However, a unique taxonomy of ATM switching architectures is very hard to find, since different keys used in different orders can be used to classify ATM switches. Very briefly, we can say that most of the ATM switch proposals rely on the adoption for the *interconnection network* (IN), which is the switch core, of multistage arrangements of very simple switching elements (SEs) each using the *packet self-routing* concept. This technique consists in allowing each SE to switch (route) autonomously the received cell(s) by only using a self-routing tag preceding the cell that identifies the addressed physical output link of the switch. Other kinds of switching architectures that are not based on multistage structures (e.g., shared memory or shared medium units) could be considered as well, even if they represent switching solutions lacking the scalability property. In fact technological limitations in the memory access speed (either the shared memory or the memory units associated with the shared medium) prevent "single-stage" ATM switches to be adopted when the number of ports of the ATM switch overcomes a given threshold. For this lack of generality such solutions will not be considered here. Since the packets to be switched (the cells) have a fixed size, the interconnection network switches all the packets from the inlets to the requested outlets in a time window called a *slot*, which are received aligned by the IN. Apparently a slot lasts a time equal to the transmission time of a cell on the input and output links of the switch. Due to this slotted type of switching operation, the non-blocking feature of the interconnection network can be achieved by adopting either a rearrangeable network (RNB) or a strict-sense non-blocking network (SNB). The former should be preferred in terms of cost, but usually requires a more complex control.

We will refer here only to the cell switching of an ATM node, by discussing the operations related to the transfer of cells from the inputs to the outputs of the switch. Thus, all the functionalities relevant to the set-up and tear-down of the virtual connections through the switch are just mentioned. The general model of a $N \times N$ switch is shown in Figure 5.2. The reference switch includes N *input port controllers* (IPC), N *output port controllers* (OPC) and an interconnection network (IN). A very important block that is not shown in the figure is the *call processor* whose task is to receive from the IPCs the virtual call requests and to apply the appropriate algorithm to decide whether to accept or refuse the call. The call processor can be connected to IPCs either directly or, with a solution that is independent from the switch size, through the IN itself. Therefore one IN outlet can be dedicated to access the call processor and one IN inlet can be used to receive the cells generated by the call processor.

The IN is capable of switching up to K_o cells to the same OPC in one slot, K_o being called an *output speed-up* since an internal bit rate higher than the external rate (or an equivalent space division technique) is required to allow the transfer of more than one cell to the same OPC. In certain architectures an *input speed-up* K_i can be accomplished, meaning that each IPC can transmit up to K_i cells to the IN. If $K_i = 1$, that is there is no input speed-up, the output speed-up will be simply referred to as *speed-up* and denoted as K. The IN is usually a multistage arrangement of very simple SEs, typically 2×2, which either are provided with internal queueing (*SE queueing*), which can be realized with input, output or shared buffers, or are unbuffered (*IN queueing*). In this last case input and output queueing, whenever adopted, take place at IPC and OPC, respectively, whereas shared queueing is accomplished by means of additional hardware associated with the IN.

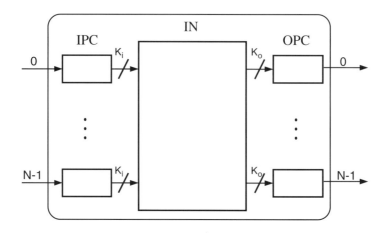

Figure 5.2. Model of ATM switch

In general two types of conflict characterize the switching operation in the interconnection network in each slot, the *internal conflicts* and the *external conflicts*. The former occur when two I/O paths compete for the same internal resource, that is the same interstage link in a multistage arrangement, whereas the latter take place when more than K packets are switched in the same slot to the same OPC (we are assuming for simplicity $K_i = 1$). An ATM interconnection network $N \times N$ with speed-up K ($K \leq N$) is said to be non-blocking (K-rearrangeable according to the definition given in Section 3.2.3) if it guarantees absence of internal conflicts for any arbitrary switching configuration free from external conflicts for the given network speed-up value K. That is a non-blocking IN is able to transfer to the OPCs up to N packets per slot, in which at most K of them address the same switch output. Note that the adoption of output queues either in an SE or in the IN is strictly related to a full exploitation of the speed-up: in fact, a structure with $K = 1$ does not require output queues, since the output interface is able to transmit downstream one packet per slot. Whenever queues are placed in different elements of the ATM switch (e.g., SE queueing, as well as input or shared queueing coupled with output queueing in IN queueing), two different internal transfer modes can be adopted:

- *backpressure* (BP), in which by means of a suitable backward signalling the number of packets actually switched to each downstream queue is limited to the current storage capability of the queue; in this case all the other head-of-line (HOL) cells remain stored in their respective upstream queue;

- *queue loss* (QL), in which cell loss takes place in the downstream queue for those HOL packets that have been transmitted by the upstream queue but cannot be stored in the addressed downstream queue.

The main functions of the port controllers are:

- rate matching between the input/output channel rate and the switching fabric rate;
- aligning cells for switching (IPC) and transmission (OPC) purposes (this requires a temporary buffer of one cell);
- processing the cell received (IPC) according to the supported protocol functionalities at the ATM layer; a mandatory task is the *routing* (switching) function, that is the allocation of a switch output and a new VPI/VCI to each cell, based on the VCI/VPI carried by the header of the received cell;
- attaching (IPC) and stripping (OPC) a self-routing label to each cell;
- with IN queueing, storing (IPC) the packets to be transmitted and probing the availability of an I/O path through the IN to the addressed output, by also checking the storage capability at the addressed output queue in the BP mode, if input queueing is adopted; queueing (OPC) the packets at the switch output, if output queueing is adopted.

An example of ATM switching is given in Figure 5.3. Two ATM cells are received by the ATM node I and their VPI/VCI labels, A and C, are mapped in the input port controller onto the new VPI/VCI labels F and E; the cells are also addressed to the output links c and f, respectively. The former packet enters the downstream switch J where its label is mapped onto the new label B and addressed to the output link c. The latter packet enters the downstream node K where it is mapped onto the new VPI/VCI A and is given the switch output address g. Even if not shown in the figure, usage of a self-routing technique for the cell within the interconnection network requires the IPC to attach the address of the output link allocated to the virtual connection to each single cell. This self-routing label is removed by the OPC before the cell leaves the switching node.

The traffic performance of ATM switches will be analyzed in the next sections by referring to an offered *uniform random traffic* in which:

- packet arrivals at the network inlets are independent and identically distributed Bernoulli processes with p ($0 < p \leq 1$) indicating the probability that a network inlet receives a packet in a generic slot;
- a network outlet is randomly selected for each packet entering the network with uniform probability $1/N$.

Note that this rather simplified pattern of offered traffic completely disregards the application of connection acceptance procedure of new virtual calls, the adoption of priority among traffic classes, the provision of different grade of services to different traffic classes, etc. Nevertheless, the uniform random traffic approach enables us to develop more easily analytical models for an evaluation of the traffic performance of each solution compared to the others. Typically three parameters are used to describe the switching fabric performance, all of them referred to steady-state conditions for the traffic:

- *Switch throughput* ρ ($0 < \rho \leq 1$): the normalized amount of traffic carried by the switch expressed as the utilization factor of its input links; it is defined as the probability that a packet received on an input link is successfully switched and transmitted by the addressed switch output; the maximum throughput ρ_{max}, also referred to as *switch capacity*, indicates the load carried by the switch for an offered load $p = 1$.

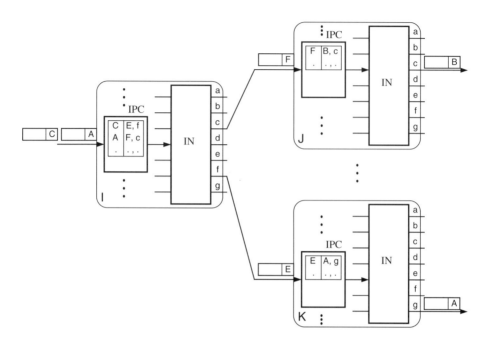

Figure 5.3. Example of ATM switching

- *Average packet delay T* ($T \geq 1$) : average number of slots it takes for a packet received at a switch input to cross the network and thus to be transmitted downstream by the addressed switch output, normalized to the number of network stages if SE queueing is adopted; thus the minimum value $T = 1$ indicate just the packet transmission time. T takes into account only the queueing delays and the packet transmission time.

- *Packet loss probability* π ($0 < \pi \leq 1$) : probability that a packet received at a switch input is lost due to buffer overflow.

Needless to say, our dream is a switching architecture with minimum complexity, capacity very close to 1, average packet delay less than a few slots and a packet loss probability as low as desired, for example less than 10^{-9} .

All the performance plots shown in the following chapters will report, unless stated otherwise, results from analytical models by continuous lines and data from computer simulation by plots. The simulation results, wherever plotted in the performance graph, have a 95% confidence interval not greater than 0.1%, 5% of the plotted values for throughput and delay figures, respectively. As far as the packet loss is concerned, these intervals are not greater than 5% (90%) of the plotted values if the loss estimate is above (below) 10^{-3} .

As far as the packet delay is concerned, we will disregard in the following the latency of packets inside multistage networks. Therefore the only components of the packet delay will be the waiting time in the buffers and the packet transmission time.

5.2. ATM Switch Taxonomy

As already mentioned, classifying all the different ATM switch architectures that have been proposed or developed is a very complicated and arduous task, as the key parameters for grouping together and selecting the different structures are too many. As a proof, we can mention the taxonomies presented in two surveys of ATM switches presented some years ago. Ahmadi and Denzel [Ahm89] identified six different classes of ATM switches according to their internal structure: banyan and buffered banyan-based fabrics, sort-banyan-based fabrics, fabrics with disjoint path topology and output queueing, crossbar-based fabrics, time division fabrics with common packet memory, fabrics with shared medium. Again the technological aspects of the ATM switch fabric were used by Tobagi [Tob90] to provide another survey of ATM switch architectures which identifies only three classes of switching fabrics: shared memory, shared medium and space-division switching fabrics. A further refinement of this taxonomy was given by Newman [New92], who further classified the space-division type switches into single-path and multiple-path switches, thus introducing a non-technological feature (the number of I/O paths) as a key of the classification.

It is easier to identify a more general taxonomy of ATM switches relying both on the functional relationship set-up between inlets and outlets by the switch and on the technological features of the switching architecture, and not just on these latter properties as in most of the previous examples. We look here at switch architectures that can be scaled to any reasonable size of input/output ports; therefore our interest is focused onto multistage structures which own the distributed switching capability required to switch the enormous amounts of traffic typical of an ATM environment.

Multistage INs can be classified as *blocking* or *non-blocking*. In the case of blocking interconnection networks, the basic IN is a banyan network, in which only one path is provided between any inlet and outlet of the switch and different I/O paths within the IN can share some interstage links. Thus the control of packet loss events requires the use of additional techniques to keep under control the traffic crossing the interconnection network. These techniques can be either the adoption of a packet storage capability in the SEs in the basic banyan network, which determines the class of *minimum-depth* INs, or the usage of deflection routing in a multiple-path IN with unbuffered SEs, which results in the class of *arbitrary-depth* INs. In the case of non-blocking interconnection networks different I/O paths are available, so that the SEs do not need internal buffers and are therefore much simpler to be implemented (a few tens of gates per SE). Nevertheless, these INs require more stages than blocking INs.

Two distinctive technological features characterizing ATM switches are the buffers configuration and the number of switching planes in the interconnection network. Three configurations of cell buffering are distinguished with reference to each single SE or to the whole IN, that is *input queueing* (IQ), *output queueing* (OQ) and *shared queueing* (SQ). The buffer is placed inside the switching element with SE queueing, whereas unbuffered SEs are used with IN queueing, the buffer being placed at the edges of the interconnection network. It is important to distinguish also the architectures based on the number of switch planes it includes, that is single-plane structures and parallel plane structures in which at least two switching planes are equipped. It is worth noting that adopting parallel planes also means that we adopt a queueing strategy that is based on, or anyway includes, output queueing. In fact the adoption of multiple switching planes is equivalent from the standpoint of the I/O functions of the overall interconnection network to accomplishing a speed-up equal to the number of planes. As already discussed in Section 5.1, output queueing is mandatory in order to control the cell loss performance when speed-up is used.

A taxonomy of ATM switch architectures, which tries to classify the main ATM switch proposals that have appeared in the technical literature can be now proposed. By means of the four keys just introduced (network blocking, network depth, number of switch planes and queueing strategy), the taxonomy of ATM interconnection network given in Figure 5.4 is obtained which only takes into account the meaningful combinations of the parameters, as witnessed by the switch proposals appearing in the technical literature. Four ATM switch classes have been identified:

- *blocking INs with minimum depth*: the interconnection network is blocking and the number of switching stages is the minimum required to reach a switch outlet from a generic switch inlet; with a single plane, SE queueing is adopted without speed-up so that only one path is available per I/O pair; with parallel planes, IN queueing and simpler unbuffered SEs are used; since a speed-up is accomplished in this latter case, output queueing is adopted either alone (OQ) or together with input queueing (IOQ);

- *blocking INs with arbitrary depth*: IN queueing and speed-up are adopted in both cases of single and parallel planes; the interconnection network, built of unbuffered SEs, is blocking but makes available more than one path per I/O pair by exploiting the principle of deflection routing; output queueing (OQ) is basically adopted;

- *non-blocking IN with single queueing*: the interconnection network is internally non-blocking and IN queueing is used with buffer being associated with the switch inputs (IQ), with the switch outputs (OQ) or shared among all the switch inlets and outlets (SQ);

- *non-blocking IN with multiple queueing*: the IN is non-blocking and a combined use of two IN queueing types is adopted (IOQ, SOQ, ISQ) with a single-plane structure; an IN with parallel planes is adopted only with combined input/output queueing (IOQ).

A chapter is dedicated in the following to each of these four ATM switch classes, each dealing with both architectural and traffic performance aspects.

Limited surveys of ATM switches using at least some of the above keys to classify the architectures have already appeared in the technical literature. Non-blocking architectures with single queueing strategy are reviewed in [Oie90b], with some performance issues better investigated in [Oie90a]. Non-blocking ATM switches with either single or multiple queueing strategies are described in terms of architectures and performance in [Pat93]. A review of blocking ATM switches with arbitrary depth IN is given in [Pat95].

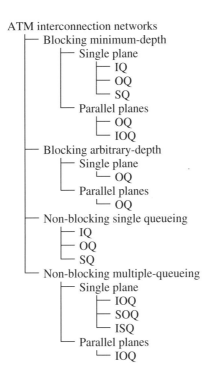

Figure 5.4. Taxonomy of ATM interconnection networks

5.3. References

[Ahm89]   H. Ahmadi, W.E. Denzel, "A survey of modern high-performance switching techniques", *IEEE J. on Selected Areas in Commun.*, Vol. 7, No. 7, Sept. 1989, pp. 1091-1103.

[New92]   P. Newman, "ATM technology for corporate networks", *IEEE Communications Magazine*, Vol. 30, No. 4, April 1992, pp. 90-101.

[Oie90a]   Y. Oie, T. Suda, M. Murata, H. Miyahara, "Survey of the performance of non-blocking switches with FIFO input buffers", *Proc. of ICC 90*, Atlanta, GA, April 1990, pp. 737-741.

[Oie90b]   Y. Oie, T. Suda, M. Murata, D. Kolson, H. Miyahara, "Survey of switching techniques in high-speed networks and their performance", *Proc. of INFOCOM 90*, San Francisco, CA, June 1990, pp. 1242-1251.

[Pat93]   A. Pattavina, "Non-blocking architectures for ATM switching", *IEEE Communications Magazine*, Vol. 31, No. 2, Feb. 1992, pp. 38-48.

[Pat95]   A. Pattavina, "ATM switching based on deflection routing", *Proc. of Int. Symp. on Advances in Comput. and Commun.*, Alexandria, Egypt, June 1995, pp. 98-104.

[Tob90]   F.A. Tobagi, "Fast packet switch architectures for broadband integrated services digital networks", *Proc. of the IEEE*, Vol. 78, No. 1, Jan 1990, pp. 133-167.

Chapter 6 *ATM Switching with Minimum-Depth Blocking Networks*

Architectures and performance of interconnection networks for ATM switching based on the adoption of banyan networks are described in this chapter. The interconnection networks presented now have the common feature of a *minimum depth* routing network, that is the path(s) from each inlet to every outlet crosses the minimum number of routing stages required to guarantee full accessibility in the interconnection network and to exploit the self-routing property. According to our usual notations this number n is given by $n = \log_b N$ for a network $N \times N$ built out of $b \times b$ switching elements. Note that a packet can cross more than n stages where switching takes place, when distribution stages are adopted between the switch inlets and the n routing stages. Nevertheless, in all these structures the switching result performed in any of these additional stages does not affect in any way the self-routing operation taking place in the last n stages of the interconnection network. These structures are inherently blocking as each interstage link is shared by several I/O paths. Thus packet loss takes place if more than one packet requires the same outlet of the switching element (SE), unless a proper storage capability is provided in the SE itself.

Unbuffered banyan networks are the simplest self-routing structure we can imagine. Nevertheless, they offer a poor traffic performance. Several approaches can be considered to improve the performance of banyan-based interconnection networks:

1. Replicating a banyan network into a set of parallel networks in order to divide the offered load among the networks;

2. Providing a certain multiplicity of interstage links, so as to allow several packets to share the interstage connection;

3. Providing each SE with internal buffers, which can be associated either with the SE inlets or to the SE outlets or can be shared by all the SE inlets and outlets;

4. Defining handshake protocols between adjacent SEs in order to avoid packet loss in a buffered SE;

5. Providing external queueing when replicating unbuffered banyan networks, so that multiple packets addressing the same destination can be concurrently switched with success.

Section 6.1 describes the performance of the unbuffered banyan networks and describes networks designed according to criteria 1 and 2; therefore networks built of a single banyan plane or parallel banyan planes are studied. Criteria 3 and 4 are exploited in Section 6.2, which provides a thorough discussion of banyan architectures suitable to ATM switching in which each switching element is provided with an internal queueing capability. Section 6.3 discusses how a set of internally unbuffered networks can be used for ATM switching if queueing is available at switch outlets with an optional queueing capacity associated with network inlets according to criterion 5. Some final remarks concerning the switch performance under offered traffic patterns other than random and other architectures of ATM switches based on minimum-depth routing networks are finally given in Section 6.4.

6.1. Unbuffered Networks

The class of unbuffered networks is described now so as to provide the background necessary for a satisfactory understanding of the ATM switching architectures to be investigated in the next sections. The structure of the basic banyan network and its traffic performance are first discussed in relation to the behavior of the crossbar network. Then improved structures using the banyan network as the basic building block are examined: multiple banyan planes and multiple interstage links are considered.

6.1.1. Crossbar and basic banyan networks

The terminology and basic concepts of crossbar and banyan networks are here recalled and the corresponding traffic performance parameters are evaluated.

6.1.1.1. Basic structures

In principle, we would like any interconnection network (IN) to provide an optimum performance, that is maximum throughput ρ and minimum packet loss probability π. Packets are lost in general for two different reasons in unbuffered networks: conflicts for an internal IN resource, or *internal conflicts*, and conflicts for the same IN outlet, or *external conflicts*. The loss due to external conflicts is independent of the particular network structure and is unavoidable in an unbuffered network. Thus, the "ideal" unbuffered structure is the *crossbar network* (see Section 2.1) that is free from internal conflicts since each of the N^2 crosspoints is dedicated to each specific I/O couple.

An $N \times N$ banyan network built out of $b \times b$ SEs includes n stages of N/b SEs in which $n = \log_b N$. An example of a banyan network with Baseline topology and size $N = 16$ is given in Figure 6.1a for $b = 2$ and in Figure 6.1b for $b = 4$. As already explained in Section 2.3.1, internal conflicts can occur in banyan networks due to the link commonality of different I/O paths. Therefore the crossbar network can provide an upper bound on throughput and loss performance of unbuffered networks and in particular of unbuffered banyan networks.

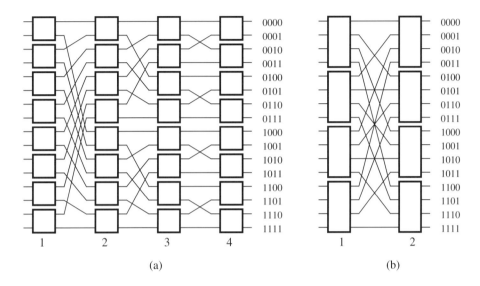

(a)

(b)

Figure 6.1. Example of banyan networks with Baseline topology

6.1.1.2. Performance

In an $N \times N$ crossbar network with random load, a specific output is idle in a slot when no packets are addressed to that port, which occurs with probability $(1 - p/N)^N$, so that the network throughput is immediately given by

$$\rho = 1 - \left(1 - \frac{p}{N}\right)^N \tag{6.1}$$

Once the switch throughput is known, the packet loss probability is simply obtained as

$$\pi = 1 - \frac{\rho}{p} = 1 - \frac{1 - \left(1 - \frac{p}{N}\right)^N}{p}$$

Thus, for an asymptotically large switch $(N \rightarrow \infty)$, the throughput is $1 - e^{-p}$ with a switch capacity $(p = 1.0)$ given by $\rho_{max} = 0.632$.

Owing to the random traffic assumption and to their single I/O path feature, banyan networks with different topologies are all characterized by the same performance. The traffic performance of unbuffered banyan networks was initially studied by Patel [Pat81], who expressed the throughput as a quadratic recurrence relation. An asymptotic solution was then provided for this relation by Kruskal and Snir. [Kru83]. A closer bound of the banyan network throughput was found by Kumar and Jump. [Kum86], who also give the analysis of replicated and dilated banyan networks to be described next. Further extensions of these results are reported by Szymanski and Hamacker. [Szy87].

The analysis given here, which summarizes the main results provided in these papers, relies on a simplifying assumption, that is the statistical independence of the events of packet arrivals at SEs of different stages. Such a hypothesis means overestimating the offered load stage by stage, especially for high loads [Yoo90].

The throughput and loss performance of the basic unbuffered $b^n \times b^n$ banyan network, which thus includes n stages of $b \times b$ SEs, can be evaluated by recursive analysis of the load on adjacent stages of the network. Let p_i $(i = 1, ..., n)$ indicate the probability that a generic outlet of an SE in stage i is "busy", that is transmits a packet (p_0 denotes the external load offered to the network). Since the probability that a packet is addressed to a given SE outlet is $1/b$, we can easily write

$$p_0 = p$$

$$p_i = 1 - \left(1 - \frac{p_{i-1}}{b}\right)^b \qquad (i = 1, ..., n) \tag{6.2}$$

Thus, throughput and loss are given by

$$\rho = p_n$$

$$\pi = 1 - \frac{p_n}{p_0}$$

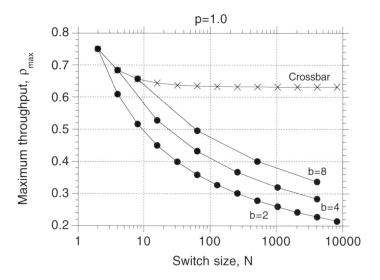

Figure 6.2. Switch capacity of a banyan network

The switch capacity, ρ_{max}, of a banyan network (Equation 6.2) with different sizes b of the basic switching element is compared in Figure 6.2 with that provided by a crossbar network (Equation 6.1) of the same size. The maximum throughput of the banyan network decreases as the switch size grows, since there are more packet conflicts due to the larger number of network stages. For a given switch size a better performance is given by a banyan network with a larger SE: apparently as the basic $b \times b$ SE grows, less stages are needed to build a banyan network with a given size N.

An asymptotic estimate of the banyan network throughput is computed in [Kru83]

$$\rho \cong \frac{2b}{(b-1)\,n + \dfrac{2b}{p}}$$

which provides an upper bound of the real network throughput and whose accuracy is larger for moderate loads and large networks. Figure 6.3 shows the accuracy of this simple bound for a banyan network loaded by three different traffic levels. The bound overestimates the real network throughput and the accuracy increases as the offered load p is lowered roughly independently of the switch size.

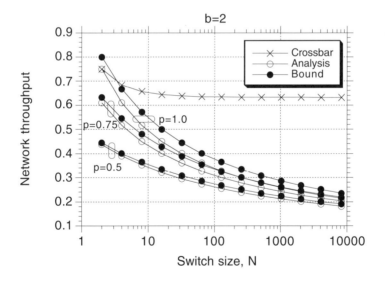

Figure 6.3. Switch capacity of a banyan network

It is also interesting to express π as a function of the loss probability $\pi_i = 1 - p_i / p_{i-1}$ $(i = 1, \ldots, n)$ occurring in the single stages. Since packets can be lost in general at any stage due to conflicts for the same SE outlet, it follows that

$$\pi = 1 - \prod_{i=1}^{n} (1 - \pi_i)$$

or equivalently by applying the theorem of total probability

$$\pi = \pi_1 + \sum_{i=2}^{n} \pi_i \prod_{h=1}^{i-1} (1 - \pi_h)$$

Therefore the loss probability can be expressed as a function of the link load stage by stage as

$$\pi = \pi_1 + \sum_{i=2}^{n} \pi_i \prod_{h=1}^{i-1} (1 - \pi_h) = 1 - \frac{p_1}{p_0} + \sum_{i=2}^{n} \left(1 - \frac{p_i}{p_{i-1}} \right) \prod_{h=1}^{i-1} \frac{p_h}{p_{h-1}} = \sum_{i=1}^{n} \frac{p_{i-1} - p_i}{p_0} \quad (6.3)$$

For the case of $b = 2$ the stage load given by Equation 6.2 assumes an expression that is worth discussion, that is

$$p_i = 1 - \left(1 - \frac{p_{i-1}}{2} \right)^2 = p_{i-1} - \frac{1}{4} p_{i-1}^2 \qquad (i = 1, ..., n) \qquad (6.4)$$

Equation 6.4 says that the probability of a busy link in stage i is given by the probability of a busy link in the previous stage $i - 1$ decreased by the probability that both the SE inlets are receiving a packet (p_{i-1}^2) and both packets address the same SE outlet $(1/4)$. So, the loss probability with 2×2 SEs given by Equation 6.3 becomes

$$\pi = \sum_{i=1}^{n} \frac{p_{i-1} - p_i}{p_0} = \sum_{i=1}^{n} \frac{\frac{1}{4} p_{i-1}^2}{p_0} \qquad (6.5)$$

6.1.2. Enhanced banyan networks

Interconnection networks based on the use of banyan networks are now introduced and their traffic performance is evaluated.

6.1.2.1. Structures

Improved structures of banyan interconnection networks were proposed [Kum86] whose basic idea is to have multiple internal paths per inlet/outlet pair. These structures either adopt multiple banyan networks in parallel or replace the interstage links by multiple parallel links.

An $N \times N$ interconnection network can be built using K parallel $N \times N$ networks (planes) interconnected to a set of N splitters $1 \times K$ and a set of N combiners $K \times 1$ through suitable input and output interconnection patterns, respectively, as shown in Figure 6.4. These structures are referred to as *replicated banyan networks* (RBN), as the topology in each plane is banyan or derivable from a banyan structure. The splitters can distribute the incoming traffic in different modes to the banyan networks; the main techniques are:

- *random loading* (RL),
- *multiple loading* (ML),
- *selective loading* (SL).

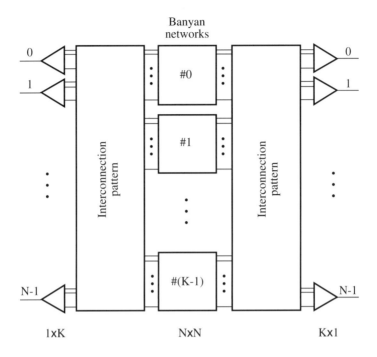

Figure 6.4. Replicated Banyan Network

RBNs with random and multiple loading are characterized by full banyan networks, the same input and output interconnection patterns, and different operations of the splitters, whereas selective loading uses "truncated" banyan networks and two different types of interconnection pattern. In all these cases each combiner that receives more than one packet in a slot discards all but one of these packets.

A replicated banyan network operating with RL or ML is represented in Figure 6.5: both interconnection patterns are of the EGS type (see Section 2.1). With random loading each splitter transmits the received packet to a randomly chosen plane out of the $K = K_r$ planes with even probability $1/K_r$. The aim is to reduce the load per banyan network so as to increase the probability that conflicts between packets for interstage links do not occur. Each received packet is broadcast concurrently to all the $K = K_m$ planes with multiple loading. The purpose is to increase the probability that at least one copy of the packet successfully reaches its destination.

Selective loading is based on dividing the outlets into $K = K_s$ disjoint subsets and dedicating each banyan network suitably truncated to one of these sets. Therefore one EGS pattern of size NK_s connects the splitters to the K_s banyan networks, whereas K_s suitable patterns (one per banyan network) of size N must be used to guarantee full access to all the combiners from every banyan inlet. The splitters selectively load the planes with the traffic addressing their respective outlets. In order to guarantee full connectivity in the interconnection network, if each banyan network includes $n - k$ stages $(k = \log_b K_s)$, the splitters transmit each packet to

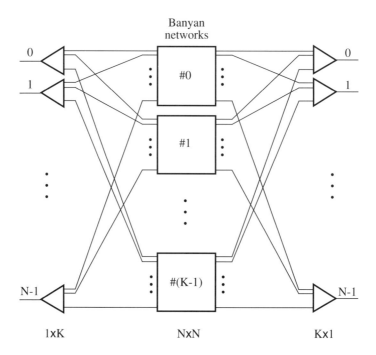

Figure 6.5. RBN with random or multiple loading

the proper plane using the first k digits (in base b) of the routing tag. The example in Figure 6.6 refers to the case of $N = 16$, $b = 2$ and $K_s = 2$ in which the truncated banyan network has the reverse Baseline topology with the last stage removed. Note that the connection between each banyan network and its combiners is a perfect shuffle (or EGS) pattern. The target of this technique is to reduce the number of packet conflicts by jointly reducing the offered load per plane and the number of conflict opportunities.

Providing multiple paths per I/O port, and hence reducing the packet loss due to conflicts for interstage links, can also be achieved by adopting a multiplicity $K = K_d$ $(K_d \geq 2)$ of physical links for each "logical" interstage link of a banyan network (see Figure 4.10 for $N = 16$, $b = 2$ and $K_d = 2$). Now up to K_d packets can be concurrently exchanged between two SEs in adjacent stages. These networks are referred to as *dilated banyan networks* (DBN). Such a solution makes the SE, whose physical size is now $2K_d \times 2K_d$, much more complex than the basic 2×2 SE. In order to drop all but one of the packets received by the last stage SEs and addressing a specific output, $K_d \times 1$ combiners can be used that concentrate the K_d physical links of a logical outlet at stage n onto one interconnection network output. However, unlike replicated networks, this concentration function could be also performed directly by each SE in the last stage.

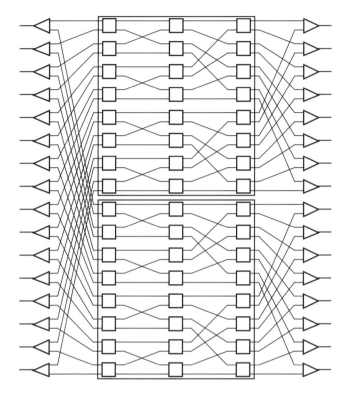

Figure 6.6. Example of RBN with selective loading

6.1.2.2. Performance

Analysis of replicated and dilated banyan networks follows directly from the analysis of a single banyan network. Operating a random loading of the K planes means evenly partitioning the offered load into K flows. The above recursive analysis can be applied again considering that the offered load per plane is now

$$p_0 = \frac{p}{K}$$

Throughput and loss in this case are

$$\rho = 1 - (1 - p_n)^K \tag{6.6}$$

$$\pi = 1 - \frac{\rho}{p} = 1 - \frac{1 - (1 - p_n)^K}{p} \tag{6.7}$$

For multiple loading it is difficult to provide simple expressions for throughput and delay. However, based on the results given in [Kum86], its performance is substantially the same as the random loading. This fact can be explained considering that replicating a packet on all planes increases the probability that at least one copy reaches the addressed output, as the choice for packet discarding is random in each plane. This advantage is compensated by the drawback of a higher load in each plane, which implies an increased number of collision (and loss) events.

With selective loading, packet loss events occur only in $n - k$ stages of each plane and the offered load per plane is still p_0 / K. The packet loss probability is again given by $\pi = 1 - \rho / p$ with the switch throughput provided by

$$\rho = 1 - (1 - p_{n-k})^K$$

since each combiner can receive up to K packets from the plane it is attached to.

In dilated networks each SE has size $bK \times bK$, but not all physical links are active, that is enabled to receive packets. SEs have 1 active inlet and b active outlets per logical port at stage 1, b active inlets and b^2 active outlets at stage 2, K active inlets and K active outlets from stage k onwards $(k = \log_b K)$. The same recursive load computation as described for the basic banyan network can be adopted here taking into account that each SE has bK physical inlets and b logical outlets, and that not all the physical SE inlets are active in stages 1 through $k - 1$. The event of m packets transmitted on a tagged link of an SE in stage i $(1 \leq i \leq n)$, whose probability is $p_i(m)$, occurs when $j \geq m$ packets are received by the SE from its b upstream SEs and m of these packets address the tagged logical outlet. If $p_0(m)$ denotes the probability that m packets are received on a tagged inlet an SE in stage 1, we can write

$$p_0(m) = \begin{cases} 1 - p & m = 0 \\ p & m = 1 \\ 0 & m > 1 \end{cases}$$

$$p_i(m) = \begin{cases} \displaystyle\sum_{j=m}^{bK} \binom{j}{m} \left(\frac{1}{b}\right)^m \left(1 - \frac{1}{b}\right)^{j-m} \sum_{m_1 + \ldots + m_b = j} p_{i-1}(m_1) \ldots p_{i-1}(m_b) & (m < K) \\ \displaystyle\sum_{j=m}^{bK} \sum_{h=m}^{j} \binom{j}{h} \left(\frac{1}{b}\right)^h \left(1 - \frac{1}{b}\right)^{j-h} \sum_{m_1 + \ldots + m_b = j} p_{i-1}(m_1) \ldots p_{i-1}(m_b) & (m = K) \end{cases}$$

The packet loss probability is given as usual by $\pi = 1 - \rho / p$ with the throughput provided by

$$\rho = 1 - p_n(0)$$

The switch capacity, ρ_{max}, of different configurations of banyan networks is shown in Figure 6.7 in comparison with the crossbar network capacity. RBNs with random and selective loading have been considered with $K_r = 2, 4$ and $K_s = 2, 4$, respectively. A dilated banyan network with link dilation factors $K_d = 2, 4$ has also been studied. RBN with random and selective loading give a comparable throughput performance, the latter behaving a little better. A dilated banyan network with dilation factor K_d behaves much better than an RBN network with replication factor $K = K_d$. The dilated banyan network with $K_d = 4$

gives a switch throughput very close to crossbar network capacity. Nevertheless the overall complexity of a dilated banyan network is much higher than in RBNs (more complex SEs are required). For example a network with size $N = 32$ and $K = K_d = 2$ includes 160 SEs of size 2×2 in an RBN and 80 SEs of size 4×4. Since we have no queueing in unbuffered banyan networks, the packet delay figure is not of interest here.

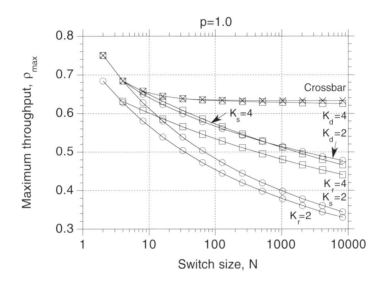

Figure 6.7. Switch capacity for different banyan networks

So far we have studied the traffic performance of unbuffered banyan networks with random offered traffic in which both internal and external conflicts among packets contribute to determine the packet loss probability. A different pattern of offered traffic consists in a set of packets that does not cause external conflicts, that is each outlet is addressed by at most one packet. A performance study of these patterns, referred to as permutations, is reported in [Szy87].

6.2. Networks with a Single Plane and Internal Queueing

In general the I/O paths along which two packets are transmitted are not link-independent in a banyan network. Thus, if two or more ATM cells address the same outlet of an SE (that is the same interstage link or the same switch output), only one of them can actually be transmitted to the requested outlet. The other cell is lost, unless a storage capability is available in the SE. We assume that each switching element with size $b \times b$ is provided with a queueing capacity per port of B cells per port and we will examine here three different types of arrangements of this memory in the SE: *input queueing*, *output queueing* and *shared queueing*. With input and output queueing b physical queues are available in the SE, whereas only one is available with shared queueing. In this latter case the buffer is said to include b logical queues, each holding the packets addressing a specific SE outlet. In all the buffered SE structure considered here we assume a FIFO cell scheduling, as suggested by simplicity requirements for hardware implementation.

Various internal protocols are considered in our study, depending on the absence or presence of signalling between adjacent stages to enable the downstream transmission of a packet by an SE. In particular we define the following internal protocols:

- **backpressure** (BP): signals are exchanged between switching elements in adjacent stages so that the generic SE can grant a packet transmission to its upstream SEs only within the current idle buffer capacity. The upstream SEs enabled to transmit are selected according to the *acknowledgment* or *grant* mode, whereas the number of idle buffer positions is determined based on the type of backpressure used, which can be either *global* (GBP) or *local* (LBP). These operations are defined as follows:

  — **acknowledgment** (ack): the generic SE in stage i ($1 \leq i < n$) issues as many *requests* as the number of SE outlets addressed by head-of-line (HOL) packets, each transmitted to the requested downstream SE. In response, each SE in stage i ($1 < i \leq n$) enables the transmission by means of *acknowledgments* to all the requesting upstream SEs, if their number does not exceed its idle buffer positions, determined according to the GBP or LBP protocol; otherwise the number of enabled upstream SEs is limited to those needed to saturate the buffer;

  — **grant** (gr): without receiving any requests, the generic SE in stage i ($1 < i \leq n$) grants the transmission to all the upstream SEs, if its idle buffer positions, n_{idle}, are at least b; otherwise only n_{idle} upstream SEs are enabled to transmit; unlike the BP-ack protocol, the SE can grant an upstream SE whose corresponding physical or logical queue is empty with the BP-gr operations;

  — **local backpressure** (LBP): the number of buffer places that can be filled in the generic SE in stage i ($1 < i \leq n$) at slot t by upstream SEs is simply given by the number of idle positions at the end of the slot $t - 1$;

  — **global backpressure** (GBP): the number of buffer places that can be filled in the generic SE in stage i ($1 < i \leq n$) at slot t by upstream SEs is given by the number of idle positions at the end of the slot $t - 1$ increased by the number of packets that are going to be transmitted by the SE in the slot t;

- **queue loss** (QL): there is no exchange of signalling information within the network, so that a packet per non-empty physical or logical queue is always transmitted downstream by each SE, independent of the current buffer status of the destination SE; packet storage in the SE takes place as long as there are enough idle buffer positions, whereas packets are lost when the buffer is full.

From the above description it is worth noting that LBP and GBP, as well as BP-ack and BP-gr, result in the same number of upstream acknowledgment/grant signals by an SE if at least b positions are idle in its buffer at the end of the preceding slot. Moreover, packets can be lost for queue overflow only at the first stage in the BP protocols and at any stage in the QL protocol. In our model the selection of packets to be backpressured in the upstream SE (BP) or to be lost (QL) in case of buffer saturation is always random among all the packets competing for the access to the same buffer. Note that such general description of the internal protocols applied to the specific type of queueing can make meaningless some cases.

The implementation of the internal backpressure requires additional internal resources to be deployed compared to the absence of internal protocols (QL). Two different solutions can be devised for accomplishing interstage backpressure, that is in the space domain or in the time domain. In the former case additional internal links must connect any couple of SEs interfaced by interstage links. In the latter case the interstage links can be used on a time division base to transfer both the signalling information and the ATM cells. Therefore an internal bit rate, C_i, higher than the link external rate, C (bit/s), is required. With the acknowledgment BP we have a two-phase signalling: the *arbitration phase* where all the SEs concurrently transmit their requests downstream and the *enable phase* where each SE can signal upstream the enabling signal to a suitable number of requesting SEs. The enable phase can be accomplished concurrently by all SEs with the local backpressure, whereas it has be a sequential operation with global backpressure. In this last case an SE needs to know how many packets it is going to transmit in the current slot to determine how many enable signals can be transmitted upstream, but such information must be first received by the downstream SEs. Thus the enable phase of the BP-ack protocol is started by SEs in stage n and ends with the receipt of enable signal by SEs in stage 1. Let l_d and l_u (bit) be the size of each downstream and upstream signalling packet, respectively, and l_c (bit) the length of an information packet (cell). Then the internal bit rate is $C_i = C$ for the QL protocol and $C_i = (1 + \eta) C$ for the BP protocol where η denotes the *switching overhead*. This factor in the BP protocol with acknowledgment is given by

$$
\eta = \begin{cases} \dfrac{l_d + (n-1) l_u}{l_c} & GBP \\[2ex] \dfrac{l_d + l_u}{l_c} & LBP \end{cases} \tag{6.8}
$$

In the BP protocol with grant we do not have any request phase and the only signalling is represented by the enable phase that is performed as in the case of the BP-ack protocol. Thus the internal rate C_i of the BP-gr protocol is given by Equation 6.8 setting $l_d = 0$.

The network is assumed to be loaded by purely random and uniform traffic; that is at stage 1:

1. A packet is received with the same probability in each time slot;
2. Each packet is given an outlet address that uniformly loads all the network outlets;
3. Packet arrival events at different inlets in the same time slots are mutually independent;
4. Packet arrival events at an inlet or at different inlets in different time slot are mutually independent.

Even if we do not provide any formal proof, assumption 2 is likely to be true at every stage, because of general considerations about flow conservation across stages. The independence assumption 3 holds for every network stage in the QL mode, since the paths leading to the different inlets of an SE in stage i cross different SEs in stage $j < i$ (recall that one path through the network connects each network inlet to each network outlet). Owing to the memory device in each SE, the assumption 4, as well as the assumption 3 for the BP protocol, no longer holds in stages other than the first. For simplicity requirements the assumption 3 is supposed to be always true in all the stages in the analytical models to be developed later. In spite of the correlation in packet arrival events at a generic SE inlet in stages 2 through n, our models assume independence of the state of SEs in different stages. Such a correlation could be taken into account by suitably modelling the upstream traffic source loading each SE inlet. Nevertheless, in order to describe simple models, each upstream source will be represented here by means of only one parameter, the average load.

We assume independence between the states of SEs in the same stage, so that one SE per stage is representative of the behavior of all the elements in the same stage (SE_i will denote such an element for stage i). For this reason the topology of the network, that is the specific kind of banyan network, does not affect in any way the result that we are going to obtain. As usual we consider $N \times N$ banyan networks with $b \times b$ switching elements, thus including $n = \log_b N$ stages.

Buffered banyan networks were initially analyzed by Dias and Jump [Dia81], who only considered asymptotic loads, and by Jenq [Jen83], who analyzed the case of single-buffered input-queued banyan networks loaded by a variable traffic level. The analysis of buffered banyan networks was extended by Kumar and Jump [Kum84], so as to include replicated and dilated buffered structures. A more general analysis of buffered banyan networks was presented by Szymanski and Shiakh [Szy89], who give both separate and combined evaluation of different SE structures, such as SE input queueing, SE output queueing, link dilation. The analysis given in this section for networks adopting SEs with input queueing or output queueing is based on this last paper and takes into account the modification and improvements described in [Pat91], mainly directed to improve the computational precision of network throughput and cell loss. In particular, the throughput is only computed as a function of the cell loss probability and not vice versa.

As far as networks with shared-queued SEs are concerned, some contributions initially appeared in the technical literature [Hlu88, Sak90, Pet90], basically aiming at the study of a single-stage network (one switching element). Convolutional approaches are often used that assume mutual independence of the packet flows addressing different destinations. Analytical models for multistage structures with shared-buffered SEs have been later developed in [Tur93] and [Mon92]. Turner [Tur93] proposed a simple model in which the destinations of the packets in the buffer were assumed mutually independent. Monterosso and Pattavina [Mon92] developed an exact Markovian model of the switching element, by introducing modelling approximation only in the interstage traffic. The former model gave very inaccurate results, whereas the latter showed severe limitation in the dimensions of the networks under study. The model described here is the simplest of the three models described in [Gia94] in which the SE state is always represented as a two-state variable. The other two more complex models therein, not developed here, take into account the correlation of the traffic received at any stage other than the first.

6.2.1. Input queueing

The functional structure of a 2×2 SE with input queueing, shown in Figure 6.8 in the solution with additional interstage links for signalling purposes, includes two (local) queues, each with capacity $B = B_i$ cells, and a controller. Each of the local queues, which interface directly the upstream SEs, performs a single read and write operation per slot. The controller receives signals from the (remote) queues of the downstream SEs and from the local queues when performing the BP protocol. With this kind of queueing there is no need for an arbitration phase with downstream signalling, since each queue is fed by only one upstream SE. Thus the BP protocol can only be of the grant type. Nevertheless, arbitration must take place slot by slot by the SE controller to resolve possible conflicts arising when more than one HOL cell of the local queues addresses the same SE outlet.

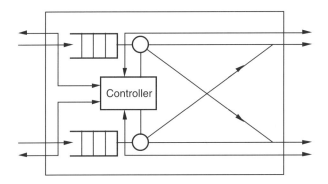

Figure 6.8. SE with input queueing

Packet transmissions to downstream SEs (or network outlets) and packet receipt from upstream SEs (or network inlets) take place concurrently in the SE at each time slot. For the sake of better understanding the protocols QL and GBP, we can well imagine for an SE that packet transmissions occur in the first half of the slot, whereas packet receipts take place in the second half of the slot based on the empty buffer space at the end of the first phase. With the LBP protocol there is no need for such decomposition as the amount of packets to be received is independent of the packets to be transmitted in the slot. In such a way we can define a virtual half of each time slot that separates transmissions from receipts.

In order to develop analytical models for the network, it turns out useful to define the following probability distributions to characterize the dynamic of the generic input queue of the SE, the *tagged* queue:

- $d_{i,t}(m)$ = Pr [the tagged queue at stage i at time t contains m packets];
- $d'_{i,t}(m)$ = Pr [the tagged queue at stage i at time t contains m packets if we consider to be removed those packets that are going to be transmitted in the slot t];
- $a_{i,t}$ = Pr [an SE at stage i at time t offers a packet to a queue at stage $i+1$]; $a_0 = p$ denoted the external offered load;
- $b_{i,t}$ = Pr [a packet offered by a queue at stage i at time t is actually transmitted by the queue];

- $c_{i,t}$ = Pr [a packet offered by a queue at stage i at time t is selected for transmission].

Note that the $d'_{i,t}(m)$ denotes the probability distribution of the tagged queue at the half-time slot if transmission and receipt of packets occur sequentially in the slot. The LBP protocol does not require the definition of the distribution $d'_{i,t}$, as the ack/grant signals depend only on the idle buffer space at the end of the last slot. Moreover, for the sake of simplicity, the following notation is used:

$$\beta(N, i, p) = \binom{N}{i} p^i (1-p)^{N-i}$$

In the following, time-dependent variables without the subscript t indicate the steady-state value assumed by the variable.

The one-step transition equations for the protocols QL and GBP describing the dynamic of the tagged queue due first to cell transmissions and then to the cell receipts are easily obtained:

$$d'_{i,t}(0) = d_{i,t-1}(1) b_{i,t} + d_{i,t-1}(0)$$

$$d'_{i,t}(h) = d_{i,t-1}(h+1) b_{i,t} + d_{i,t-1}(h) (1-b_{i,t}) \qquad (1 \le h \le B_i - 1)$$

$$d'_{i,t}(B_i) = d_{i,t-1}(B_i) (1-b_{i,t})$$

$$d_{i,t}(0) = d'_{i,t}(0) (1-a_{i-1,t})$$

$$d_{i,t}(h) = d'_{i,t}(h-1) a_{i-1,t} + d'_{i,t}(h) (1-a_{i-1,t}) \qquad (1 \le h \le B_i - 1)$$

$$d_{i,t}(B_i) = d'_{i,t}(B_i) + d'_{i,t}(B_i - 1) a_{i-1,t}$$

The analogous equations for the LBP protocol with $B_i \ge 3$ are

$$d_{i,t}(0) = d_{i,t-1}(1) (1-a_{i-1,t}) b_{i,t} + d_{i,t-1}(0) (1-a_{i-1,t})$$

$$d_{i,t}(1) = d_{i,t-1}(2) (1-a_{i-1,t}) b_{i,t} + d_{i,t-1}(1) [a_{i-1,t} b_{i,t} + (1-a_{i-1,t}) (1-b_{i,t})]$$
$$+ d_{i,t-1}(0) a_{i-1,t}$$

$$d_{i,t}(h) = d_{i,t-1}(h+1) (1-a_{i-1,t}) b_{i,t} + d_{i,t-1}(h) [a_{i-1,t} b_{i,t} + (1-a_{i-1,t}) (1-b_{i,t})]$$
$$+ d_{i,t-1}(h-1) a_{i-1,t} (1-b_{i,t}) \qquad (1 < h < B_i - 1)$$

$$d_{i,t}(B_i - 1) = d_{i,t-1}(B_i) b_{i,t} + d_{i,t-1}(B_i - 1) [a_{i-1,t} b_{i,t} + (1-a_{i-1,t}) (1-b_{i,t})]$$
$$+ d_{i,t-1}(B_i - 2) a_{i-1,t} (1-b_{i,t})$$

$$d_{i,t}(B_i) = d_{i,t-1}(B_i) (1-b_{i,t}) + d_{i,t-1}(B_i - 1) a_{i-1,t} (1-b_{i,t})$$

which for $B_i = 2$ reduce to

$$d_{i,t}(0) = d_{i,t-1}(1)(1 - a_{i-1,t})b_{i,t} + d_{i,t-1}(0)(1 - a_{i-1,t})$$

$$d_{i,t}(1)$$
$$= d_{i,t-1}(2)b_{i,t} + d_{i,t-1}(1)[a_{i-1,t}b_{i,t} + (1 - a_{i-1,t})(1 - b_{i,t})] + d_{i,t-1}(0)a_{i-1,t}$$

$$d_{i,t}(2) = d_{i,t-1}(2)(1 - b_{i,t}) + d_{i,t-1}(1)a_{i-1,t}(1 - b_{i,t})$$

and for $B_i = 1$ to

$$d_{i,t}(0) = d_{i,t-1}(1)b_{i,t} + d_{i,t-1}(0)(1 - a_{i-1,t})$$
$$d_{i,t}(1) = d_{i,t-1}(1)(1 - b_{i,t}) + d_{i,t-1}(0)a_{i-1,t}$$

Based on the independence assumption of packet arrivals at each stage, the distribution probability of $a_{i,t}$ is immediately obtained:

$$a_{i,t} = 1 - \left(1 - \frac{1 - d_{i,t-1}(0)}{b}\right)^b \qquad (1 \le i \le n) \qquad (6.9)$$

with the boundary condition

$$a_{0,t} = p$$

Since the probability $c_{i,t}$ that a HOL packet is selected to be transmitted to the downstream SE is

$$c_{i,t} = \sum_{j=0}^{b-1} \beta\left(b - 1, j, \frac{1 - d_{i,t}(0)}{b}\right)\frac{1}{j+1} \qquad (1 \le i \le n)$$

the distribution probability of $b_{i,t}$ is given by

$$b_{i,t} = \begin{cases} c_{i,t} & (1 \le i \le n) & QL \\ c_{i,t}[1 - d_{i+1,t}(B_i)] & (1 \le i \le n-1) & LBP \\ c_{i,t}[1 - d'_{i+1,t}(B_i)] & (1 \le i \le n-1) & GBP \\ c_{n,t} & (i = n) & BP \end{cases}$$

An iterative approach is used to solve this set of equations in which we compute all the state variables from stage 1 to stage n using the values obtained in the preceding iteration for the unknowns. A steady state is reached when the relative variation in the value assumed by the variables is small enough. Assuming that a suitable and consistent initial value for these variables is assigned, we are so able to evaluate the overall network performance.

Packet losses take place only at stage 1 with backpressure, whereas in the QL mode a packet is lost at stage i $(i \geq 2)$ only if it is not lost in stages 1 through $i-1$, that is

$$\pi = \begin{cases} d'_1(B_i) + \sum_{i=2}^{n} d'_i(B_i) \prod_{j=1}^{i-1} (1 - d'_j(B_i)) & QL \\ d_1(B_i) & LBP \\ d'_1(B_i) & GBP \end{cases}$$

Moreover the switch throughput, ρ, is the traffic carried by the last stage

$$\rho = a_n \tag{6.10}$$

and the average packet delay, T, is straightforwardly obtained through the Little's formula

$$T = \begin{cases} \dfrac{1}{n} \sum_{i=1}^{n} \dfrac{1}{a_i} \sum_{h=0}^{B_i} h d_i(h) & QL \\ \dfrac{1}{n\rho} \sum_{i=1}^{n} \sum_{h=0}^{B_i} h d_i(h) & BP \end{cases} \tag{6.11}$$

The accuracy of the analytical model in terms of packet loss probability is assessed in Figures 6.9-6.11 by comparing data obtained from the model with results given by computer simulation for a network with $N = 256$ and $b = 2$ (hence the network includes eight stages). In these figures the overall buffer capacity per SE $B_t = bB_i$ has been chosen ranging from $B_t = 4$ to $B_t = 32$ cells. The best accuracy is attained with the GBP protocol especially if low offered loads are considered, whereas the model for LBP and QL turns out to be less accurate.

The loss performance given by the analytical model for three protocols GBP, LBP and QL for the same buffer size is shown in Figure 6.12. As one might expect, the GBP protocol gives the best performance and behaves significantly better than the other two protocols especially for small buffers. Apparently, if the buffer is quite large the performance improvement enabled by the exploiting of the buffer positions (at most one with IQ) being emptied in the same slot (GBP over LBP) becomes rather marginal.

6.2.2. Output queueing

With output queueing, the (local) queues of the SE, each with capacity $B = B_o$ cells, interface the SE outlets, as represented in Figure 6.13 for a 2×2 SE in the space division solution for the inter-stage signalling. Now switching precedes rather than following queueing so that each queue must be able to perform up to b write and 1 read operations per slot. The SE controller exchanges information with the SEs in the adjacent stages and with the local queues when the BP protocol is operated. In case of possible saturation of any local queues, it is a task of the SE controller to select the upstream SEs enabled to transmit a packet without overflow-

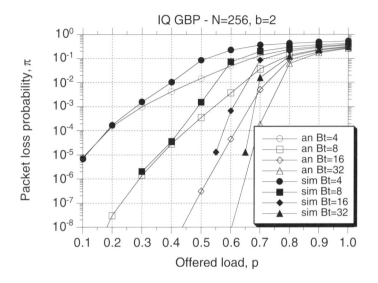

Figure 6.9. Loss performance with IQ and GBP

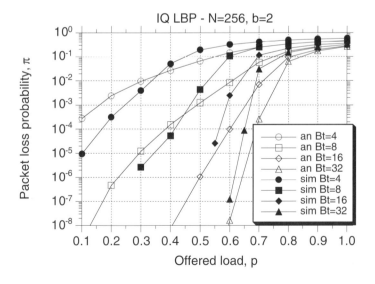

Figure 6.10. Loss performance with IQ and LBP

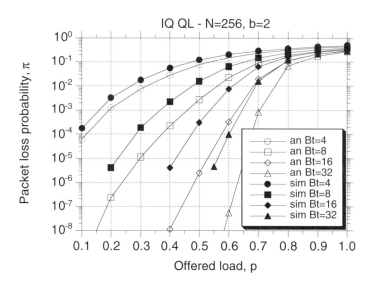

Figure 6.11. Loss performance with IQ and QL

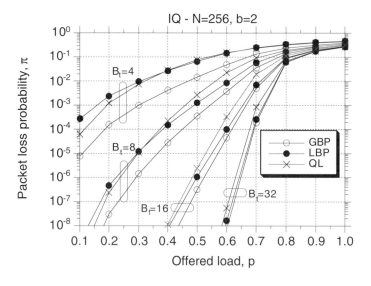

Figure 6.12. Loss performance with IQ and different protocols

ing the local queue capacity. Note that now there is no need of arbitration by the SE controller in the downstream packet transmission as each local queue feeds only one downstream SE.

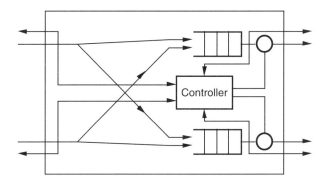

Figure 6.13. SE with output queueing

In general output-queued SEs are expected to perform better than input-queued SEs. In fact, in the latter structure the HOL blocking can take place, that is a HOL cell is not transmitted owing to a conflict with the HOL cell of other local queue(s) for the same SE outlet, thus reducing the SE throughput. With output-queued SEs each local queue has exclusive access to a SE outlet and eventual multiple cell arrivals from upstream SEs are handled through suitable hardware solutions. Thus, SEs with output queueing are much more complex than SEs with input queueing.

With output queueing the one-step transitions equations for the protocols QL and GBP describing the dynamics of the tagged output queue due to packet transmissions are then given by

$$d'_{i,t}(0) = d_{i,t-1}(1)\, b_{i,t} + d_{i,t-1}(0)$$

$$d'_{i,t}(h) = d_{i,t-1}(h+1)\, b_{i,t} + d_{i,t-1}(h)\,(1-b_{i,t}) \qquad (1 \le h \le B_o - 1)$$

$$d'_{i,t}(B_o) = d_{i,t-1}(B_o)\,(1-b_{i,t})$$

A different behavior characterizes the SE dynamics under BP protocol with acknowledgment or grant, when the number of idle places in the buffer is less than the SE size b. Therefore the evolution of the tagged output queue due to packet receipt under QL or BP-ack is described by

$$d_{i,t}(s) = \begin{cases} \displaystyle\sum_{h=\max(0,\,s-b)}^{s} d'_{i,t}(h)\, \beta\!\left(b, s-h, \frac{a_{i-1,t}}{b}\right) & (0 \le s \le B_o - 1) \\[2em] \displaystyle\sum_{h=\max(0,\,B_o-b)}^{B_o} d'_{i,t}(h) \sum_{j=B_o-h}^{b} \beta\!\left(b, j, \frac{a_{i-1,t}}{b}\right) & (s = B_o) \end{cases}$$

and under the BP-gr protocol by

$$
d_{i,t}(s) = \begin{cases}
\displaystyle\sum_{h=\max(0,\,s-b)}^{s} d'_{i,t}(h)\,\beta\!\left(\min(b, B_o - h),\, s-h, \dfrac{a_{i-1,t}}{b}\right) & (0 \le s \le B_o - 1) \\[3ex]
\displaystyle\sum_{h=\max(0,\,B_o-b)}^{B_o} d'_{i,t}(h) \sum_{j=B_o-h}^{\min(b,\,B_o-h)} \beta\!\left(\min(b, B_o-h),\, j, \dfrac{a_{i-1,t}}{b}\right) & (s = B_o)
\end{cases}
$$

After defining the function

$$
X(h) = \begin{cases}
b_{i,t} & (h > 0) \\
0 & (h = 0)
\end{cases}
$$

which represents the probability that a queue holding h packets transmits a packet, the one-step transition equations in the case of LBP-ack protocol are

$$
d_{i,t}(s) = \sum_{h=\max(0,\,s-b)}^{s} d_{i,t-1}(h)\,[1 - X(h)]\,\beta\!\left(b, s-h, \dfrac{a_{i-1,t}}{b}\right)
$$

$$
+ \sum_{h=\max(0,\,s-b+1)}^{s+1} d_{i,t-1}(h)\,X(h)\,\beta\!\left(b, s-h+1, \dfrac{a_{i-1,t}}{b}\right) \quad (0 \le s \le B_o - 2)
$$

$$
d_{i,t}(B_o - 1) = \sum_{h=\max(0,\,B_o-b-1)}^{B_O-1} d_{i,t-1}(h)\,[1 - X(h)]\,\beta\!\left(b, s-h, \dfrac{a_{i-1,t}}{b}\right)
$$

$$
+ \sum_{h=\max(0,\,B_O-b)}^{B_O} d_{i,t-1}(h)\,X(h) \sum_{j=B_O-h}^{b} \beta\!\left(b, j, \dfrac{a_{i-1,t}}{b}\right)
$$

$$
d_{i,t}(B_o) = \sum_{h=\max(0,\,B_O-b)}^{B_O} d_{i,t-1}(h)\,[1 - X(h)] \sum_{j=B_O-h}^{b} \beta\!\left(b, j, \dfrac{a_{i-1,t}}{b}\right)
$$

The analogous equations for the LBP-gr protocol are obtained by simply replacing b with $\min(b, B_o - h)$ when b appears as first parameter in the function β and as superior edge in a sum.

The distributions $a_{i,t}$ and $b_{i,t}$ for the GBP protocol are given by

$$
a_{i,t} = 1 - d_{i,t-1}(0) \qquad (1 \le i \le n)
$$

$$
b_{i,\,t} =
\begin{cases}
\displaystyle\sum_{h=0}^{B_o} d'_{i+1,\,t}(h) \sum_{j=0}^{b-1} \beta\!\left(b-1, j, \frac{1-d_{i,\,t}(0)}{b}\right) \min\!\left(\frac{B_o-h}{j+1}, 1\right) & (1 \le i \le n-1) \quad \text{ack} \\[4mm]
\displaystyle\sum_{h=0}^{B_o} d'_{i+1,\,t}(h) \min\!\left(\frac{B_o-h}{b}, 1\right) & (1 \le i \le n-1) \quad \text{gr} \\[4mm]
1 & (i = n)
\end{cases}
$$

Note that $min\,(x/y,\,1)$ denotes the probability that the HOL packet of the tagged queue in stage i is actually granted the transmission given that x positions are available in the downstream buffer and y SEs in stage i compete for them. These y elements are the tagged queue together with other $y-1$ non-empty queues addressing the same SE outlet in stage $i+1$ under the acknowledgment protocol, just the b SEs in stage i interfacing the same SE of stage i+1 as the tagged queue under the grant protocol. This probability value becomes 1 for $x > y$, since all the contending packets, including the HOL packet in the tagged queue, are accepted downstream.

The analogous equations for the LBP protocol are obtained by simply replacing $d'_{i+1,\,t}(h)$ with $d_{i+1,\,t}(h)$, whereas for the QL mode we obviously have

$$
b_{i,\,t} = 1 \qquad (1 \le i \le n)
$$

After applying the iterative computation of these equations already described in the preceding Section, the steady-state network performance measures are obtained. Throughput and delay figures are expressed as in the case of input queueing, so that the throughput value is given by Equation 6.10 and the delay by Equation 6.11. The packet loss probability is now

$$
\pi =
\begin{cases}
\displaystyle\pi = \pi_1 + \sum_{i=2}^{n} \pi_i \prod_{j=1}^{i-1} (1 - \pi_j) & QL \\[4mm]
\displaystyle\sum_{h=0}^{B_o} d_1(h)\,\theta_1\,(B_o - h) & LBP \\[4mm]
\displaystyle\sum_{h=0}^{B_o} d'_1(h)\,\theta_1\,(B_o - h) & GBP
\end{cases}
$$

where π_i is the loss probability at stage i and $\theta_i(x)$ represents the probability that a packet offered to a memory with x idle positions in stage i is refused. These variables are obtained as

$$
\pi_i = \sum_{h=0}^{B_o} d'_i(h)\,\theta_i\,(B_o - h)
$$

$$\theta_i(x) = \begin{cases} \displaystyle\sum_{r=x}^{b-1} \beta\left(b-1, r, \frac{a_{i-1,t}}{b}\right)\frac{r+1-x}{r+1} & (0 \le x \le b-1) & QL, BP-\text{ack} \\ 0 & (x \ge b) & QL, BP-\text{ack} \\ 1 - \min\left(1, \dfrac{x}{b}\right) & & BP-\text{gr} \end{cases}$$

As with IQ, we assess now the accuracy of the analytical model by considering a network with $N = 256$ and $b = 2$ with a total buffer capacity per SE $B_t = bB_o$ in the range of cells 4–32 cells. Now the GBP protocol with acknowledgment gives a very good matching with simulation data as with input queueing (Figure 6.14), whereas the same is no more true when grant is used (Figure 6.15). The degree of accuracy in evaluating loss probabilities by the GBP-gr protocols applies also to the LBP protocols, in both acknowledgment and grant versions. In the case of the QL protocol, the model accuracy with output queueing is comparable with that shown in Figure 6.11 for input queueing.

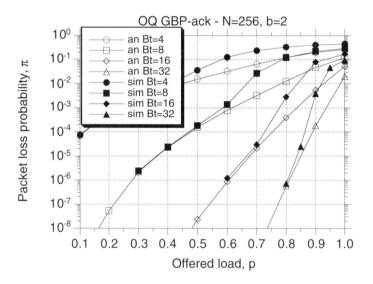

Figure 6.14. Loss performance with OQ and GBP-ack

The packet loss probability of the five protocols with output queueing given by the analytical model is plotted in Figure 6.16. As with IQ, the GBP significantly improves the performance of the LBP only for small buffers. The same reasoning applies to the behavior of the acknowledgment protocols compared to the grant protocols. In both cases the better usage of the buffer enabled by GBP and by ack when the idle positions are less than b is appreciable only when the buffer size is not much larger than b.

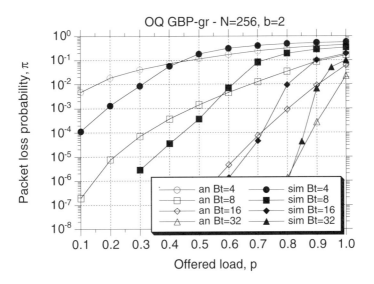

Figure 6.15. Loss performance with OQ and GBP-gr

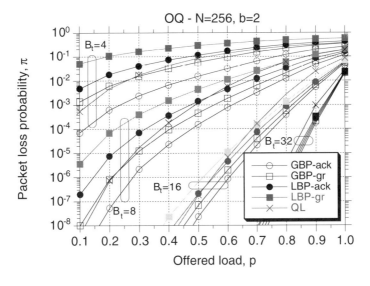

Figure 6.16. Loss performance with OQ and different protocols

6.2.3. Shared queueing

An SE with internal shared queueing is provided with a total buffer capacity of $B = B_s$ (cells) that is shared among all the SE inlets and outlets (see Figure 6.17 in which additional interstage links have been used for signalling purposes). The buffer is said to include b *logical queues* each holding the packets addressing a specific SE outlet. On the SE inlet side, the cells offered by the upstream SEs are stored concurrently in the buffer that holds all the cells independently of the individual destinations. According to our FIFO assumption, the controller must store sequentially the received packets in each logical queue, so as to be able to transmit one packet per non-empty logical queue in each slot. Thus the queue must be able to perform up to b write and b read operations per slot. As in the case of SEs with output queueing, HOL blocking cannot take place since there is no contention among different queues for the same SE outlets. The SE controller, which exchanges information with the SEs in adjacent stages and with the local queue, performs the arbitration for the concurrent access to the local queue by the upstream switching element when buffer overflow is going to occur.

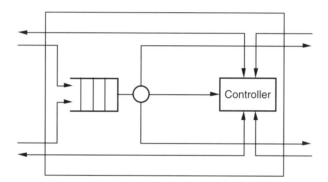

Figure 6.17. SE with shared queueing

Based on the model assumptions defined at the beginning of Section 6.2 and analogously to the two previous queueing models only one SE per stage, the tagged SE, is studied as representative of the behavior of all the elements in the same stage (SE_i will denote such an element for stage i). Let us define the following state variables for the tagged SE:

- $s_{i,t}$: state of an SE in stage i (SE_i) at time t;
- $s'_{i,t}$: state that SE_i at time t would assume if the packets to be transmitted during slot t are removed from its buffer (that is the state assumed by the SE at half time slot if transmission and receipt of packets are considered to occur sequentially in the slot);

whose probability distributions are

- $d_{i,t}(m) = \Pr$ [the state of SE_i at time t is $s_{i,t} = m$];
- $d'_{i,t}(m) = \Pr$ [the state of SE_i at time t is $s'_{i,t} = m$];

Note that the LBP protocol does not require the definition of the variable $s'_{i,t}$ and the corresponding distribution $d'_{i,t}(m)$ as the ack/grant signals depend only on the idle buffer space at the end of the last slot. Observe that now the state variable $s_{i,t}$ is much more general than the scalar representing the queue state with the previous queueing types. Now the following additional variables are needed:

- $g_{i,t}(h)$ = Pr [the buffer of SE_i at time t holds h packets];
- $g'_{i,t}(h)$ = Pr [the buffer of SE_i at time t holds h packets, if the packets to be transmitted during slot t are removed from the buffer];

For the sake of convenience we redefine here also the variable describing the interstage traffic, that is

- $a_{i,t}$ = Pr [a link outgoing from SE_i offers a packet at time t]; $a_0 = p$ denotes the external offered load;
- $b_{i,t}$ = Pr [a packet offered by SE_i at time t is actually transmitted by SE_i].

If S denotes the set of all the SE states, the SE_i dynamics can be expressed as

$$d'_{i,t}(m) = \sum_{j=0}^{|S|-1} d_{i,t}(j) \, u_{i,t}(j,m)$$
$$\qquad\qquad (0 \le m \le |S|-1) \qquad\qquad (6.12)$$
$$d_{i,t+1}(m) = \sum_{j=0}^{|S|-1} d'_{i,t}(j) \, e_{i,t}(j,m)$$

where

- $u_{i,t}(j,m)$ = Pr [a transition occurs from state $s_{i,t} = j$ to state $s'_{i,t} = m$];
- $e_{i,t}(j,m)$ = Pr [a transition occurs from state $s'_{i,t} = j$ to state $s_{i,t+1} = m$].

Different approaches have been proposed in the technical literature. We simply recall here the main assumptions and characteristics of two basic models by referring to the original papers in the literature for the analytical derivations of the performance results. In the first proposal by Turner [Tur93], which here will be referred to as a *scalar model*, the state $s_{i,t} = j$ simply represents the number of packets in the SE_i buffer at time t. With this model the destinations of packets in the buffer are assumed to be mutually independent. In the second proposal by Monterosso and Pattavina [Mon92], which here will be called a *vectorial model*, the independence assumption of the addresses of packets sitting in the same buffer is removed. The state $\mathbf{s}$ is a vector of b components $s(m) = (s_{c1}(m), s_{c2}(m), ..., s_{cb}(m))$ $(s_{ci}(m) \le B_s)$, in which s_{ci} represents the number of packets addressing one specific SE output and the different components are sorted in decreasing order. The shortcomings of these two models are very poor results for the former and a large state space growth with buffer or SE size in the latter.

Our approach to make the analysis feasible and accurate is to represent the buffer content by means of only two variables, one being the content of a specific logical queue (the tagged queue) and the other being the cumulative content of the other $b-1$ logical queues [Gia94]. Thus, if S denotes the set of the SE states, the buffer content in the generic SE state $m \in S$ will be represented by the two state variables

- $s_q(m)$: number of packets in the tagged logical queue when the SE state is m;
- $s_Q(m)$: cumulative number of packets in the other $b-1$ logical queues when the SE state is m;

with the total buffer content indicated by $s_0(m)$ and the obvious boundary conditions

$$s_q(m) \leq B_s$$

$$s_Q(m) \leq B_s$$

$$s_0(m) = s_q(m) + s_Q(m) \leq B_s$$

This model is called here a *bidimensional model*, since only two variables characterize the SE. Two other more complex (and accurate) models are also described in [Gia94] which use one (*tridimensional model*) or two (*four-dimensional model*) additional state variables to take into account the correlation in the interstage traffic. A different kind of bidimensional model is described in [Bia93]. In our bidimensional model this traffic is assumed to be strictly random and thus characterized by only one parameter. The two distributions $g_{i,t}(h)$ and $d_{i,t}(m)$, as well as $g'_{i,t}(h)$ and $d'_{i,t}(m)$, can now be related by

$$g_{i,t}(h) = \sum_{s_0(m) = h} d_{i,t}(m)$$

$$g'_{i,t}(h) = \sum_{s_0(m) = h} d'_{i,t}(m) \tag{6.13}$$

(note that m is a state index, while h is an integer).

In order to solve Equation 6.12 it is useful to split $e_{i,t}$ into two factors:

$$e_{i,t} = v_{i,t}(j, m) f(j, m) \tag{6.14}$$

where

- $v_{i,t}(j, m) = \Pr[SE_i$ at time t receives the number of packets necessary to reach state m from state j];
- $f(j, m) = \Pr[$the transition from state j to state m takes place, given that the SE has received the number of packets required by this transition].

The two factors of $e_{i,t}$ are given by

$$v_{i,t}(j, m) = \begin{cases} \beta(b, s_0(m) - s_0(j), a_{i-1,t}) & (s_0(m) \neq B_s) \\ \displaystyle\sum_{x = s_0(m) - s_0(j)}^{b} \beta(b, x, a_{i-1,t}) & (s_0(m) = B_s) \end{cases} \tag{6.15}$$

$$f(j, m) = \binom{s_0(m) - s_0(j)}{s_q(m) - s_q(j)} \left(\frac{1}{b}\right)^{s_q(m) - s_q(j)} \left(1 - \frac{1}{b}\right)^{s_Q(m) - s_Q(j)}$$

where the average interstage load is

$$a_{i,t} = 1 - \sum_{s_q(m) = 0} d_{i,t}(m) \qquad (1 \leq i \leq n) \tag{6.16}$$

with the usual boundary condition $a_0 = p$.

The function $u_{i,t}(j, m)$ describing the packet transmission process by SE_i is given by

$$u_{i,t}(j, m) = \beta(\min(1, s_q(j)), s_q(j) - s_q(m), b_{i,t}) \tag{6.17}$$

$$\sum_{v = s_Q(j) - s_Q(m)}^{b-1} P_{i,t}(v|s_Q(j))\beta(v, s_Q(j) - s_Q(m), b_{i,t})$$

in which $P_{i,t}(v|l)$ represents the probability that v non-tagged logical queues of SE_i at time t hold a packet ready to be transmitted, given that the SE_i buffer holds l packets addressed to the $b-1$ non-tagged outlets. We can compute this function as follows:

$$P_{i,t}(v|l) = \frac{\displaystyle\sum_{k_j = I_{v,l}} \Pr[s_{Q(1), i, t} = k_1, \ldots, s_{Q(b-1), i, t} = k_{b-1}]}{\displaystyle\sum_{k_j = I_l} \Pr[s_{Q(1), i, t} = k_1, \ldots, s_{Q(b-1), i, t} = k_{b-1}]}$$

with

$$I_{v,l} = \left\{ k_j : \sum_{j=1}^{b-1} \min(1, k_j) = v; \sum_{j=1}^{b-1} k_j = l \right\}$$

$$I_l = \left\{ k_j : \sum_{j=1}^{b-1} k_j = l \right\}$$

where $s_{Q(j), i, t}$ indicates the number of packets in the j-th (non-tagged) logical queue of SE_i at time t.

In order to calculate the joint distribution

$$\Pr[s_{Q(1), i, t} = k_1, \ldots, s_{Q(b-1), i, t} = k_{b-1}]$$

some approximations are introduced. We assume all the logical queues to be mutually independent and their distribution to be equal to that of the tagged logical queue $\Pr[s_{q, i, t} = k]$. Therefore we have

$$\Pr[s_{Q(1), i, t} = k_1, \ldots, s_{Q(b-1), i, t} = k_{b-1}] \cong \Pr[s_{q, i, t} = k_1] \ldots \Pr[s_{q, i, t} = k_{b-1}]$$

where $\Pr[s_{q, i, t} = k]$ is given by

$$\Pr[s_{q, i, t} = k] = \sum_{m \in I_k} d_{i,t}(m) \qquad (0 \le k \le B_s)$$

with

$$I_k = \{ m : (s_q(m) = k) \}$$

Completing the calculation of $u_{i,t}(j,m)$ requires the specification of the probability $b_{i,t}$ that a packet is transmitted downstream by a non–empty logical queue. This function is given by

$$
b_{i,t} = \begin{cases}
1 & (1 \le i \le n) \quad QL \\
& (i = n) \quad BP \\
\displaystyle\sum_{h=0}^{B_s} G_{i+1,t-1}(h)\,\Omega_{i,t-1}(B_s-h) & (1 \le i < n) \quad BP
\end{cases}
\tag{6.18}
$$

in which G indicates the function g with LBP and g' with GBP. The function $\Omega_{i,t}(x)$, representing the probability that a packet offered by SE_i is accepted by SE_{i+1}, given that x idle positions are available in the SE_{i+1} buffer, is given by

$$
\Omega_{i,t}(x) = \begin{cases}
1 & (x = b) \\
\dfrac{x}{b} & (x < b) \quad gr \\
\displaystyle\sum_{r=0}^{b-1} \beta(b-1,r,a_{i,t})\min\left(1,\dfrac{x}{r+1}\right) & (x < b) \quad ak
\end{cases}
\tag{6.19}
$$

The overall network performance can now be evaluated. Cell loss can take place at any stage, so that the total packet loss probability is given by

$$
\pi = \begin{cases}
\displaystyle \pi = \pi_1 + \sum_{i=2}^{n}\pi_i\prod_{j=1}^{i-1}(1-\pi_j) & QL \\
\pi_1 & BP
\end{cases}
\tag{6.20}
$$

The packet loss probability π_i at stage i is computed as the ratio between the average number of lost packets and the average number of offered packets, that is

$$
\pi_i = \frac{\displaystyle\sum_{x=1}^{b} x \sum_{h=0}^{B_s} G_i(h)\,\beta(b,B_s+x-h,a_{i-1})}{b\,a_{i-1}}
\tag{6.21}
$$

in which G is given by g if there is local backpressure, otherwise by g'. The switch throughput ρ is given by the load carried by SE_n, that is

$$
\rho = a_n
$$

whereas the average packet delay can be computed using Little's formula by

$$
T = \begin{cases}
\dfrac{1}{nb} \displaystyle\sum_{i=1}^{n} \dfrac{1}{a_i} \sum_{h=0}^{B_s} h g_i(h) & QL \\[3ex]
\dfrac{1}{nb\rho} \displaystyle\sum_{i=1}^{n} \sum_{h=0}^{B_s} h g_i(h) & BP
\end{cases}
$$

Figures 6.18–6.20 show the model accuracy with shared queueing for the packet loss probability in a network with eight stages and 2×2 SEs with a shared buffer capacity $B_t = B_s$ ranging from 4 to 32 cells. Unlike OQ, both acknowledgment and grant versions of the GBP protocol are characterized by a very good accuracy of the model, whereas in the case of the QL protocol the accuracy is qualitatively similar to that obtained with IQ and OQ. The loss parameter for these three protocols given by the model is shown in Figure 6.21: basically if the buffer size is above $B_t = 8$ cells, there is no advantage in using the more complex acknowledgment protocol rather than the grant protocol. The advantage given by the use of backpressure can be qualified in terms of a gain of one order of magnitude in the packet loss probability for a given offered load.

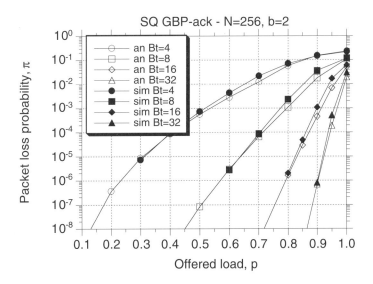

Figure 6.18. Loss performance with SQ and GBP-ack

6.2.4. Performance

In all the analytical models for banyan networks with internal queueing the existence of a steady state is assumed so that, after choosing an arbitrary initial state for the system, the evolution is computed slot by slot until a satisfactory steady state is reached. A better understanding of the overall procedure can be provided by Figure 6.22, showing a pseudo-code of the iterative procedure followed for the analysis of a banyan network with shared queueing. For each

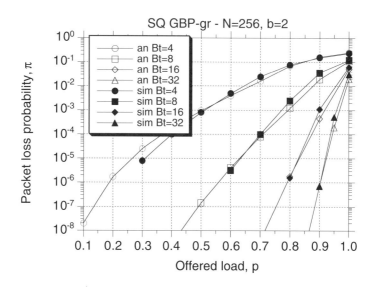

Figure 6.19. Loss performance with SQ and GBP-gr

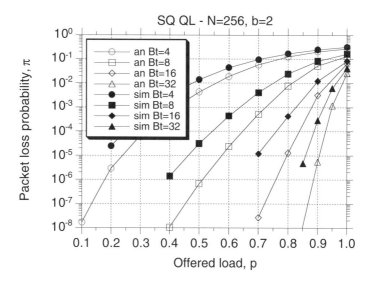

Figure 6.20. Loss performance with SQ and QL

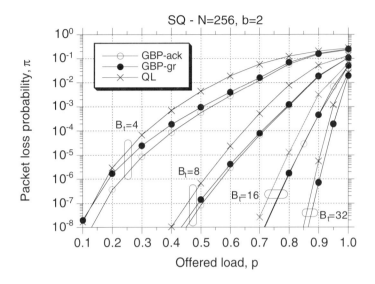

Figure 6.21. Loss performance with SQ and different protocols

iteration step, the probability distributions characterizing all the network stages are computed, so that the overall network performance figures can be evaluated. Note that the iteration goes from stage 1 to stage n, so that the backpressure information received by SE_i at time t from the downstream stage must be based on the SE_{i+1} state computed at the previous iteration, i.e. at time $t-1$. In order to determine the steady-state conditions for the procedure, the most critical performance parameter is selected, that is the network loss probability, by referring to its relative variation between consecutive iterations. In particular a tolerance value such as $\varepsilon = 10^{-5}$ could be assumed. It is worth observing that the procedure is simplified when no backpressured is applied. In fact there is no upstream signalling with the QL protocol, so that the steady-state distribution is found stage after stage for increasing stage index.

After describing and analyzing the three modes of internal queueing in a banyan network, that is input, output and shared queueing, and assessing at the same time the model accuracy for each of them, we give now some comparative results. In particular we draw attention to the switch capacity, that is the carried load when the offered load is maximum $(p = 1)$, the average packet delay, and the cell loss performance. The maximum throughput that can be obtained with the various queueing and internal protocols as a function of the total buffer size B_t per SE is shown in Figure 6.23. The switch capacity increases with B_t and shows the asymptotic values $\rho_{max} = 0.75$ for input queueing and $\rho_{max} = 1.0$ for output queueing. In fact in the latter case there is no head–of–line blocking and all SE outlets are always busy. In the former case the switch capacity is determined by the first stage that is characterized by a throughput lower than one due to the HOL blocking. Its carried load is given by Equation 6.9, in which the probability that the buffer is empty is zero (the computation of the switch capacity assumes that the offered load is $p = 1$). For finite buffer sizes the BP protocol always gives switch capacities slightly higher than the QL protocol. When the buffer size is

STEP	EQUATION
loop 1 (counter t)	
for i=1 to n	
$b_{i,t} = f(g_{i+1,t-1}, \Omega_{i,t-1})$	6.18
$u_{i,t} = f(b_{i,t}, P_{i,t})$	6.17
$v_{i,t} = f(a_{i-1,t})$	6.15
$e_{i,t} = f(v_{i,t}, f)$	6.14
$d'_{i,t} = f(d_{i,t}, u_{i,t})$	6.12
$d_{i,t+1} = f(d'_{i,t}, e_{i,t})$	6.12
$G_{i,t} = d(d_{i,t}, d'_{i,t})$	6.13
$\pi_{i,t} = f(G_{i,t}, a_{i-1,t})$	6.21
$a_{i,t} = f(d_{i,t})$	6.16
$\Omega_{i,t} = f(a_{i,t})$	6.19
end cycle for	
$\pi_t = f(\pi_{i,t})$	6.20
$if \; \lvert \pi_t - \pi_{t-1} \rvert \le \varepsilon\pi_{t-1}$	
then end loop 1	
$t = t+1$	
end loop 1	

Figure 6.22. Pseudo-code showing the iterative solution of the model

rather small the switch capacity is much lower than the asymptotic value as the number of switch stages increases, as is shown in Figure 6.24 for a total buffer size per SE $B_t = 4$.

An example of the delay performance of a banyan network computed by computer simulation is given in Figure 6.25 for an eight-stage network with $B_t = 16$. For offered loads up to $p = 0.4$ all the protocols give an average packet delay T very close to 1 slot. Above this value the IQ curves grow much faster than the OQ and SQ ones, since the switch capacity with input queueing is much lower than with output or shared queueing.

Finally Figures 6.26–6.28 compare the packet loss performance of the three queueing strategies under GBP-ack, GBP-gr and QL, whose results are obtained simulating banyan networks with three stages and a total buffer $B_t = 8b$ per SE. By means of this assumption the three networks (their sizes are $N = 8, 64, 512$) have the same amount of buffer per SE port, independent of the SE size that ranges from $b = 2$ to $b = 8$. Shared queueing is clearly much better than the other two types of queueing under any offered load and internal protocol. Interestingly enough, when the offered load is quite low and for small network sizes, input queueing does not behave much worse than output queueing.

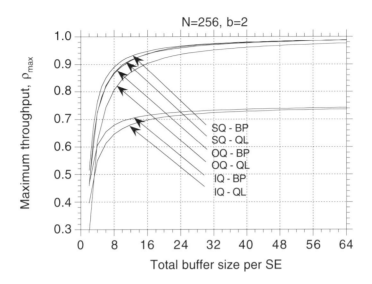

Figure 6.23. Switch capacity with different queueings and protocols

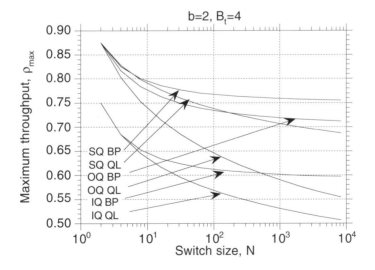

Figure 6.24. Switch capacity with different queueings and protocols

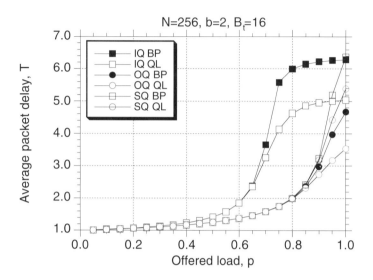

Figure 6.25. Delay performance with different queueings and protocols

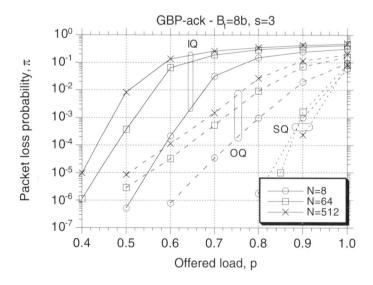

Figure 6.26. Loss performance with GBP-ack and different queueings

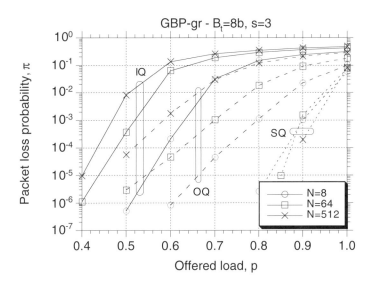

Figure 6.27. Loss performance with GBP-gr and different queueings

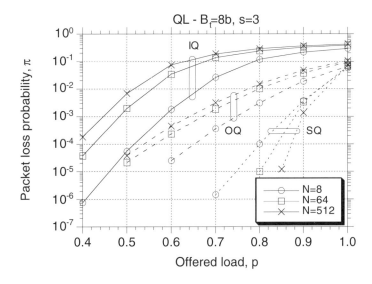

Figure 6.28. Loss performance with QL and different queueings

6.3. Networks with Unbuffered Parallel Switching Planes

In the previous section it has been shown how adding a queueing capability to each SE of a banyan network, the performance typical of an ATM switch can be easily obtained. An alternative approach for improving the traffic performance of a basic banyan network consists in using unbuffered parallel planes coupled with external queueing. The adoption of multiple unbuffered parallel switching planes implies that output queueing is mandatory in order to control the packet loss performance. In fact more than one cell per slot can be received at each network output interface but just one cell per slot can be transmitted downstream by the switch. In this section we will examine how output queueing alone or combined with input queueing can be adopted in a *buffered replicated banyan network* (BRBN) in which different arrangements and operations of the banyan networks are considered.

6.3.1. Basic architectures

The general architecture of a buffered replicated banyan network is represented by the general scheme of Figure 6.4 in which each splitter is preceded by a buffer (the input queue) and each combiner is replaced by a buffer (the output queue). Note that at most one read and one write operation per slot are required in each input queue, whereas multiple writes, up to K, occur at each output queue. The output interconnection pattern is always a shuffle pattern that guarantees full accessibility of each banyan network to all the output buffers. The input interconnection pattern, together with the operation mode of the splitters, determines the managing technique of the K banyan networks in the BRBN structure, that is the way of distributing the packet received at the switch inlets to the K banyan networks.

Analogously to the operation in the (unbuffered) RBN, three basic modes of packet distribution techniques can be defined, in which all of them guarantee now full accessibility to any output queue from all banyan networks:

- *random loading* (RL),
- *multiple loading* (ML),
- *alternate loading* (AL).

The input interconnection pattern is a shuffle pattern with random and multiple loading, so that each BRBN inlet can access all the $K = K_r$ (RL) or $K = K_m$ (ML) banyan networks (Figure 6.29). Each packet is randomly routed by the splitters with even probability $1/K_r$ to any of the K_r banyan planes with RL, whereas the splitter generates K_m copies of the same packet (one per plane) with ML. So, the ML technique requires additional hardware in association with the output interconnection pattern to store in the output buffers only one instance of the multiple copies (up to K_m) of the same packet that crossed successfully the banyan networks.

Each splitter degenerates into a 1×1 connection with BRBN alternate loading and the input interconnection pattern is selected so as to minimize the conflicts between packets in the $K = K_a$ banyan networks (Figure 6.30). This kind of packet distribution technique, originally proposed in [Cor93] for replicated banyan networks adopting also input queue and internal backpressure, consists in ensuring that the packets do not conflict with each other in the first

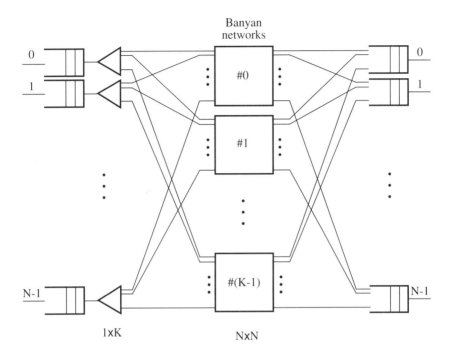

Figure 6.29. BRBN with random or multiple loading

$\log_b K_a$ stages of each banyan network, independent of their respective destinations. This result is achieved by properly selecting the input interconnection pattern between the N BRBN inlets and the N "active" inlets in the K_a banyan networks. For example if $N = 16$, $K_a = 4$, $b = 2$ each banyan network has $N/K_a = 4$ active inlets out of its 16 inlets. Now, if the banyan topology is a an Omega network [Law75] (see Figure 2.16a), it is very easy to show that packet conflicts can take place only in stages 3 and 4 if the 4 active inlets per banyan network are adjacent (for example they interface the two top SEs in stage 1). An analogous property is obtained for isomorphic banyan networks, as defined in Section 2.3.1.2, by suitably defining the input interconnection pattern of the BRBN.

6.3.2. Architectures with output queueing

Our attention is now focused on BRBN architectures where only output queueing is adopted. Therefore packets are lost if they cannot enter the addressed queue in the slot of their arrival. In this case the packet loss performance is highly dependent on the capability of the cell to arrive at the output queue avoiding conflicts in the banyan network being crossed. For this reason other specific architectures of BRBN with only output queueing will be defined now that are capable of improving the loss performance of the basic architectures with random, multiple or alternate loading.

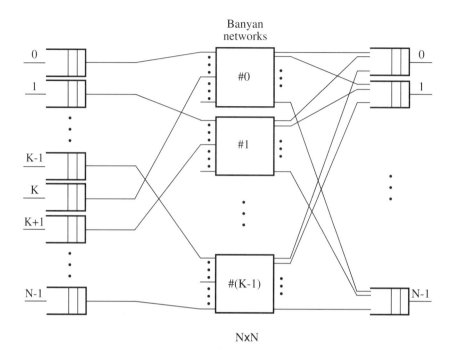

Figure 6.30. BRBN with alternate loading

6.3.2.1. Specific architectures

Two different packet distribution techniques derived from the ML technique are defined [Fer93]:

- *multiple topology loading* (MTL),
- *multiple priority loading* (MPL),

With multiple topology loading each banyan network is preceded by a block whose function is to perform a random permutation of the set of packets entering the network (see the example of Figure 6.31 for $N = 8$, $K_{mt} = 4$). The permutations applied in each slot by these K_{mt} blocks are selected independently from each other. This kind of operation makes the BRBN network with MTL loading of Figure 6.31 equivalent to a BRBN with ML and a "random topology" in each of its K_{mt} planes.

We found out that if the banyan topologies of the different planes are selected so that the sets of permutations set up by any two of them differ significantly, the traffic performance of such BRBN and the BRBN with MTL provide very similar traffic performance. An example of a BRBN network is given in Figure 6.32, in which each plane adopts a specific topology, that is Baseline, Omega, reverse 3-cube and SW-banyan.

With all the basic packet distribution techniques (RL, ML, AL) we have implicitly assumed that in case of conflict between two packets addressing the same SE outlet, the packet to be

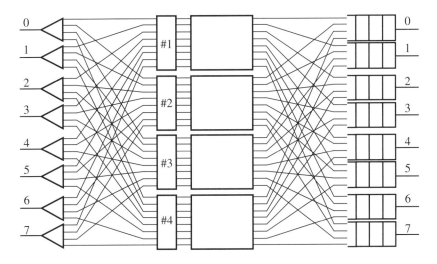

Figure 6.31. BRBN with multiple topology loading

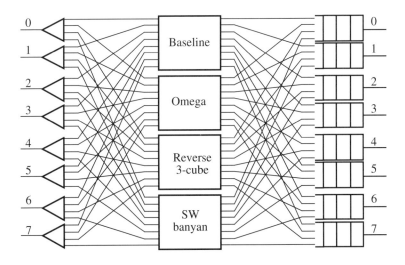

Figure 6.32. BRBN with multiple "random" banyan topologies

discarded is randomly selected. Apparently, such a choice has no effect on the average traffic performance of random and alternate loading, since it does not matter which specific packet is lost. When multiple loading is applied, how to choose the particular packet to be discarded in each plane affects the overall performance, since it can vary the number of packets for which all the K_m copies are lost, thus affecting the network throughput. Unlike the MTL technique where each plane is different from the other, we have the same topology in all planes in the multiple priority loading, but the selection of the packet to be discarded in case of conflict in each plane is no longer random. By altering the collision resolution mechanism in each banyan network, each incoming packet has a different loss probability in the different planes.

The MPL technique is based on the adoption of different priority rules in the BRBN planes for selecting the winner packets in case of conflicts for an interstage link. MPL is defined in topologies with 2×2 SEs ($b = 2$) and is thus based on three types of SEs: *random winner* (RW), where the conflict winner is selected random, *top winner* (TW) and *bottom winner* (BW), in which the conflict winner is always the packet received on the top (0) and bottom (1) SE inlet, respectively.

To explain how this result is accomplished, consider first the case in which MPL is applied only on the first stage of each banyan network (one priority level), while SEs randomly choose the winner packet in case of contention in all the other stages. To guarantee a fair treatment to all the packets we need at least two banyan networks: in one of them priority in case of collision at the first stage is given to packets entering the SE on the top inlet, whereas packets entering the bottom SE inlet receive priority in the other plane. Applying the same concept to different priorities in the first r stages requires a stack of 2^r banyan networks, in order to implement all the combinations of priority collision resolution patterns that grant a fair treatment to all packets. Such a result is obtained by applying the following algorithm. Let the planes be numbered 0 through $K_{mp} - 1$ with plane j having the binary representation $j = j_{r-1}, j_{r-2}, ..., j_0$. Then SEs in stage i of plane j are RW elements for $r < i \leq n$, TW (BW) elements if the bit j_{r-i} is 0 (1) for $1 \leq i \leq r$. An example is represented in Figure 6.33 for $N = 8$, $r = 2$, $K_{mp} = 4$. Note that the BRBN becomes non-blocking if $r = n$ (or $K_{mp} = N$). If a number of planes $K_{mp} = h2^r$ (h integer) is selected with $h > 1$, the network includes h banyan networks of each type.

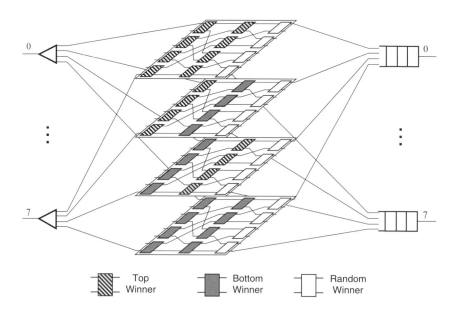

Figure 6.33. Example of BRBN with multiple priority loading

6.3.2.2. Performance

The traffic performance of a BRBN with output queueing using 2×2 switching elements is now described using computer simulation. In order to emphasize the effect on the traffic handling capacity of the different loading strategies, an infinite capacity for the output queues is assumed. This means that multiple packet arrivals at each output queue are handled without cell loss: due to the infinitely large buffer space, all the packets are buffered to be eventually transmitted to the switch output link. Apparently the traffic performance results provided under this assumption are optimistic compared to the real case of output queues with finite capacity.

How the number of banyan planes in the buffered replicated banyan network affects the overall loss performance is first investigated. Figure 6.34 shows the packet loss probability of a BRBN with size $N = 1024$ under multiple loading for an increasing number of planes K_m. A larger number of planes improves the loss performance but even in the case of $K_m = 16$ the load must be lower than $p = 0.1$ to obtain a loss probability smaller than 10^{-3}. Random and alternate loadings give a loss probability worse than multiple loading for a large number of planes, e.g. $K = 16$, comparable for small values of K, e.g. $K = 2$.

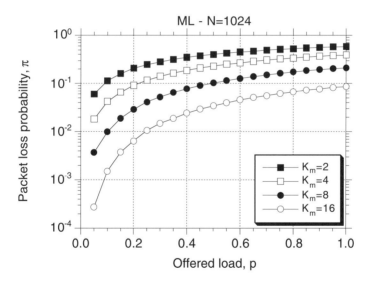

Figure 6.34. Loss performance of a BRBN with output queueing and multiple loading

The size of the switch has a very limited effect on the loss performance. Figure 6.35 gives the cell loss probability of a BRBN under multiple loading with $K_m = 8$ planes for a switch size ranging from $N = 16$ to $N = 1024$. Compared to the largest network size, the smaller switches gain very little in packet loss performance in spite of the smaller number of network stages and hence of the smaller number of conflict opportunities for a packet. Independently from the offered load value, the smallest and largest switches are characterized by loss probability differing by less than one order of magnitude.

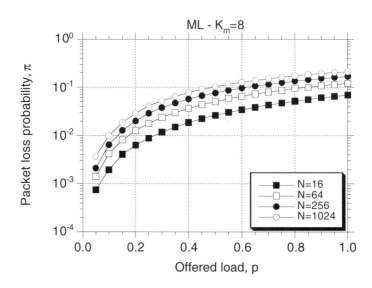

Figure 6.35. Loss performance of a BRBN with output queueing and multiple loading

The loss performance of a BRBN under multiple priority loading is now presented. A switch of size $N = 1024$ has been considered with a varying number of switching planes, from $K_{mp} = 2$ to $K_{mp} = 16$, in which the number r of stages where priority is applied grows with K_{mp}, that is $r = \log_2 K_{mp}$. The effect of having a reduced number of stages where all the copies of a packet can be lost due to contention is the reduction of the cell loss probability, especially for low offered loads, as is shown in Figure 6.36. Nevertheless, even with the minimum offered load considered in the figure, $p = 0.05$, the cell loss probability is not lower than 10^{-4}.

The traffic performance of the different loading strategies, that is RL, AL, MPL and MTL, is now compared for a large switch ($N = 1024$), for the two cases of $K = 4$ and $K = 16$ whose results are given in Figures 6.37 and 6.38, respectively. Random and alternate loading give comparable loss probabilities independently of the number of banyan planes. The multiple priority loading improves the loss figure only if K is sufficiently large, whereas the multiple topology loading behaves significantly better than all the other techniques for any number of banyan planes. Nevertheless, as a general consideration, BRBNs with output queueing do not seem suitable to provide the very low loss probabilities that are expected from an ATM switch. It will be shown in the next section that the adoption of input queueing combined with output queueing improves greatly the loss performance, thus making a BRBN switch suitable to ATM switching.

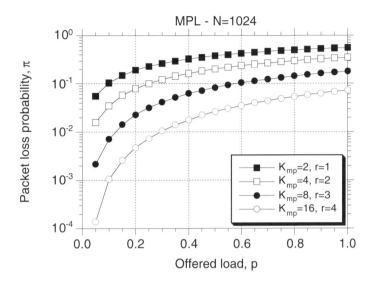

Figure 6.36. Loss performance of a BRBN with output queueing and multiple priority loading

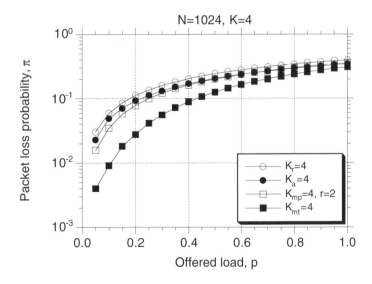

Figure 6.37. Comparison of loss performance of a BRBN with output queueing and different loadings

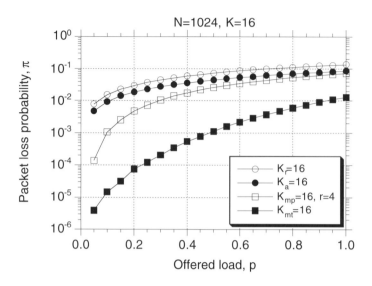

Figure 6.38. Comparison of loss performance of a BRBN with output queueing and different loadings

6.3.3. Architectures with combined input–output queueing

The general structure of a BRBN is now considered where both input and output queues are equipped, thus achieving a combined input–output queueing[1]. Input and output queues reside in the input (IPC) and output (OPC) port controller, respectively, where all the other functions related to the interface with the switch input and output links are performed. The availability of the input queues makes much less critical the choice of the packet distribution technique. For this reason only the two simplest techniques will now be considered here, that is the random loading (RL) and the alternate loading (AL), whose corresponding network architectures are shown in Figure 6.29 and Figure 6.30, respectively. It is worth noting that the input queues could be moved between the splitters and the input interconnection pattern, thus configuring NK input queues, one per banyan network inlet. This technique could provide some performance advantages as it allows more than N packets (actually up to NK) to be transmitted at the same time through the K banyan networks. However the price to pay is that packets may become out of sequence; this can take place on the cell flow crossing the switch on a given inlet/outlet couple.

In a BRBN a packet remains stored in the head-of-line (HOL) position of its input queue as long as its successful transmission to the requested output queue is not accomplished. This result can be obtained by having each IPC transmit in each slot a probe packet carrying the switch outlet requested by the HOL packet in the input queue. An acknowledgment packet is

---

1. A simpler switch architecture analogous to the BRBN in which the output queues are not equipped is described in [Cor93b].

returned by the OPC for each probe packet that successfully crosses the interconnection network using the same path traced by the probe packet. The same path is kept alive for the transmission of the HOL packets by those IPC receiving back the acknowledgment packet. If a backpressure protocol is applied, the acknowledgment packets are sent only within the current storage capability of the addressed output queue.

6.3.3.1. Models for performance analysis

Two analytical models are now described that enable the performance evaluation of BRBN with input–output queueing under both random and alternate loading. The first model, which will be referred to as *model 1-d* (monodimensional), has been proposed in [Cor93a] and [Awd94] with reference to the AL technique and is based on the study of a tagged input queue and a tagged output queue as independent from all other queues. Note that the model developed in [Awd94] uses an interconnection network including just one EGS network of size $NK \times NK$ with n stages rather than K banyan networks of size $N \times N$ as in our BRBN shown in Figure 6.30. However, it is possible to show that these two networks are isomorphic (see Problem 6.11). The second model, called *model 2-d* (bidimensional), has been proposed in [Cat96]; it removes the independence assumption of the input queues by thus improving the model accuracy. In both models the destinations of the HOL packets are mutually independent due to the external random traffic assumption.

The model 1-d is developed for both network operation modes, that is queue loss and backpressure. The behavior of a tagged input queue is studied as representative of any other input queue. This queue is modelled as $Geom/G/1/B_i$ in which B_i is the capacity in cells of the tagged input queue. The arrival process is clearly geometric due to our general assumption of an uncorrelated random input traffic and its mean value is denoted by p $(0 \le p \le 1)$. As far as the service process is concerned, it is assumed that the transmission attempts of the HOL packet across consecutive time slots are mutually independent. Therefore the probability distribution of the service time θ in the input queue is geometrical, that is

$$\Pr[\theta = j] = q(1-q)^{j-1}$$

where q is the probability that a probe packet is successful and is therefore the probability that the HOL packet is actually transmitted in the current slot. By means of the procedure described in the Appendix, the queue $Geom/G/1/B_i$ can be analyzed and its performance measures, throughput ρ_i, cell loss probability π_i, and average waiting time $E[\eta_i]$, are computed.

In order to compute the success probability q we need to analyze the behavior of the stack of the K banyan networks. The load offered to each inlet of a banyan network can be easily expressed as a function of the probability that the corresponding input queue is empty considering the specific loading strategy. With random loading the load of each queue is divided into K planes with even probability, whereas each input queue feeds just one banyan network inlet with alternate loading, so that

$$p_0 = \begin{cases} \dfrac{1 - Pr[n_i = 0]}{K} & RL \\[2ex] 1 - Pr[n_i = 0] & AL \end{cases}$$

The computation of the load per stage in the banyan network is immediately given by Equation 6.2 for random loading, that is

$$p_i = 1 - \left(1 - \frac{p_{i-1}}{b}\right)^b \qquad (1 \le i \le n) \tag{6.22}$$

In the case of alternate loading we have to take into account that the first $\log_b K_s$ stages are conflict-free and hence

$$p_i = \begin{cases} \dfrac{p_{i-1}}{b} & (1 \le i \le \log_b K_s) \\[2ex] 1 - \left(1 - \dfrac{p_{i-1}}{b}\right)^b & (\log_b K_s < i \le n) \end{cases} \tag{6.23}$$

Now the success probability q can be computed as the ratio between the output load and the load offered to the first conflicting stage, that is

$$q = \begin{cases} p_n/p_0 & RL \\ p_n/p_{\log_b K_s} & AL \end{cases} \tag{6.24}$$

So we have been able to express the evolution of the input queue as a function of the success probability which depends on the input and output loads of the banyan network which in turn depend on the occupancy probability of the input queue. For this reason an iterative solution is adopted, which is explained by the flow chart in Figure 6.39 for the QL mode (backpressure is not applied and hence packets can be lost due to the output queue overflow).

An initial value of the probability distribution of the input queue occupancy is assigned and hence also an initial value for the loss probability π_i. Afterwards each cycle requires to compute the load per stage, which gives the success probability q enabling the analysis of the input queue. The new value of the packet loss probability is thus computed and the procedure is iterated as long as the relative difference between two consecutive values of π_i exceeds a given threshold ε, for example 10^{-5}.

The load offered to the tagged output queue, p_o, is given by the traffic carried by the tagged input queue and is given by

$$p_o = \rho_i = p(1 - \pi_i) \tag{6.25}$$

where p is the input load to the switch. This value is used as the total average load offered by the K Bernoulli sources feeding the tagged output queue that is modelled as $Geom(K)/D/1/B_o$, B_o being the capacity of each output queue. By means of the proce-

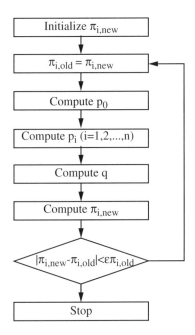

Figure 6.39. Iterative solution of the input queue

dure described in the Appendix, this queue can be analyzed by thus obtaining its performance measures, namely throughput ρ_o, cell loss probability π_o, and average waiting time $E[\eta_o]$.

The throughput, loss and delay performance values of the overall switch are then given by

$$\rho = \rho_o$$

$$\pi = \pi_i + (1 - \pi_i)\pi_o$$

$$T = 1 + E[\eta_i] + E[\eta_o] \qquad (6.26)$$

The previous procedure is slightly modified if backpressure is adopted meaning that the success probability (q_{BP}) of probe packets is also conditioned to the actual availability of buffer capacity in the addressed output queue, that is

$$q_{BP} = q_{QL}(1 - Pr[n_o = B_o]) \qquad (6.27)$$

where q_{QL} is the success probability without backpressure given by Equation 6.24. The previous iterative procedure must now include also the evaluation of the behavior of the output queue. The flow chart in Figure 6.39 remains valid if the execution of the block "compute q" is preceded by the evaluation of the probability distribution in the output queue whose offered load per source is given by the value p_n just computed. The load carried by the output queue, which also represents the switch throughput ρ, equals the load carried by the input queue and is given by

$$\rho = \rho_i = p\,(1 - \pi_i) \tag{6.28}$$

The overall performance measures of the switch under backpressure operation are now provided by Equation 6.28 for the switch throughput, by

$$\pi = \pi_i$$

for packet loss probability and by Equation 6.26 for the average packet delay.

In order to overcome the limits of the previous model 1-d, the bidimensional model 2-d is defined which keeps track of both the number of packets in the tagged queue, n_i, and the number of packets waiting to be transmitted in the HOL position of all the input queues, n_h (n_h also represents the number of non-empty input queues). Therefore the number of states identified by this model is $N(B_i + 1)$.

The evaluation of the transition probabilities between states requires the knowledge of two distributions: the number of HOL packets successfully transmitted through the interconnection network, n_t, and the number of packet arrivals to the HOL position of the input queues, n_a.

We assume that, for both random and alternate loading the traffic offered to the first stage when n_h input queues are non-empty is

$$p_0 = \begin{cases} n_h / (KN) & RL \\ n_h / N & AL \end{cases} \tag{6.29}$$

Let q_{nh} denote the probability that a HOL packet is successfully transmitted in a time slot when there are n_h requests. Then q_{nh} can be evaluated using Equation 6.24 (QL) or Equation 6.27 (BP) together with Equation 6.22 for random loading and Equation 6.23 for alternate loading (recall that now p_0 is a function of n_h). Note that the assumption about the distribution of the offered load to the interconnection network, as expressed by Equation 6.29, means underestimating the real value of q_{nh} when the number of requesting packets is small. If, for example, there is only one HOL packet, this does not experience any conflict and its success probability is equal to one, whereas it is a little less than one using Equation 6.29. For example even assuming $K = 1$, $q_1 = 0.94$ for $N = 16$ and $q_1 = 0.99$ for $N = 256$.

Since all the HOL packets are equally likely to be successful, the distribution of successful packets n_t in a slot when the number of requests is n_h is binomial. The distribution of the number of packet arrivals to the HOL position of the input queues as well as the transition probabilities in the bidimensional chain (n_i, n_h) are computed in [Cat96].

6.3.3.2. Performance results

The accuracy of the monodimensional and bidimensional models is now assessed and the different network operations compared. Unless stated otherwise, results refer to a switch with $N = 64$, $b = 2$, $B_i = 4$ and $B_o = 16$.

The switch capacity, ρ_{max}, that is the load carried by each output queue when the external load is $p = 1.0$, is given in Table 6.1 for a switch size $N = 64$ for different values of B_o

and K. These numerical values have been obtained through computer simulation since the model does not accept an offered load greater than the maximum throughput. Apparently the switch capacity is independent of the input buffer capacity. It is noted that the QL policy performs better than BP for small output capacity. The reason is the same as that which gives a crossbar switch a capacity larger than a non-blocking input-queued switch (see Section 7.1.2.1). In other words, discarding the conflicting packets in most situations represents a statistically better condition than keeping all the HOL losing packets. We expect both QL and BP to give the same switch capacity when the output queues do not saturate.

Table 6.1. Switch capacity for different network configurations

		$K = 2$		$K = 4$		$K = 8$		$K = 16$	
		RL	AL	RL	AL	RL	AL	RL	AL
$B_o = 2$	QL	0.518	0.548	0.649	0.690	0.718	0.775	0.753	0.779
	BP	0.513	0.531	0.608	0.632	0.645	0.663	0.661	0.673
$B_o = 4$	QL	0.532	0.556	0.698	0.751	0.794	0.836	0.843	0.872
	BP	0.531	0.555	0.674	0.706	0.727	0.743	0.745	0.757
$B_o = 8$	QL	0.533	0.556	0.711	0.773	0.828	0.886	0.889	0.925
	BP	0.533	0.556	0.708	0.759	0.793	0.818	0.820	0.830
$B_o = 16$	QL	0.533	0.557	0.712	0.775	0.836	0.905	0.908	0.952
	BP	0.533	0.557	0.712	0.775	0.825	0.873	0.871	0.890
$B_o = 32$	QL	0.533	0.557	0.712	0.775	0.837	0.908	0.913	0.960
	BP	0.533	0.557	0.712	0.775	0.837	0.905	0.910	0.930

Figure 6.40 shows the packet loss probability performance given by the two models 1-d and 2-d for increasing offered loads and speed-ups in the QL mode for random loading, together with data from computer simulation. The bidimensional model underestimates the HOL success probability when the number of packets requesting transmission is small and provides rather precise results when this number is large. Thus the results obtained by this analysis are closer to the simulation data ones when the blocking characteristic is not negligible and the evaluation of q_{nh} is more accurate. The monodimensional analysis provides optimistic results because it assumes that the traffic offered to the interconnection network is more uniform than in reality. More accurate matching in this case is thus provided when large values of the internal speed-up are used. It is remarkable to notice that, for opposite reasons, the two analyses provide an upper and a lower bound to the simulation results. Similar considerations apply to the analytical results obtained with QL under alternate loading as well as with BP under both RL and AL.

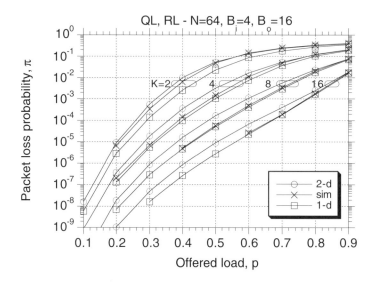

Figure 6.40. Loss performance of a QL BRBN switch with random loading

Figure 6.41 shows the loss performance for increasing loads and input queue sizes without backpressure under alternate loading. A very good matching is found between the bidimensional model and simulation results whereas precision deteriorates in the monodimensional model. The reason is that in this latter model the underestimation of the time spent by the HOL packet in the server of the input queue increases as the average number of packets in the queue increases and this value is proportional to B_i. Analogous results are obtained under alternate loading or with backpressure.

The average packet delay of a BRBN switch is plotted in Figure 6.42 under QL and RL operation. Results from the model 2-d are much more accurate than those from the model 1-d and maintain their pessimistic estimate as long as the offered load is lower than the maximum value (see Table 6.1 for the asymptotic throughput values).

The two different loading modes, RL and AL, are compared now in terms of loss performance they provide in Figure 6.43 and Figure 6.44 without and with backpressure, respectively, using the model 2-d. Alternate loading always provides a better loss performance than random loading: more than one order of magnitude is gained by AL over RL when the speed-up is enough, for example for $K \geq 8$ and a proper output queue size is adopted. The better performance is clearly due to the higher probability of success in the interconnection network by the HOL packets.

We would like now to compare the switch performance with and without backpressure for a given total budget of total queueing capacity per switch port $B_t = B_i + B_o$. The adoption of the backpressure mechanism is not complex at all since the probe phase is substantially the same as in QL; on the other hand the backpressure introduces several advantages. As is shown in Figure 6.45, whose data have been obtained through the model 2-d, the BP loss probability is lower than the QL loss when the input queue size is greater than a fixed value. This can be

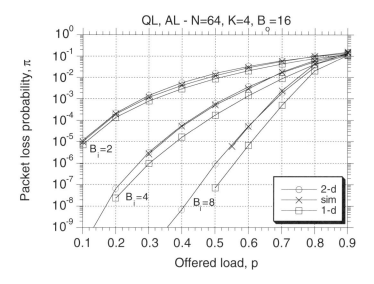

Figure 6.41. Loss performance of a QL BRBN switch with alternate loading

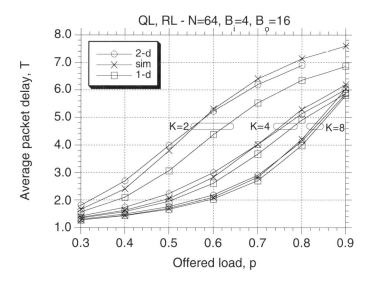

Figure 6.42. Delay performance of a QL BRBN switch with random loading

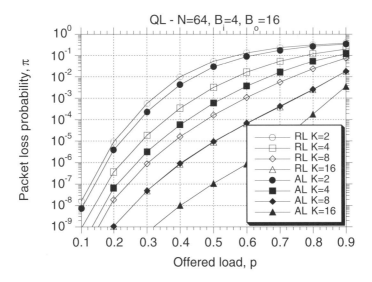

Figure 6.43. Loss performance comparison between RL and AL in a QL BRBN switch

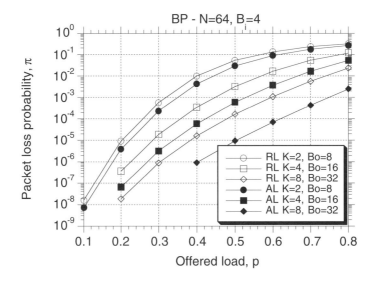

Figure 6.44. Loss performance comparison between RL and AL in a BP BRBN switch

explained with a better queueing resource usage that this mechanism allows. If the output buffer is full the packet will not be lost and wait for the next slot. The results plotted in the figure also show that there is an optimal buffer allocation for both BP and QL. In the latter case this can be explained with the loss shift from the input to the output queue when B_i grows. The former case is more complex. When the output queue size reduces and the loss probability increases, packets remain in their input queues which thus leave less space for new incoming packets. Let B_i^* be the input queue capacity providing the minimum loss probability. The input buffer saturates and the cell loss grows very quickly when $B_i > B_i^*$. The results shown here for the loss probability, given a total buffer budget, are analogous to those obtained for a switch with the same queueing strategy (input and output) with a non-blocking interconnection network (see Figure 8.25).

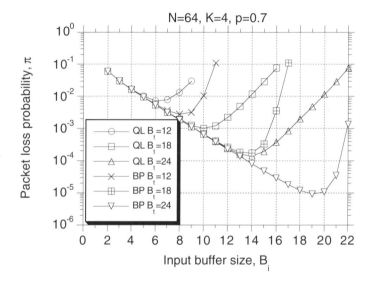

Figure 6.45. Loss performance comparison between BP and QL in a BRBN switch

6.4. Additional Remarks

The traffic performance of all the architectures in this chapter has been evaluated relying on the random and uniform traffic assumption in which the switch inlets are each loaded by the same kind of Bernoulli source and all switch outlets are equally likely to be addressed. Nevertheless, the random traffic assumption is only an oversimplifying hypothesis that, although allowing a performance analysis of the switching architecture, is far from modelling the real traffic patterns that are likely to be offered to an ATM switch. In particular, traffic correlation and non-uniformity in traffic loading should be considered. Performance analyses of minimum-depth banyan networks with correlated traffic patterns have been reported in [Xio93], [Mor94] for output queueing and in [Pat94] for shared queueing. Non-uniformity in traffic patterns has been studied for minimum–depth banyan networks in [Kim90] with input queueing and in [Pat94] with shared queueing.

Other ATM switch architectures have been described in which the basic banyan network is enhanced so as to include other features such as the capability of partially sharing internal buffers [Kim94, Wid94] or the availability of interstage link dilation without [Wid95] or with recirculation [You93]. By still using a minimum–depth routing network it is also possible to build an ATM switch [Kim93] in which a $(\log_2 N)$ -stage distribution network precedes the $(\log_2 N)$ -stage routing network. Both networks have the same topology (they are banyan networks) but only the latter requires internal queueing. The former network is just used to distribute the traffic entering from each port of the switch onto different input ports of the routing network. This feature can become important to provide fairness in the presence of non–uniform traffic patterns (remember that a banyan network supports only one path per inlet/outlet pair and different inlet/outlet paths share interstage links). Another proposal consists of parallel banyan planes one of which acting as a control plane to resolve the conflicts for all the other data planes so that multiple packet transmissions can take place without conflicts to the same output port [Won95]. Different approaches to manage the queues have also been studied. One technique consists in operating a non–FIFO queueing, analogous to the windowing technique described in Section 7.1.3.2, on a BRBN architecture [Su94]. Another approach, referred to as "cut–through switching", consists in allowing a packet to cross a SE without spending a slot in the buffer if queueing is not strictly needed (for example to cope with conflicts) [Wid93].

6.5. References

[Awd94]   R.Y. Awdeh, H.T. Mouftah, "The expanded delta fast packet switch", *Proc. of ICC 94*, New Orleans, LA, May 1994, pp. 397-401.

[Bia93]   G. Bianchi, J.S. Turner, "Improved queueing analysis of shared buffer switching networks", *IEEE/ACM Trans. on Networking*, Vol. 1, No. 4, Aug. 1993, pp.482-490.

[Cat96]   C. Catania, A. Pattavina, "Analysis of replicated banyan networks with input and output queueing for ATM switching", *Proc. of ICC 96*, Dallas, TX, June 1996, pp. 1685-1689.

[Cor93a]   G. Corazza, C. Raffaelli, "Input/output buffered replicated banyan networks for broadband switching applications", *Eur. Trans. on Telecommun.*, Vol. 4, No. 1, Jan.-Feb. 1993, pp. 95-105.

[Cor93b]   G. Corazza, C. Raffaelli, "Performance evaluation of input-buffered replicated banyan networks", *IEEE Trans. on Commun.*, Vol. 41, No. 6, June 1993, pp. 841-4845.

[Dia81]   D.M. Dias, R. Jump, "Analysis and simulation of buffered delta networks", *IEEE Trans. on Comput.*, Vol. C-30, No. 4, Apr. 1981, pp. 273-282.

[Fer93]   G. Ferrari, M. Lenti, A. Pattavina, "Distributed routing techniques for internally unbuffered interconnection networks", *Eur. Trans. on Telecommun.*, Vol. 4, No. 1, Jan.-Feb. 1993, pp. 85-94.

[Gia94]   S. Gianatti, A. Pattavina, "Performance analysis of ATM banyan networks with shared queueing: Part I - Random offered traffic", *IEEE/ACM Trans. on Networking*, Vol. 2, No. 4, Aug. 1994, pp. 398-410.

[Hlu88] M.G. Hluchyj, K.J. Karol, "Queueing in high-performance packet switching", *IEEE J. on Selected Areas in Commun.*, Vol. 6, No. 9, Dec. 1988, pp. 1587-1597.

[Jen83] Y. Jenq, "Performance analysis of a packet switch based on single-buffered banyan networks", *IEEE J. on Selected Areas in Commun.*, Vol. SAC-1, No. 6, Dec. 1983, pp. 1014-1021.

[Kim90] H.S. Kim, A.L. Garcia, "Performance of buffered banyan networks under nonuniform traffic patterns", *IEEE Trans. on Commun.*, Vol. 38, No. 5, May 1990, pp. 648-658.

[Kim93] Y.M. Kim, K.Y. Lee, "PR-banyan: a packet switch with a pseudorandomizer for nonuniform traffic", *IEEE Trans. on Commun.*, Vol. 41, No. 7, July 1993, pp. 1039-1042.

[Kim94] H.S. Kim, "Design and performance of MULTINET switch: a multistage ATM switch architecture with partially shared buffers", *IEEE/ACM Trans. on Networking*, Vol. 2, No. 6, Dec. 1994, pp. 571-580.

[Kru83] C.P. Kruskal, M. Snir, "The performance of multistage interconnection networks for multiprocessors", *IEEE Trans. on Comput.*, Vol. C-32, No. 12, Dec. 1983, pp. 1091-1098.

[Kum84] M. Kumar, J.R. Jump, "Performance enhancement in buffered delta networks using crossbar switches", *J. of Parallel and Distributed Computing*, Vol. 1, 1984, pp. 81-103.

[Kum86] M. Kumar, J.R. Jump, "Performance of unbuffered shuffle-exchange networks", *IEEE Trans. on Comput.*, Vol. C-35, No. 6, June 1986, pp. 573-578.

[Law75] D.H. Lawrie, "Access and alignment of data in an array processor", *IEEE Trans. on Comput.*, Vol. C-24, No. 12, Dec. 1975, pp. 1145-1155.

[Mon92] A. Monterosso, A. Pattavina, "Performance analysis of multistage interconnection networks with shared-buffered switching elements for ATM switching", *Proc. of INFOCOM 92*, Florence, Italy, May 1992, pp. 124-131.

[Mor94] T.D. Morris, H.G. Perros, "Performance modelling of a multi-buffered banyan switch under bursty traffic", *IEEE Trans. on Commun.*, Vol. 42, No. 2-4, Feb.-Apr. 1994, pp. 891-895.

[Pat81] J.H. Patel, "Performance of processor-memory interconnections for multiprocessors", *IEEE Trans. on Computers*, Vol C-30, No. 10, pp. 771-780.

[Pat91] A. Pattavina, "Broadband switching systems: first generation", *European Trans. on Telecommun. and Related Technol.*, Vol. 2, No. 1, Jan.-Feb. 1991, pp. 75-87.

[Pat94] A. Pattavina, S. Gianatti, "Performance analysis of ATM banyan networks with shared queueing - Part II: correlated/unbalanced traffic", *IEEE/ACM Trans. on Networking*, Vol. 2, No. 4, Aug. 1994, pp. 411-424.

[Pet90] G.H. Petit, E.M. Desmet, "Performance evaluation of shared buffer multiserver output queue used in ATM", *Proc of 7th ITC Seminar*, Morristown, NJ, Oct. 1990.

[Sak90] Y. Sakurai, N. Ido, S. Gohara, N. Endo, "Large scale ATM multistage network with shared buffer memory switches", *Proc. of ISS 90*, Stockholm, Sweden, May 1990, Vol. IV, pp.121-126.

[Szy87] T. Szymanski, V.C. Hamacker, "On the permutation capability of multistage interconnection networks", *IEEE Trans on Computers*, Vol. C-36, No. 7, July 1987, pp. 810-822.

[Szy89] T. Szymanski, S. Shaikh, "Markov chain analysis of packet-switched banyans with arbitrary switch sizes, queue sizes, link multiplicities and speedups", *Proc. of INFOCOM 89*, Ottawa, Canada, April 1989, pp. 960-971.

[Su94] Y.-S. Su, J.-H. Huang, "Throughput analysis and optimal design of banyan switches with bypass queues", *IEEE Trans. on Commun.*, Vol. 42, No. 10, Oct. 1994, pp. 2781-2784.

[Tur93]   J.S. Turner, "Queueing analysis of buffered switching networks", *IEEE Trans. on Commun.*, Vol. 41, No. 2, Feb. 1993, pp. 412-420.

[Wid93]   I. Widjaja, A. Leon-Garcia, H.T. Mouftah, "The effect of cut-through switching on the performance of buffered banyan networks", *Comput. Networks and ISDN Systems*, Vol. 26, 1993, pp. 139-159.

[Wid94]   I. Widjaja, A. Leon-Garcia, "The Helical switch: a multipath ATM switch which preserves cell sequence", *IEEE Trans. on Commun.*, Vol. 42, No. 8, Aug. 1994, pp. 2618-2629.

[Wid95]   I. Widjaja, H.S. Kim, H.T. Mouftah, "A high capacity broadband packet switch architecture based on a multilink approach", *Int. J. of Commun Systems*, Vol. 8, 1995, pp.69-78.

[Won95]   P.C. Wong, "Design and analysis of a novel fast packet switch - Pipeline Banyan", *IEEE/ ACM Trans. on Networking*, Vol. 3, No. 1, Feb. 1995, pp. 63-69.

[Xio93]   Y. Xiong, H. Bruneel, G. Petit, "On the performance evaluation of an ATM self-routing multistage switch with bursty and uniform traffic", *Proc. of ICC 93*, Geneva, Switzerland, May 1993, pp. 1391-1397.

[Yoo90]   H. Yoon, K.Y. Lee, M.T. Liu, "Performance analysis of multibuffered packet switching network", *IEEE Trans. on Comput.*, Vol. 39, No. 3, Mar. 1990, pp. 319-327.

[You93]   Y.S. Youn, C.K. Un, "Performance of dilated banyan network with recirculation", *Electronics Letters*, Vol. 29, No. 1, Jan. 1993, pp.62-63.

6.6. Problems

6.1   Draw the RBN with selective loading with parameters $N = 16$, $b = 2$, $K_s = 2$, in which the truncated banyan network is: (a) a reverse Baseline network with the first stage removed; (b) a reverse Omega network with the last stage removed. For both cases specify formally the permutation required between each banyan network and the combiners to guarantee full accessibility.

6.2   Explain why the capacity of a single-stage switch with input queueing and $b = 2$, $B_i = 2$ (see Figure 6.24) is the same as a crossbar network with the same size.

6.3   Compute the capacity of a single-stage switch with $b = 4$, $B_i = 2$ and explain why this value is different from that of a crossbar network of the same size.

6.4   Compute the capacity of a single-stage switch with input queueing, $b = 2$ and generic buffer size B_i.

6.5   Compute the capacity of a single-stage switch with output queueing, $b = 2$, $B_o = 2$ and compare its value with that given in Figure 6.24.

6.6   Express the capacity of a single-stage switch with output queueing and $b = 2$ as a function of the buffer size B_o.

6.7   Draw the BRBN with selective loading with parameters $N = 16$, $b = 2$, $K_s = 4$ in which the banyan networks have: (a) the reverse Baseline topology; (b) the n-cube topology.

6.8   Compute the network throughput and the packet loss probability for a BRBN with output queueing in the case of multiple loading assuming that output queues have infinite capacity and that the collision events in the single banyan planes are mutually independent. Determine whether the performance results underestimate or overestimate those obtained through computer simulation (use, e.g., data in Figures 6.34–6.35) and justify the result.

6.9   Compute the network throughput and the packet loss probability for a BRBN with output queueing in the case of multiple priority loading assuming that output queues have infinite capacity. Verify if the performance results underestimate or overestimate those obtained through computer simulation (use, e.g., data in Figure 6.36) and justify the result.

6.10   Compute the network throughput and the packet loss probability for a BRBN with output queueing in the two cases of random loading and selective loading assuming that output queues have infinite capacity. Check if the performance results match well those obtained through computer simulation (use, e.g., data in Figures 6.37–6.38).

6.11   Prove that an EGS network of size $NK \times NK$ with $\log_2 N$ stages, K equal to a power of 2 and N active inlets is isomorphic to a set of K banyan networks of size $N \times N$ each with N/K active inlets. Identify the corresponding active inlets in the two architectures.

6.12   Assume that the packet loss probability of a BRBN switch with output queueing is known, which takes into account the loading strategy adopted. Write the equations expressing the network throughput, the average cell delay and the overall packet loss probability by suitably modelling the output queues for which a finite size B_o is considered.

Chapter 7   *ATM Switching with Non-Blocking Single-Queueing Networks*

A large class of ATM switches is represented by those architectures using a non-blocking interconnection network. In principle a non-blocking interconnection network is a crossbar structure that guarantees absence of switching conflicts (internal conflicts) between cells addressing different switch outlets. Non-blocking multistage interconnection networks based on the self-routing principle, such as sorting–routing networks, are very promising structures capable of running at the speed required by an ATM switch owing to their self-routing property and their VLSI implementation suitability. It has been shown in Section 6.1.1.2 that a non-blocking interconnection network (e.g., a crossbar network) has a maximum throughput $\rho_{max} = 0.63$ per switch outlet due to external conflicts, that is multiple cells addressing the same outlet in the same slot. Even more serious than such low utilization factor is the very small load level that guarantees a cell loss beyond significant limits.

Queueing in non-blocking multistage networks is adopted for improving the loss performance and whenever possible also for increasing the maximum throughput of the switch. Conceptually three kinds of queueing strategies are possible:

- *input queueing* (IQ), in which cells addressing different switch outlets are stored at the switch input interfaces as long as their conflict-free switching through the interconnection network is not possible;
- *output queueing* (OQ), where multiple cells addressing the same switch outlet are first switched through the interconnection network and then stored in the switch output interface while waiting to be transmitted downstream;
- *shared queueing* (SQ), in which a queueing capability shared by all switch input and output interfaces is available for all the cells that cannot be switched immediately to the desired switch outlet.

Figure 7.1 shows a general model for an $N \times M$ ATM switch: it is composed of N *input port controllers* (IPC), a non-blocking interconnection network and M *output port controllers* (OPC). Usually, unless required by other considerations, the IPC and OPC with the same index are implemented as a single port controller (PC) interfacing an input and an output channel. In this case the switch becomes squared and, unless stated otherwise, N and M are assumed to be powers of 2.

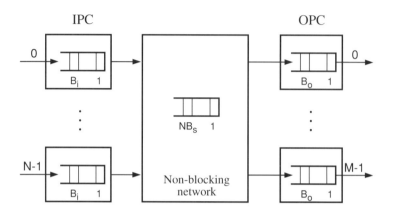

Figure 7.1. Model of non-blocking ATM switch

Each IPC is provided with a queue of B_i cells, whereas a queue of B_o cells is available in each OPC. Moreover, a shared queue of B_s cells per switch inlet (a total capacity of NB_s cells is available) is associated with the overall interconnection network. The buffer capacity B takes 1 as the minimum value and, based on the analytical models to be developed later in this section and in the following one for input and output queueing, it is assumed that the packet is held in the queue as long as its service has not been completed. With single queueing strategy, we assume that each IPC is able to transmit at most 1 cell per slot to the interconnection network whereas each OPC can concurrently receive up to K cells per slot addressing the interface, K being referred to as *(output) speed-up*. The interconnection network is implemented, unless stated otherwise, as a multistage network that includes as basic building blocks a *sorting network* and, if required, also a *routing network*. As far as the former network is concerned, we choose to adopt a Batcher network to perform the sorting function, whereas the *n*-cube or the Omega topology of a banyan network is selected as routing network; in fact, as shown in Section 3.2.2, such network configuration is internally non-blocking (that is free from internal conflicts). The specific models of ATM switches with non-blocking interconnection network that we are going to describe will always be mapped onto the general model of Figure 7.1 by specifying the values of the queue capacity and speed-up factor of the switch. Unless stated otherwise, a squared switch is considered ($N = M$) and each queue operates on a FIFO basis.

This chapter is devoted to the study of the switching architectures adopting only one of the three different queueing strategies just mentioned. Adoption of multiple queueing strategies within the same switching fabric will be discussed in the next chapter. ATM switching architectures and technologies based on input, output and shared queueing are presented in Sections 7.1, 7.2 and 7.3, respectively. A performance comparison of ATM switches with single queueing strategy is discussed in Section 7.4 and additional remarks concerning this class of ATM switching architectures are given in Section 7.5.

7.1. Input Queueing

By referring to the general switch model of Figure 7.1, an ATM switch with pure input queueing (IQ) is characterized by $B_i > 0$, $B_o = B_s = 0$ and $K = 1$; the general model of a squared $N \times N$ switch is shown in Figure 7.2. Cells are stored at switch input interfaces so that in each slot only cells addressing different outlets are switched by the multistage interconnection network. Thus a contention resolution mechanism is needed slot by slot to identify a set of cells in different input queues addressing different network outlets. Two basic architectures will be described which differ in the algorithm they adopt to resolve the output contentions. It will be shown how both these structures suffer from a severe throughput limitation inherent in the type of queueing adopted. Enhanced architectures will be described as well that aim at overcoming the mentioned throughput limit by means of a more efficient handling of the input queues.

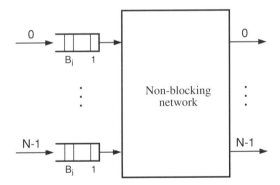

Figure 7.2. Model of non-blocking ATM switch with input queueing

7.1.1. Basic architectures

With input queueing, two schemes have been proposed to perform the outlet contention resolution that are likely to be compatible with the rate requirements of the switch I/O links: the *three-phase algorithm* and the *ring-reservation algorithm*. The ATM switching architectures adopting these schemes, referred to as *Three-Phase switch* and *Ring-Reservation switch,* are now described.

7.1.1.1. The Three-Phase switch

The block structure of the Three-Phase switch [Hui87] is represented in Figure 7.3: it includes N port controllers PC_i $(i = 0, ..., N-1)$ each interfacing an input channel, I_i, and an output channel, O_i, a Batcher sorting network (SN), a banyan routing network (RN) and a channel allocation network (AN). The purpose of network AN is to identify winners and losers in the contention for the switch outlets by means of the *three-phase algorithm* [Hui87]. This scheme has been conceived to exploit the sorting and routing capability of the multistage network in order to resolve the contentions for the switch outlets. The algorithm, which is run every slot, evolves according to three phases:

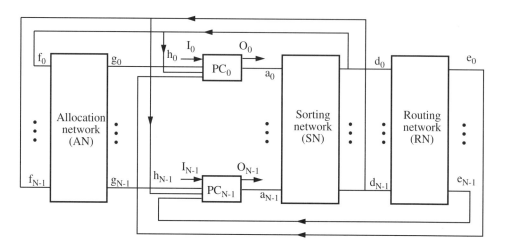

Figure 7.3. Architecture of the Three-Phase switch

I   *Probe phase*: port controllers request the permission to transmit a cell stored in their queue to a switch outlet; the requests are processed in order to grant at most one request per addressed switch outlet.

II   *Acknowledgment (ack) phase*: based on the processing carried out in Phase I, acknowledgment signals are sent back to each requesting port controller.

III   *Data phase*: the port controllers whose request is granted transmit their cell.

The algorithm uses three types of control packets (for simplicity, we do not consider other control fields required by hardware operations, e.g., an activity bit that must precede each packet to distinguish an idle line from a line transporting a packet with all fields set to "0")[1]:

- Packet $REQ(j,i)$ is composed of the *destination* address j of the switch outlet requested by the HOL cell in the input queue of PC_i and the *source* address i of the transmitting port controller. Both addresses are $n = \log_2 N$ bit long.

- Packet $ACK(i,a)$ includes the *source* address i, which is n bits long, to whom the acknowledgment packet is addressed and the *grant* bit a carrying the contention result.

- Packet $DATA(j,cell)$ contains the n–bit *destination* address j of the HOL cell and the cell itself.

---

1. All the fields of the control packets used in the three-phase algorithm are transmitted with the most significant bit first.

In the *probe phase* (see Figure 7.4) each port controller with a non–empty input queue sends a request packet REQ(j,i) through the interconnection network. The packets REQ(j,i) are sorted in non-decreasing order by network SN using the destination and source fields as primary and secondary sorting key, respectively, so that the requests for the same switch outlets are adjacent at the outputs of network SN. The sorted packets REQ(j,i) enter network AN which grants only one request per addressed switch outlet, that is the one received on the lowest-index AN inlet. Thus network AN generates for each packet REQ(j,i) a grant field a indicating the contention outcome ($a = 0$ winner, $a = 1$ loser).

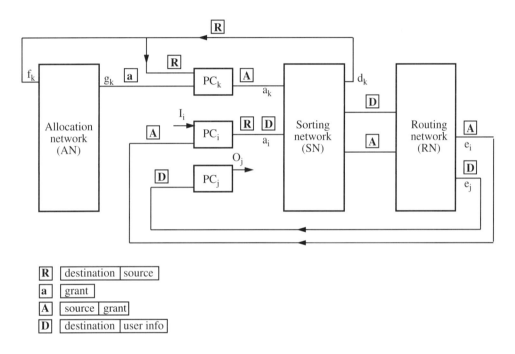

R	destination	source
a	grant
A	source	grant
D	destination	user info

Figure 7.4. Packet flow in the Three-Phase switch

In the *acknowledgment phase* (see Figure 7.4) the port controllers generate packets ACK(i,a) including the field source just received from the network SN within the packet REQ(j,i) and the grant field computed by the network AN. The packet ACK(i,a) is delivered through the sorting and routing networks to its due "destination" i in order to signal to PC$_i$ the contention outcome for its request. Packets ACK(i,a) cannot collide with each other because all the destination address i are different by definition (each port controller cannot issue more than one request per slot).

In the *data phase* (see Figure 7.4) the port controller PC$_i$ receiving the packet ACK($i,0$) transmits a data packet DATA(j,cell) carrying its HOL cell to the switch outlet j, whereas upon receiving packet ACK($i,1$) the HOL cell is kept in the queue and the same request REQ(j,i) will be issued again in the next slot.

An example of packet switching according to the three-phase algorithm for $N = 8$ is shown in Figure 7.5. Only four out of the seven requests are granted since two network outlets are addressed by more than one request.

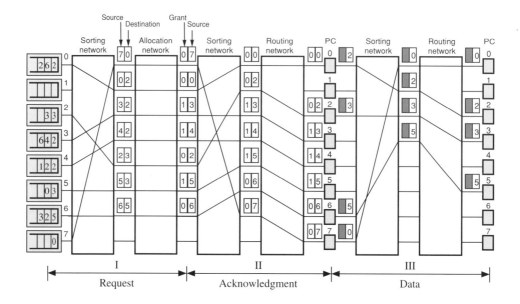

Figure 7.5. Example of switching with the three-phase algorithm

The structure of networks SN and RN is described in Section 3.2.2. The sorting Batcher network includes $n(n+1)/2$ stages of $N/2$ sorting elements 2×2, whereas n stages of $N/2$ switching elements 2×2 compose the banyan network, with $n = \log_2 N$.

The hardware implementation of the channel allocation network is very simple, owing to the sorting operation of the packets $REQ(j,i)$ already performed by the network SN. In fact, since all the requests for the same outlet appear on adjacent inlets of the network AN, we can simply compare the destination addresses on adjacent AN inlets and grant the request received on the AN inlet f_k, if the AN inlet f_{k-1} carries a request for a different outlet. The logic associated with port g_{k+1} $(k = 0, ..., N-2)$ of network AN is given in Figure 7.6. The

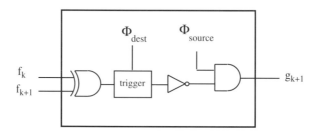

Figure 7.6. Logic associated with each port of AN

destination address of packets REQ(.,.) received on inputs f_k and f_{k+1} are compared bit by bit by an EX-OR gate, whose output sets the trigger by the first mismatching bit in f_k and f_{k+1}. The trigger keeps its state for a time sufficient for packet ACK(i,a) to be generated by port controller PC$_i$. The trigger is reset by the rising edge of signal Φ_{dest} at the start of the address comparison.

Port controllers generate packets ACK(.,.) by transmitting field *source* of packet REQ(.,.) being received from network SN, immediately followed by field a received from network AN. The AND gate in network AN synchronizes the transmission of field a with the end of receipt of field *source* by the port controller. The signal on outlet g_0 is always low ($a = 0$), independent of the input signals on f_k ($k = 1, ..., N-1$), as the request packet received on inlet f_0 (if any) is always granted (it is the request received on the lowest-index AN inlet for the requested switch outlet). The structure of network AN is so simple that it can be partitioned and its function can be performed by each single port controller, as suggested in the original proposal of three-phase algorithm [Hui87]: the hardware associated to outlet g_k ($k = 1, ..., N-1$) is implemented within port controller PC$_k$, which thus receives signals both from b_{k-1} and from b_k.

Since the networks SN and RN are used to transfer both the user information (the cells within packets DATA(.,.)) and the control packets, the internal rate C_i of the switch must be higher than the external rate C, so that the time to complete the three-phase algorithm equals the cell transmission time. The transfer of user information takes place only in Phase III of the three-phase algorithm, as Phases I and II represent switching overhead (we disregard here the additional overhead needed to transmit the destination field of packet DATA(j,cell)). Let η denote the *switching overhead*, defined as the ratio between the total duration, $T_{I, II}$, of Phases I and II and the transmission time, T_{III}, of packet DATA(.,.) in Phase III ($T_{I, II}$ and T_{III} will be expressed in bit times, where the time unit is the time it takes to transmit a bit on the external channels). Then, $C(1 + \eta)$ bit/s is the bit rate of each digital pipe inside the switch that is required to allow a flow of C bit/s on the input and output channels.

The number of bit times it takes for a signal to cross a network will be referred to as *signal latency* in the network and each sorting/switching stage is accounted for with a latency of 1 bit. The duration of Phase I is given by the latency $n(n+1)/2$ in the Batcher network and the transmission time n of the field destination in packet REQ(.,.) (the field source in packet REQ(.,.) becomes the field source in packet ACK(.,.) and its transmission time is summed up in Phase II). The duration of Phase II includes the latency $n(n+1)/2$ in the Batcher network, the latency n in the banyan network, the transmission time $n+1$ of packet ACK(i,a). Hence, $T_{I, II}$ is given by

$$T_{I-II} = n(n+4) + 1 = \log_2 N(\log_2 N + 4) + 1$$

For $N = 1024$ and T_{III} given by the standard cell length ($T_{III} = 53 \cdot 8$), we obtain $\eta \cong 0.333$ (the n bit time needed to transmit the destination field of packet DATA(.,.) has been disregarded). Thus, if $C = 150$ Mbit/s is the external link rate, the switch internal rate must be about 200 Mbit/s.

The reservation algorithm has been described assuming for the sake of simplicity that all the requests for a given switch outlet are equally important. Actually, in order to avoid unfairness in the selection of the contention winners (the requests issued by lower-index PCs are always given implicit priority by the combined operations of the sorting and allocation networks), a priority must be associated with each request. This type of operation in which each request packet includes now three fields will be described in Section 7.1.3.1 where the three-phase algorithm is applied to an enhanced switch architecture with input queueing. Other solutions for providing fairness could be devised as well that do not necessarily require an additional field in the request packet (see, e.g., [Pat91]).

7.1.1.2. The Ring-Reservation switch

The Ring-Reservation switch [Bin88a, Bin88b] includes a non-blocking self-routing interconnection network, typically a Batcher-banyan network and a ring structure that serially connects all the port controllers of the switching fabric (see Figure 7.7). Contentions among port controllers for seizing the same switch outlets are resolved exchanging control information on the ring: the port controller PC_i ($i = 0, ..., N-1$) receives control information from $PC_{(i-1) \bmod N}$ and transmits control information to $PC_{(i+1) \bmod N}$.

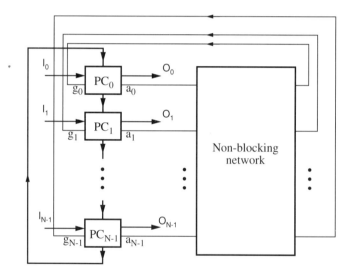

Figure 7.7. Architecture of the Ring-Reservation switch

Port controller PC_0 generates a frame containing N fields, each 1 bit long initially set to 0, that crosses all the downstream port controllers along the ring to be finally received back by PC_0. The field i ($i = 0, ..., N-1$) of the frame carries the idle/reserved status (0/1) for the switch outlet O_i. A port controller holding a HOL packet in its buffer with destination address j sets to 1 the j-th bit of the frame if that field is received set to 0. If the switch outlet has already been seized by an upstream reservation, the port controller will repeat the reservation process in the next slot. The port controllers that have successfully reserved a switch outlet can transmit their HOL packet preceded by the self-routing label through the interconnection network. Such a reservation procedure, together with the non-blocking property of the Batcher-banyan interconnection network, guarantees absence of internal and external conflicts between data packets, as each outlet is reserved by at most one port controller.

Note that the interconnection network bit rate in the Ring-Reservation architecture needs not be higher than the external bit rate, as in the case of the Three-Phase switch (we disregard again the transmission time of the packet self-routing label). However the price to pay here is a contention resolution algorithm that is run serially on additional hardware. In order to guarantee that the interconnection network is not underutilized, the reservation cycle must be completed in the time needed to transmit a data packet by port controllers, whose length is $T_{DATA} = 53 \cdot 8$ time units. Apparently, the reservation phase and the user packet transmission phase can be pipelined, so that the packets transmitted through the interconnection network in slot n are the winners of the contention process taking place in the ring in slot $n-1$. Thus the minimum bit rate on the ring is $2^{n}C/T_{DATA}$, if C is the bit rate in the interconnection network. Therefore, a Ring-Reservation switch with $N \geq 512$ requires a bit rate on the ring for the contention resolution algorithm larger than the bit rate in the interconnection network.

The availability of a ring structure for resolving the output contentions can suggest a different implementation for the interconnection network [Bus89]. In fact, using the control information exchanged through the ring it is possible in each reservation cycle not only to allocate the addressed switch outlets to the requesting PCs winning the contention (*busy* PCs), but also to associate each of the non-reserved switch outlets with a non-busy port controller (*idle* PC). A port controller is idle if its queue is empty or it did not succeed in reserving the switch outlet addressed by its HOL cell. In such a way N packets with different addresses j ($j \in \{0, \ldots, N-1\}$) can be transmitted, each by a different port controller, which are either the HOL packets of the busy PCs or empty packets issued by idle PCs. Based on the operation of a sorting network described in Section 2.3.2.2, such arrangement makes it is possible to use only a sorting Batcher network as the interconnection network, since all the switch outlets are addressed by one and only one packet. Apparently only the non-empty packets received at the switch outlets will be transmitted downstream by the switching fabric.

The allocation of the non-reserved switch outlets to idle PCs can be carried out making the reservation frame round twice across the port controllers. In the first round the switch outlet reservation is carried out as already described, whereas in the second round each port controller sets to 1 the first idle field of the reservation frame it receives. Since the number of non-reserved outlets at the end of the first round equals the number of idle port controllers, this procedure guarantees a one-to-one mapping between port controllers and switch outlets in each slot.

Compared to the basic Ring-Reservation switch architecture, saving the banyan network in this implementation has the drawback of doubling the minimum bit rate on the ring which is now equal to $2 \cdot 2^{n}C/T_{DATA}$ (two rounds of the frame on the ring must be completed in a slot). Thus a switching fabric with $N \geq 256$ requires a bit rate on the ring for the two-round contention resolution algorithm larger than the bit rate in the Batcher interconnection network.

7.1.2. Performance analysis

The performance of the basic architecture of an $N \times M$ ATM switch with input queueing (IQ) is now analyzed. The concept of *virtual queue* is now introduced: the virtual queue VQ_i $(i = 0, ..., M-1)$ is defined as the set of the HOL positions in the different input queues holding a cell addressed to outlet i. The server of the virtual queue VQ_i is the transmission line terminating the switch outlet i. A cell with outlet address i entering the HOL position also enters the virtual queue VQ_i. So, the capacity of each virtual queue is N (cells) and the total content of all the M virtual queues never exceeds N. A graphical representation of the virtual queue VQ_j is given in Figure 7.8. The analysis assumes a first-in-first-out (FIFO) service in the input queues and a FIFO or random order (RO) in the virtual queue. Under the hypothesis of random traffic offered to the switch we first evaluate the asymptotic throughput and then the average delay. Cell loss probability is also evaluated for finite values B_i of the input queue.

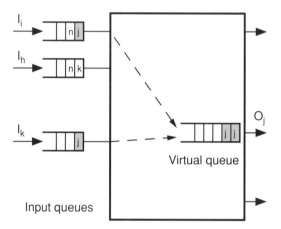

Figure 7.8. Representation of the virtual queue

In the analysis it is assumed that $N \to \infty$, $M \to \infty$ while keeping a constant ratio $M/N = E$. Thus the number of cells entering the virtual queues in a slot approaches infinity and the queue joined by each such cell is independently and randomly selected. Furthermore, since the arrival process from individual inlets to a virtual queue is asymptotically negligible, the interarrival time from an input queue to a virtual queue becomes sufficiently long. Therefore virtual queues, as well as input queues, form a mutually-independent discrete-time system. Owing to the random traffic and complete fairness assumption, the analysis will be referred to the behavior of a generic "tagged" input (or virtual) queue, as representative of any other input (or virtual) queue. This operation is an abstraction of the real behavior of the Three-Phase switch, since the hardware sorting operation determines implicitly a biased selection of the HOL packets to transmit, thus making different the behavior of the different queues. Also the Ring-Reservation switch is affected by a similar unfairness, since the first port controllers to perform the reservation have a higher probability of booking successfully the addressed network outlet.

7.1.2.1. Asymptotic throughput

The asymptotic throughput analysis is carried out following the procedure defined in [Bas76] for a completely different environment (a multiprocessor system with multiple memory modules). Such an approach is based on the analysis of a synchronous $M/D/1$ queue with internal server, as defined in the Appendix. A more recent and analogous analysis of the asymptotic throughput of ATM switches with input queueing is described in [Kar87], in which a synchronous $M/D/1$ queue with external server is used.

In order to evaluate the limiting throughput conditions we assume that each switch inlet receives a cell in a generic slot with probability $p = 1$ and that $N, M \rightarrow \infty$ with a constant ratio $M/N = E$, referred to as an *expansion ratio*. For a generic "tagged" queue we define

Q_n : number of cells stored in the tagged virtual queue at the end of slot n;

A_n : number of cells entering the tagged virtual queue at the beginning of slot n.

The evolution of the tagged virtual queue is expressed by

$$Q_n = \max(0, Q_{n-1} - 1) + A_n \tag{7.1}$$

This analysis assumes steady-state conditions, thus the index n is omitted. When N and M are finite, the distribution function of the new cells entering the virtual queue can be found considering that the total number of new cells in all virtual queues in a slot equals the number M_b of busy virtual queue servers in the preceding slot. Since $1/M$ is the probability that a new cell enters the tagged virtual queue, the probability a_i of i new cells in the tagged virtual queue in a generic slot is

$$a_i = \Pr[A = i] = \binom{M_b}{i}\left(\frac{1}{M}\right)^i\left(1 - \frac{1}{M}\right)^{M_b - i}$$

When $N \rightarrow \infty$, $M \rightarrow \infty$ the number of busy virtual servers will become a constant fraction of the total number of servers, denoted as ρ_M, that represents the maximum utilization factor of each server (we are assuming $p = 1$). Thus the cell arrival distribution becomes Poisson, that is

$$a_i = \frac{e^{-\rho_M}\rho_M^i}{i!} \tag{7.2}$$

As is shown in the Appendix, the virtual queue under study described by Equations 7.1 and 7.2 is a synchronous $M/D/1$ queue with internal server whose average queue length is

$$E[Q] = \rho_M + \frac{\rho_M^2}{2(1-\rho_M)}$$

Assuming $p = 1$ means that all input queues are never empty. Since each HOL cell is queued in only one of the M virtual queues, the average number of cells in each virtual queue is also expressed by N/M. Thus

$$\frac{N}{M} = \rho_M + \frac{\rho_M^2}{2(1 - \rho_M)}$$

which provides

$$\rho_M = \frac{N}{M} + 1 - \sqrt{\left(\frac{N}{M}\right)^2 + 1}$$

Since ρ_M also represents the utilization factor of each switch outlet, the switch asymptotic throughput ρ_{max} (the switch capacity) which is always referred to each switch inlet, becomes

$$\rho_{max} = \frac{M}{N}\rho_M = \frac{M}{N} + 1 - \sqrt{\left(\frac{M}{N}\right)^2 + 1}$$

The switch capacities for different expansion ratios $E = M/N$ are given in Table 7.1. A squared $N \times N$ switch with input queueing has a capacity $\rho_{max} = 2 - \sqrt{2} = 0.586$, whereas a higher throughput $\rho_{max} = 0.632$ characterizes a pure crossbar system. The reason is that in the former case only some HOL positions are filled with new cells, whereas in the latter case all the HOL cells are generated slot by slot. This partial memory of the HOL cell pattern in a switch with input queueing results in a higher number of conflicts for the switch outlets than in a memoryless crossbar switch. The term head-of-line (HOL) blocking is used to denote such a low capacity of an IQ switch compared to the ideal value of $\rho_{max} = 1$. When the expansion ratio M/N is larger (smaller) than 1, ρ_{max} correspondingly increases (decreases) quite fast.

Table 7.1. Switch capacity for different expansion ratios

$\frac{M}{N}$	ρ_{max}	$\frac{M}{N}$	ρ_{max}
1	0.586	1	0.586
1/2	0.382	2	0.764
1/4	0.219	4	0.877
1/8	0.117	8	0.938
1/16	0.061	16	0.969
1/32	0.031	32	0.984

When $M = N$ and N is finite, the maximum throughput can be obtained through straightforward Markovian analysis as reported in [Bha75]. Nevertheless, since the state space grows exponentially, this approach is feasible only for small N. The throughput values so obtained are contained in Table 7.2, together with values obtained through computer simulation. It is interesting to note that the throughput asymptotic value $\rho_{max} = 0.58$ is approached quite fast as N increases. We could say that $N = 64$ is a switch size large enough to approximate quite well the behavior of an infinitely large switch.

An interesting observation arises from Table 7.2: the switch capacity for $N = 2$ is the same ($\rho_{max} = 0.75$) as for the basic crossbar network 2×2, (see Section 6.1.1.2) which by

Table 7.2. Switch capacity for different network sizes

N	ρ^{AN}_{max}	ρ^{SIM}_{max}
2	0.7500	0.7500
4	0.6553	0.6552
8	0.6184	0.6184
16		0.6013
32		0.5934
64		0.5898
128		0.5876
∞	.5858	

definition has no buffers. With any other switch size, input queueing degrades the performance of the pure crossbar network. The explanation is that at least one packet is always switched in each slot in a 2×2 switch, so that at most one packet is blocked in one of the two input queues, while the other HOL packet is new. Therefore the output addresses requested by the two HOL packets are mutually independent (the system behaves as if both addresses were newly assigned slot by slot).

7.1.2.2. Packet delay

The packet delay of a $N \times M$ IQ switch can be evaluated according to the procedure described in [Kar87] (a different approach is described in [Hui87]). Owing to the Bernoulli arrival process of a packet to each switch inlet with probability p, the tagged input queue behaves as a $Geom/G/1$ queue; its service time θ_i is the time it takes for the HOL packet to win the output channel contention and thus to be transmitted, or, equivalently, the time spent inside the tagged virtual queue. As shown in [Hui87], [Li90], the steady-state number of packets entering the tagged virtual queue becomes Poisson with rate $p_v = pN/M$, when $N, M \to \infty$ (each switch outlet is addressed with the same probability $1/M$). Thus, the tagged virtual queue behaves as a synchronous $M/D/1$ with internal server (see Appendix) with average arrival rate p_v in which the waiting time η_v is the time it takes for a packet to win the output channel contention and the service time θ_v equals 1 slot. Consequently, the queueing delay $\eta_v + \theta_v$ represents the service time θ_i of the input queue $Geom/G/1$. Using the results found in Appendix for this queue, the average time spent in the input queue, which equals the average total delay T, is

$$E[\delta_i] = E[\eta_i] + E[\theta_i] = \frac{pE[\theta_i(\theta_i - 1)]}{2(1 - pE[\theta_i])} + E[\eta_v] + 1 \tag{7.3}$$

$$= \frac{p(E[\eta_v^2] + E[\eta_v])}{2[1 - p(E[\eta_v] + 1)]} + E[\eta_v] + 1$$

The first and second moment of the waiting time in the $M/D/1$ queue can be found again in the Appendix, assuming either a FIFO or a RO type of service in the virtual queue.

The former case corresponds to a switch that gives priority in the channel contention to older cells, that is cells that spent more time in the HOL positions. The latter case means assuming that the cell winning a channel contention is always selected random in the virtual queue. An implementation example of the FIFO service in the virtual queue is provided while describing the enhanced architecture with channel grouping in Section 7.1.3.1. Note that the p that makes the denominator vanish also gives the maximum throughput ρ_{max} already found in the preceding section with a different approach.

The accuracy of the analytical model is assessed in Figure 7.9 for both service disciplines FIFO and RO in the virtual queue (a continuous line represents analytical results, whereas simulation data are shown by "×" and "+" signs for FIFO and RO, respectively). The model assumes an infinite size for the squared switch, whereas the simulation refers to a switch size $N = 256$ in which a complete fairness is accomplished. The model provides very accurate results since it does not introduce any approximation on the real switch behavior and, as discussed in the previous section, the asymptotic throughputs of the two switches are almost the same. As one might expect, Figure 7.9 shows that the average packet delay grows unbounded as the offered load approaches the switch capacity.

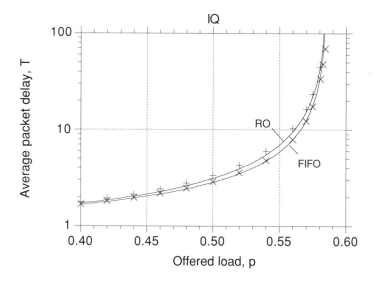

Figure 7.9. Delay performance of an ATM switch with input queueing

7.1.2.3. Packet loss probability

It has been shown that a switch with pure input queueing has a maximum throughput lower than a crossbar switch. Then a question could arise: what is the advantage of adding input queueing to a non–blocking unbuffered structure (the crossbar switch), since the result is a decrease of the switch capacity? The answer is rather simple: input queueing enables control the packet loss performance for carried loads $\rho < \rho_{max}$ by suitably sizing the input queues.

The procedure to evaluate the cell loss probability when the input queues have a finite size of B_i cells is a subcase of the iterative analysis described in Section 8.1.2.2 for an infinite-size non-blocking switch with combined input–output queueing in which $B_o = 1$ (the only packet being served can sit in the output queue) and $K = 1$ (at most one packet per switch outlet can be transmitted in a slot). The results are plotted in Figure 7.10 for a buffer B_i ranging from 1 to 32 (simulation results for a switch with $N = 256$ are shown by dots) together with the loss probability $1 - (1 - e^{-p}) / p$ of a crossbar switch. If our target is to guarantee a packet loss probability less than 10^{-9}, we simply limit the offered load to $p = 0.3$ with $B_i = 8$ or to $p = 0.5$ for $B_i = 32$, whereas the packet loss probability of the crossbar switch is above 10^{-2} even for $p = 0.05$. So, input queueing does control the packet loss performance.

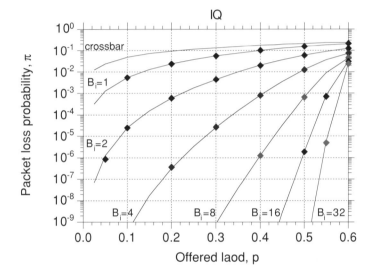

Figure 7.10. Loss performance of a non-blocking ATM switch with input queueing

A simple upper bound on the packet loss probability has also been evaluated [Hui87], by relying on the distribution of the buffer content in a switch with infinitely large input queues. The loss probability with a finite buffer B_i is then set equal to the probability that the content of the infinite queue exceeds B_i and gives

$$\pi = \frac{p(2-p)}{2(1-p)} \left[\frac{p^2}{2(1-p)^2} \right]^{B_i - 1} \tag{7.4}$$

7.1.3. Enhanced architectures

The throughput limitations of input queueing architectures shown in Section 7.1.2.1 and due to the HOL blocking phenomenon can be partially overcome in different ways. Two techniques are described here that are called *Channel grouping* and *Windowing*.

7.1.3.1. Architecture with channel grouping

With the traditional *unichannel bandwidth allocation*, an amount of bandwidth is reserved on an output channel of the switching node for each source at call establishment time. A *multichannel bandwidth allocation* scheme is proposed in [Pat88] in which input and output channels of a switch are organized in *channel groups*, one group representing the physical support of *virtual connections* between two routing entities, each residing either in a switching node or in a user–network interface. More than one channel (i.e., a digital pipe on the order of 150/600 Mbit/s) and more than one channel group can be available between any two routing entities. Consider for example Figure 7.11 where four channels are available between the network interface NI_1 (NI_2) and its access node SN_1 (SN_2) and six channels connect the two nodes SN_1 and SN_2 (Figure 7.11a). The same network in which multichannel bandwidth is available includes for example two groups of three channels between the two network nodes, two groups of two channels between NI_1 and SN_1 and one group of four channels between NI_2 and SN_2 (Figure 7.11b). With the multichannel bandwidth allocation the virtual connections are allocated to a channel group, not to a single channel and the cells of the connections can be transmitted on any channel in the group. Then we could say that the switch bandwidth is allocated according to a two-step procedure: at *connection set-up time* connections are allocated to a channel group, whereas at *transmission time* cells are allocated to single channels within a group.

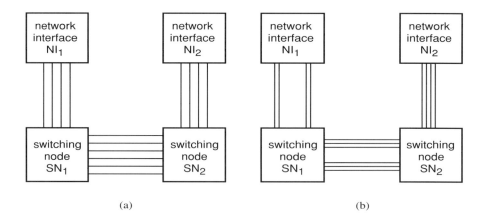

(a)   (b)

Figure 7.11. Arrangement of broadband channels into groups

The bandwidth to be allocated at connection set-up time is determined as a function of the channel group capacity, the traffic characteristics of the source, and the delay performance expected. The criterion for choosing such bandwidth, as well as the selection strategy of the specific link in the group, is an important engineering problem not addressed here. Our interest is here focused on the second step of the bandwidth allocation procedure.

At transmission time, before packets are switched, specific channels within a group are assigned to the packets addressed to that group, so that the channels in a group behave as a set of servers with a shared waiting list. The corresponding statistical advantage over a packet switch with a waiting list per output channel is well known. This "optimal" bandwidth assignment at transmission time requires coordination among the port controllers which may be achieved by designing a fast hardware "channel allocator". This allocator, in each slot, collects the channel group requests of all port controllers and optimally allocates them to specific channels in the requested channel groups. The number of channels within group j assigned in a slot equals the minimum between the number of packets requesting group j in the slot and the number of channels in group j. Packets that are denied the channel allocation remain stored in the buffer of the input port.

This multichannel bandwidth allocation has two noteworthy implications for the kind of service provided by the switching node. On the one hand, it enables the network to perform a "super-rate switching": virtual connections requiring a bandwidth greater than the channel capacity are naturally provided, as sources are assigned to a channel group, not a single channel. On the other hand, packets could be delivered out-of-sequence to the receiving user, since there is no guarantee that the packets of a virtual connection will be transmitted on the same channel. It will be shown how the implementation of the multichannel bandwidth allocation scheme is feasible in an ATM switch with input queueing and what performance improvements are associated with the scheme.

Switch architecture. An ATM switch with input queueing adopting the multichannel bandwidth allocation called the MULTIPAC switch, is described in detail in [Pat90]. Figure 7.3, showing the basic architecture of the Three-Phase switch, also represents the structure of the MULTIPAC switch if the functionality of the allocation network is properly enriched. Now the contention resolution algorithm requires the addresses of the output ports terminating the channels of the same group to be consecutive. This requirement could seriously constrain a change of the configuration of the interswitch communication facilities, e.g., following a link failure or a change in the expected traffic patterns. For this reason, a *logical addressing scheme* of the output channels is defined, which decouples the channel address from the physical address of the output port terminating the channel.

Each channel is assigned a *logical address*, so that a channel group is composed of channels with consecutive logical addresses, and a one-to-one mapping is defined between the channel logical address and the physical address of the port terminating the channel. The channel with the lowest logical address in a group is the *group leader*. The group leader's logical address also represents the *group address*. A specific channel in a group is identified by a *channel offset* given by the difference between the channel's logical address and the group leader's logical address. Each port controller is provided with two tables, K_a and K_c. K_a maps the logical address to the physical address (i.e., the port address) of each channel and K_c specifies the maximum value, *maxoff(i)*, allowed for the channel offset in group i. Tables K_a and K_c are changed only when the output channel group configuration is modified.

If the N switch outlets are organized into G groups where R_i is the number of channels, or *capacity*, of group i, then $N = R_1 + ... + R_G$. Therefore $R_i = maxoff(i) + 1$. Let R_{max} be the maximum capacity allowed for a channel group and, for simplicity, N be a power of 2. Let $n = \log_2 N$ and d denote the number of bits needed to code the logical address of a channel (or the physical address of a port) and the channel offset, respectively.

The procedure to allocate bandwidth at transmission time, which is referred to as a *multi-channel three-phase algorithm*, is derived from the three-phase algorithm described in Section 7.1.1, whose principles are assumed now to be known. In addition to solving the output channel contention, the algorithm includes the means to assign optimally the channels within a group to the requesting packets slot by slot. Compared to the control packet formats described in Section 7.1.1.1, now the control packets are slightly different:

- Packet REQ(j,v,i) is composed of the identifier j of the *destination channel group* (i.e., the logical address of the group leader), the *request packet priority* v and the physical address i of the *source port controller*. Field *priority*, which is n_p bit long, is used to give priority in the channel allocation process to the older user packets.
- Packet ACK($i,actoff(j)$) includes the PC *source* address i and the *actual offset* field $actoff(j)$. The actual offset field is n_a bit long and identifies the output channel within the requested group assigned to the corresponding request.
- Packet DATA(m,cell) includes the switch outlet address m allocated to the PC and the cell to be switched.

In Phase I, port controller PC$_i$ with a cell to be transmitted to channel group j sends a request packet REQ(j,v,i). The value of v is given by $2^{n_p} - 1$ decreased by the number of slots spent in the HOL position by the user packet. When the priority range is saturated, that is the user packet has spent at least $2^{n_p} - 1$ slots in the switch, the corresponding request packet will always maintain the priority value $v = 0$. The "channel allocator" assigns an actual offset $actoff(j)$ to each request for group j, within its capacity limit R_j, to spread the requests over all the channels of group j. Note that there is no guarantee that the number of requests for group j does not exceed the number of channels in the group. Each offset belonging to the interval $[0, R_{max} - 1]$ is assigned only once to the requests for channel group j, while other requests for the same group are given an offset $actoff(j) \geq R_{max}$. Since $maxoff(j) \leq R_{max} - 1$, each channel of group j is allocated to only one request for group j. So, the number of bits needed to code the actual offset is not less than $n_a = \lceil \log_2 R_{max} \rceil$.

In Phase II, field *source* of packet REQ(j,v,i) and $actoff(j)$ assigned to this request are received by a port controller, say PC$_k$, that transmits them in packet ACK($i,actoff(j)$) to PC$_i$ through the interconnection network. It is easy to verify that the interconnection network structure of the MULTIPAC switch is such that any two packets ACK($i,actoff(j)$) are generated by different port controllers and all packets ACK($i,actoff(j)$) are delivered to the addressed port controllers without path contention, i.e., by disjoint paths within the interconnection network.

In Phase III, if $actoff(j) \leq maxoff(j)$, PC$_i$ sends its data packet DATA(m,cell) to a specific output channel of group j with physical address m. Table K_a maps the allocated logical channel $j + actoff(j)$ to the corresponding physical address m. Packets DATA(m,cell) cross the Batcher-banyan section of the interconnection network without collisions, since the winning requests have been assigned different output logical addresses and, hence, different physical addresses of output channels. If $actoff(j) > maxoff(j)$, the port controller will issue a new request REQ(j,v,i) the next slot for its head–of–line cell, which remains stored in the input port queue.

The packet flow through interconnection network is the same shown in Figure 7.4 for the basic IQ Three-Phase switch, in which the signal a now represents the offset $actoff(.)$ of the acknowledgment packet transmitted on a_k and associated to the request packet emerging on outlet d_k. The channel allocation network is now more complex than in the unichannel three-phase algorithm. It is composed of subnetworks A and B (Figure 7.12). Subnetwork A receives a set of adjacent packets with non-decreasing destination addresses and identifies the requests for the same channel group. Subnetwork B, which includes *s* stages of adders, assigns an actual offset *actoff(j)* to each packet addressed to channel group *j*, so that the offsets corresponding to each member of the group are assigned only once.

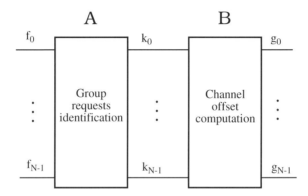

Figure 7.12. Channel allocation network

An example of the packet switching in a multichannel Three-Phase switch with $N = 8$ is shown in Figure 7.13. In particular the operation of the allocation network is detailed in the process of generating the acknowledgment packets starting from the request packets. In the example three requests are not accepted, that is one addressing group 0 that is given *actoff*(0) = 1 (the group maximum offset stored in K_c is 0) and two addressing groups 4, which are given *actoff*(4) = 3, 4 (the group maximum offset is 2). The following data phase is also shown in which the five PCs that are winners of the channel contention transmit a data packet containing their HOL cell to the switch outlets whose physical addresses are given by table K_a.

As already mentioned, the adoption of the channel grouping technique has some implications on cell sequencing within each virtual call. In fact since the ATM cells of a virtual call can be switched onto different outgoing links of the switch (all belonging to the same group), calls can become out of sequence owing to the independent queueing delays of the cells belonging to the virtual call in the downstream switch terminating the channel group. It follows that some additional queueing must be performed at the node interfacing the end-user, so that the correct cell sequence is restored edge-to-edge in the overall ATM network.

Implementation guidelines. The hardware structure of subnetwork A is the same as presented in Figure 7.6, which now represents the hardware associated with outlet k_{k+1} of subnetwork A. The EX-OR gate performs the same function. Now, port controllers generate packet ACK(.,.) by transmitting field *source* of the packet REQ(.,.,.) being received, immediately followed by the computed *actoff*(.). When the first bit of packet REQ(.,.,.) is transmitted on outlet d_k, it takes $n + s$ bit times to generate the first bit of *actoff*(.) by the channel alloca-

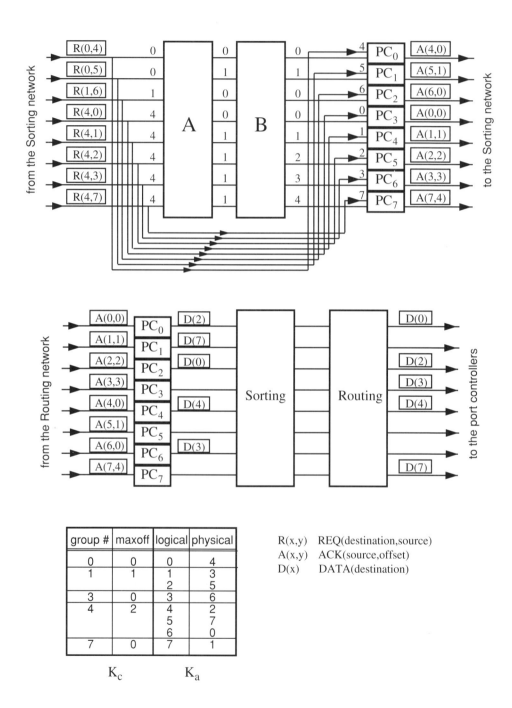

Figure 7.13. Example of packet switching in the MULTIPAC switch

tion network (n bit in subnetwork A and s bit in subnetwork B), while it takes $2n$ bit times to complete the transmission of the first field of packet ACK(.,.) by a port controller. The AND gate in subnetwork A suitably delays the start of computation of the channel offset in subnetwork B, so as to avoid the storage of $actoff(.)$ in the port controller. Also in this case the signal on outlet k_0 is always low, as is required by subnetwork B, which always gives the value 0 to the $actoff(.)$ associated to the packet REQ(.,.,.) transmitted on outlet d_0.

Subnetwork B is a running sum adder of $s = \lceil \log_2 R_{max} \rceil$ stages computing the n_a digits of the $actoff(.)$ assigned to each packet REQ(.,.). Subnetwork B is represented in Figure 7.14 for the case of $N = 16$ and $5 \leq R_{max} \leq 8$. The AND gate A_2 is enabled by signal Φ_1 for one bit time, so that the adder of the first stage at port k_{k+1} only receives the binary number 1 (if k_{k+1} is high) or the binary number 0 (if k_{k+1} is low). Based on the structure of subnetwork B, the output of each adder of stage z ($z = 1, ..., s$) is a binary stream smaller than or equal to 2^z with the least significant bit first. Hence, $n_a = 1 + s = 1 + \lceil \log_2 R_{max} \rceil$ bit are needed to code $actoff(j)$ that emerges from stage s of subnetwork B. The AND gates A_1 allow an independent computation of the running sums for each requested channel group, by resetting the running sum on each inlet k_k with low signal. This is made possible by subnetwork A that guarantees that at least one inlet k_k with low signal separates any two sets of adjacent inlets with high signal. In fact, the complete set of requests for channel group j transmitted on outlets $D(j, k, l) = \{d_k, d_{k+1}, ..., d_{k+l}\}$ determines a low signal on inlet k_k and a high signal on inlets $k_{k+1}, ..., k_{k+l}$.

The binary stream on output g_k ($k = 0, ..., N-1$) of subnetwork B represents the offset $actoff(j)$ assigned to packet REQ(j,v,i) transmitted on outlet d_k. The offset for packet REQ(.,.,.) transmitted on outlet d_0 is always 0, because any other requests for the same channel group will be given an offset greater than 0. Thus, the offset allocated by subnetwork B to the request transmitted on output $d_i \in D(j, k, l)$ is

$$actoff(j) = \begin{cases} i - k & \text{if } i \leq k + 2^s - 1 \\ 2^s & \text{if } i > k + 2^s - 1 \end{cases}$$

Note that with this implementation R_{max} is constrained to be a power of 2. An example of operation of the channel allocation network is also shown in Figure 7.14. According to the offset allocation procedure, in this case two out of ten requests for the channel group 7 are given the same $actoff(7) = 8$. If $R_7 = 6$ then four requests for channel group j, i.e. those transmitted by ports $f_{10} - f_{13}$, lose the contention, since they receive an $actoff(7) > 5$. The running sum operations in subnetwork B for this example can also be traced stage by stage in Figure 7.14.

With the same assumptions and procedure adopted in the (unichannel) Three-Phase switch described in Section 7.1.1.1, the switching overhead $\eta = T_{I-II}/T_{III}$ required to perform the three-phase algorithm is now computed. The duration of Phase I is given by the latency $n(n+1)/2$ in the Batcher network and the transmission time $n + p$ of the first two fields in packet REQ(.,.,.) (the field source in packet REQ(.,.,.) becomes field source in packet ACK(.,.) and its transmission time is summed up in Phase II). The duration of Phase II includes the latency $n(n+1)/2$ in the Batcher network, the latency n in the banyan network, the

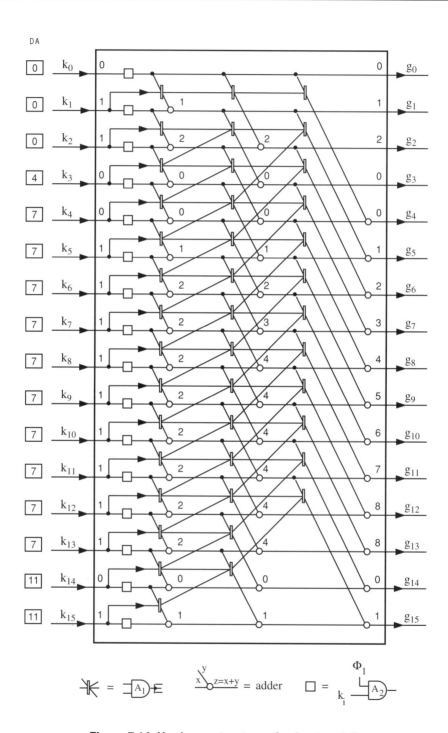

Figure 7.14. Hardware structure of subnetwork B

transmission time $n + n_a$ of packet ACK($i, actoff(j)$) and the eventual additional time needed by the port controller to check if $actoff(j) \leq maxoff(j)$ and to sum j and $actoff(j)$. It can be shown [Pat88] that these tasks can be completed in 1 bit time after the complete reception of packets ACK(.,.) by the port controllers. Note that the subnetwork A of AN has no latency (the result of the address comparison is available at the EX-OR gate output when the receipt of the destination field in packet REQ(.,.,.) is complete). Furthermore, the latency of subnetwork B of AN must not be summed up as the running sum lasts s bit times and this interval overlaps the transmission time of fields priority and source in packet ACK(.,.) (this condition holds as long as $s \leq n + p$). We further assume that the channel logical address $j + actoff(j)$ is mapped onto the corresponding channel physical address in a negligible time. Hence, the total duration of Phases I and II for a multichannel switch is given by

$$T_{I-II} = n(n+4) + n_a + p + 1 = (\log_2 N)(\log_2 N + 4) + \lceil \log_2 R_{max} \rceil + p + 2$$

whereas $T_{I-II} = (\log_2 N)(\log_2 N + 4) + 1$ in a unichannel switch.

Thus providing the multichannel capability to a Three-Phase switch implies a small additional overhead that is a logarithmic function of the maximum channels group capacity. For a reasonable value $R_{max} = 64$, a switch size $N = 1024$, no priority ($p = 0$) and the standard cell length of 53 bytes that implies $T_{III} = 53 \cdot 8$, we have $\eta \cong 0.333$ in the unichannel switch and $\eta \cong 0.349$ in the multichannel switch.

In order to reduce the switching overhead the multichannel three-phase algorithm can be run more efficiently by pipelining the signal transmission through the different networks so as to minimize their idle time [Pat91]. In this pipelined algorithm it takes at least two slots to successfully complete a reservation cycle from the generation of the request packet to the transmission of the corresponding cell. By doing so, the minimum cell switching delay becomes two slots but the switching overhead reduces from $\eta \cong 0.333$ to $\eta \cong 0.191$. Thus, with an external channel rate $C = 150$ Mbit/s, the switch internal rate $C_i = C(1 + \eta)$ is correspondingly reduced from 200 Mbit/s to 179 Mbit/s.

Performance evaluation. The performance of the multichannel MULTIPAC switch will be evaluated using the same queueing model adopted for the basic input queueing architecture (see Section 7.1.2). The analysis assumes $N, M \to \infty$, while keeping a constant expansion ratio $E = M/N$. The input queue can be modelled as a $Geom/G/1$ queue with offered load p, where the service time θ_i is given the queueing time spent in the virtual queue that is modelled by a synchronous $M/D/R$ queue with average arrival rate $p_v = pRN/M$ (each virtual queue includes R output channels and has a probability R/M of being addressed by a HOL packet). A FIFO service is assumed both in the input queue and in the virtual queue. Thus, Equation 7.3 provides the maximum switch throughput (the switch capacity) ρ_{max}, equal to the p value that makes the denominator vanish, and the average cell delay $T = E[\delta_i]$. The moments of the $M/D/R$ queue are provided according to the procedure described in [Pat90].

Table 7.3 gives the maximum throughput ρ_{max} for different values of channel group capacity and expansion ratio. Channel grouping is effective especially for small expansion ratios: with a group capacity $R = 16$, the maximum throughput increases by 50% for $M/N = 1$ (from 0.586 to 0.878), and by 30% for $M/N = 2$ (it becomes very close to 1).

This group capacity value is particularly interesting considering that each optical fiber running at a rate of 2.4 Gbit/s, which is likely to be adopted for interconnecting ATM switches, is able to support a group of 16 channels at a rate $C = 150$ Mbit/s. Less significant throughput increases are obtained for higher expansion ratios.

Table 7.3. Switch capacity for different channel group sizes and expansion ratios

	$\frac{M}{N} = 1$	2	4	8	16	32
$R = 1$	0.586	0.764	0.877	0.938	0.969	0.984
2	0.686	0.885	0.966	0.991	0.998	0.999
4	0.768	0.959	0.996	1.000	1.000	1.000
8	0.831	0.991	1.000	1.000	1.000	1.000
16	0.878	0.999	1.000	1.000	1.000	1.000
32	0.912	1.000	1.000	1.000	1.000	1.000
64	0.937	1.000	1.000	1.000	1.000	1.000

The average packet delay T given by the model for $N = M$, a channel group capacity $R = 2^i$ ($i = 0, ..., 5$) is plotted in Figure 7.15 and compared to results obtained by computer simulation of a switch with $N = 256$ and very large input queues. The analytical results are quite accurate, in particular close to saturation conditions. The mismatching between theoretical and simulation delay for small group capacities in a non–saturated switch is due to the approximate computation of the second moment of η_v in the virtual queue.

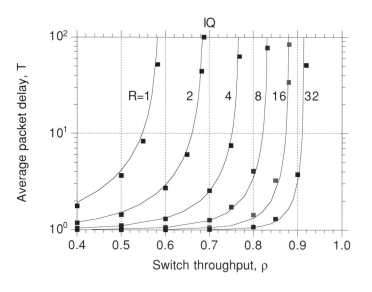

Figure 7.15. Delay performance of an IQ switch

The priority mechanism described in multichannel architecture (*local priority*) that gives priority to older cells in the HOL position of their queue has been correctly modelled by a FIFO service in the virtual queue. Nevertheless, a better performance is expected to be provided by a priority scheme (*global priority*) in which the cell age is the whole time spent in the input queue, rather than just the HOL time. The latter scheme, in fact, aims at smoothing out the cell delay variations taking into account the total queueing time. These two priority schemes have been compared by computer simulation and the result is given in Figure 7.16. As one might expect, local and global priority schemes provide similar delays for very low traffic levels and for asymptotic throughput values. For intermediate throughput values the global scheme performs considerably better than the local scheme.

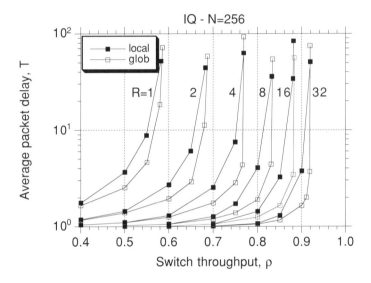

Figure 7.16. Delay performance of an IQ switch with local and global priority

Figure 7.17 shows the effect of channel grouping on the loss performance of a switch with input queueing, when a given input queue is selected (the results have been obtained through computer simulation). For an input queue size $B_i = 4$ and a loss performance target, say 10^{-6}, the acceptable load is only $p = 0.2$ without channel grouping ($R = 1$), whereas it grows remarkably with the channel group size: this load level becomes $p = 0.5$ for $R = 4$ and $p = 0.8$ for $R = 8$. This improvement is basically due to the higher maximum throughput characterizing a switch with input queueing. Analogous improvements in the loss performance are given by other input queue sizes.

7.1.3.2. Architecture with windowing

A different approach for relieving the throughput degradation due to the HOL blocking in switches with input queueing is *windowing* [Hlu88]. It consists in allowing a cell other than the HOL cell of an input queue to be transmitted if the HOL cell is blocked because of a contention for the addressed switch outlet. Such a technique assumes that a non-FIFO queueing

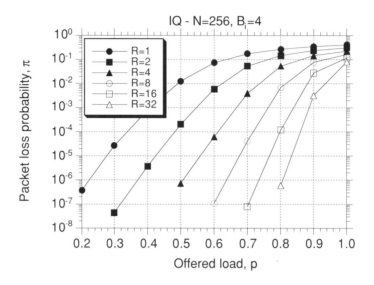

Figure 7.17. Loss performance with different channel group sizes

capability is available in the input queue management. The switch outlets are allocated using W reservation cycles per slot rather than the single cycle used in the basic input queueing architecture. In the first cycle the port controllers will request a switch outlet for their HOL cell: the contention winners stop the reservation process, while the losers will contend for the non-reserved outlets in the second cycle. The request will be for the switch outlet addressed by a packet younger than the HOL packet. Such a younger packet is identified based on either of two simple algorithms [Pat89]:

(a) it is the oldest packet in the buffer for which a reservation request was not yet issued in the current slot;

(b) it is the oldest packet in the buffer for which, in the current slot, a reservation request was not yet issued and whose outlet address has not yet been requested by the port controller.

Then in the generic cycle i $(1 < i \leq W)$, each port controller who has been a loser in the previous cycle $i-1$ will issue a new request for another younger packet in its queue. Since up to W adjacent cells with algorithm (a), or W cells addressing different switch outlets with algorithm (b) contend in a slot for the N switch outlets, the technique is said to use a *window* of size W. Apparently the number of reserved outlets in a slot is a non-decreasing function of the window size.

The adoption of a non-FIFO queueing implies that cell out-of-sequence on a virtual call can take place if algorithm (a) is adopted with a priority scheme different from that based on cell age in the switch adopted in Section 7.1.3.1. For example cell sequencing is not guaranteed if a service priority rather than a time priority is used: priority is given to cells belonging to certain virtual circuits independent of their queueing delay. As with channel grouping, a queueing capability at the ATM network exit should be provided to guarantee cell sequencing edge-to-edge in the network.

Switch architectures. The windowing technique can be implemented in principle with both the Three-Phase and the Ring-Reservation switch. An architecture analogous to the former switch is described in [Ara90], where a Batcher sorting network is used. Also in this case the output contention among PCs is resolved by means of request packets that cross the sorting network and are then compared on adjacent lines of an arbiter network. However, the contention result is now sent back to the requesting PCs by using backwards the same physical path set up in the sorting network by the request packets. In any case both Batcher-banyan based and Ring-Reservation switches have the bottleneck of a large internal bit rate required to run W reservation cycles.

Windowing technique can be accomplished more easily by a pipelined architecture of the Three-Phase switch architecture (see Section 7.1.1.1 for the basic architecture) that avoids the hardware commonality between the reservation phase (probe-ack phases) and the data phase. Such an architecture for an $N \times N$ switch, referred to as a *Pipelined Three-Phase* switch (P3-Ph), is represented in Figure 7.18. The N I/O channels are each controlled by a port controller PC_i $(i = 0, ..., N-1)$. The interconnection network includes a Batcher sorting network (SN_1) and a banyan routing network (RN), both dedicated to the data phase, while another Batcher sorting network (SN_2) and a port allocation network (AN) are used to perform the windowing reservation. Note that here a routing banyan network is not needed for the reservation phase as the same sorting network SN_2 is used to perform the routing function. In fact all the N port controllers issue a reservation request in each cycle, either true or idle, so that exactly N acknowledgment packets must be delivered to the requesting port controllers. Since each of these packets is addressed to a different port controller by definition (exactly N requests have been issued by N different PCs), the sorting network is sufficient to route correctly these N acknowledgment packets. The hardware structure of these networks is the same as described for the basic Three-Phase switch. It is worth noting that doubling the sorting network in this architecture does not require additional design effort, as both sorting networks SN_1 and SN_2 perform the same function.

The request packets contain the outlet address of the HOL packets in the first reservation cycle. In the following cycles, the request packet of a winner PC contains the same outlet address as in the previous cycle, whereas a loser PC requests the outlet address of a younger packet in the queue selected according to algorithm (a) or (b). In order to guarantee that a packet being a contention winner in cycle i does not lose the contention in cycle j $(j > i)$, starting from cycle $i + 1$ the request packet must always contain a priority field (at least one bit is needed) whose value is set to a conventional value guaranteeing its condition of contention winner in the following cycles. This priority field will be placed just between fields DA and SA in the request packet.

An example of the algorithm allocating the output ports is given in Figure 7.19 for $N = 8$ showing the packet flow for a single reservation cycle. In this example the use of a priority field in the requests is intentionally omitted for the sake of simplicity. In the request phase (Figure 7.19a) each port controller sends a request packet to network SN_2 containing, in order of transmission, an activity bit AC, the requested destination address DA and its own

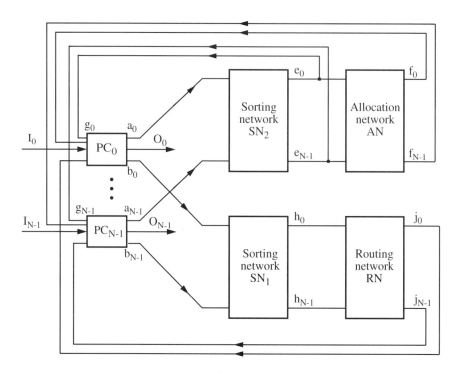

Figure 7.18. Architecture of the Pipelined Three-Phase switch

source address SA. The AC bit is 1 if the PC is actually requesting an output port, 0 if its queue is empty. Network SN_2 sorts the request packets received on inlets $a_0, ..., a_{N-1}$ according to ascending keys, the key being the whole request packet, so that requests for the same output port DA emerge on adjacent outlets of the network SN_2. Network AN, which receives request packets on inlets e_i $(i = 0, ..., N-1)$, generates on outlets f_i $(i = 0, ..., N-1)$, the grant allocation bit GR. GR is set to 0 by the network AN if the fields DA received on inlets e_{i-1} and e_i are different, otherwise is set to 1. The generic port controller receives field SA from SN outlets e_i and GR from AN outlets f_i. These two fields are combined into an acknowledgment packet, thus starting the ack phase. Acknowledgment packets are re-entered into the network SN_2, which sorts them according to field SA. Since each PC has issued a request packet, the ack packet with $SA = i$ will be transmitted to outlet e_i and, hence, received by PC_i through lead g_i without requiring a banyan routing network. In the data phase (Figure 7.19b), PC_i transmits during slot j the user packet for which the request packet was issued in slot $j-1$, if the corresponding acknowledgment packet has been received by PC_i with $GR = 0$. If $GR = 1$, a new reservation will be tried in the next slot.

Performing W reservation cycles in the Ring-Reservation switch simply consists in making W cycles of the reservation frame. In the first cycle the reservation is the same as in the basic architecture. In the generic cycle i $(1 < i \le W)$ only a port controller which has been a contention loser in the former cycle $i-1$ attempts a new reservation of the switch outlet

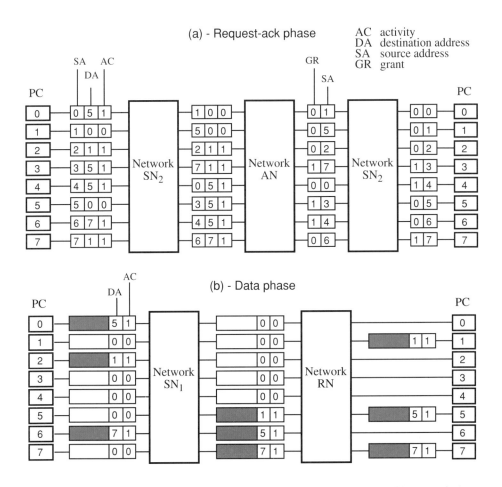

Figure 7.19. Example of packet switching in the Pipelined Three-Phase switch

addressed by the packet selected according to algorithm (a) or (b). With this structure the frame is expected to reduce the unreserved fields cycle after cycle.

It is interesting to compare the bit rate required in these two architectures to implement the windowing technique. The bit rate required in the networks SN_2 and AN can be computed with the same reasoning developed in Section 7.1.1 for the basic structure, considering that now the network latency of the probe phase must be decreased by n bits (the latency of the removed banyan network) and that the request packet now includes also the 1-bit field AC and the priority field whose minimum size is chosen here (1 bit). Thus the duration T_{I-II} of Phases I and II is given by

$$T_{I-II} = W[n(n+3) + 3] = W[\log_2 N(\log_2 N + 3) + 3]$$

Since W cycles must be completed in a time period equal to the transmission time of a data packet, then the minimum bit rate on networks SN_2 and AN is $CW[n(n+3)+3]/(53 \cdot 8)$, whereas the bit rate on the ring is $CW2^n/(53 \cdot 8)$, C being the bit rate in the interconnection network.

Figure 7.20 compares the bit rate required in the $N \times N$ interconnection network ($N = 2^n$) for the Pipelined Three-Phase switch (P3-Ph), for the Ring-Reservation switch (RR) and for the basic Three-Phase switch (3-Ph) described in Section 7.1.1. The clock rate grows very fast for the Ring-Reservation switch due to its serial reservation process, thus making it unsuitable for non-small switches. The pipelined switch with $W \leq 3$ requires a bit rate always smaller than the basic Three-Phase switch for $n \leq 13$ and is thus convenient unless very large switches with $W > 3$ are considered. The rate in this latter switch grows rather slowly with n, even if the architecture does not give the throughput improvements expected from the windowing scheme of the pipelined switch.

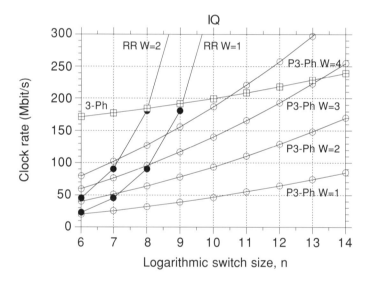

Figure 7.20. Bit rate required for accomplishing the reservation process

Performance evaluation. The reasoning developed in Section 6.1.1.2 for the analysis of a crossbar network can be extended here to evaluate the switch throughput when windowing is applied [Hui90]. Therefore we are implicitly assuming that the destinations required by the contending cells are always mutually independent in any contention cycle (each cell addresses any switch outlet with the constant probability $1/N$). If ρ_i denotes the total maximum throughput (the switch capacity) due to i reservation cycles, we immediately obtain

$$\rho_i = \begin{cases} p_1 & (i = 1) \\ \rho_{i-1} + (1 - \rho_{i-1})p_i & (i > 1) \end{cases} \tag{7.5}$$

where p_i indicates the probability that an outlet not selected in the first $i-1$ cycles is selected in the i-th cycle. Since the number of unbooked outlets after completion of cycle $i-1$ is $N(1-p_{i-1})$, the probability p_i is given by

$$p_i = \begin{cases} 1 - \left(1 - \dfrac{1}{N}\right)^N & (i = 1) \\ 1 - \left(1 - \dfrac{1}{N}\right)^{N(1-p_{i-1})} & (i > 1) \end{cases}$$

Note that the case $i = 1$ corresponds to a pure crossbar network. For an infinitely large switch $(N \to \infty)$ the switch capacity with i reservation cycles becomes

$$p_i = p_{i-1} + (1-p_{i-1})\,[1 - e^{-(1-p_{i-1})}] = 1 - (1-p_{i-1})\,e^{-(1-p_{i-1})} \tag{7.6}$$

The throughput values provided by Equations 7.5 and 7.6 only provide an approximation of the real throughput analogously to what happens for $i = 1$ (no windowing) where the crossbar model gives a switch capacity $\rho_{max} = 0.632$ versus a real capacity $\rho_{max} = 0.586$ provided by the pure input queueing model. In fact the input queues, which are not taken into account by the model, make statistically dependent the events of address allocation to different cells in different cycles of a slot or in different slots.

The maximum throughput values obtained by means of the windowing technique have also been evaluated through computer simulation and the corresponding results for algorithm (a) are given in Table 7.5 for different network and window sizes. The switch capacity increases with the window size and goes beyond 0.9 for $W = 16$. Nevertheless, implementing large window sizes $(W = 8, 16)$ in the pipelined architecture has the severe drawback of a substantial increase of the internal clock rate.

Table 7.4. Maximum throughput for different switch and window sizes

	N = 2	4	8	16	32	64	128	256
W = 1	0.750	0.655	0.618	0.601	0.593	0.590	0.588	0.587
2	0.842	0.757	0.725	0.710	0.702	0.699	0.697	0.697
4	0.910	0.842	0.815	0.803	0.798	0.795	0.794	0.794
8	0.952	0.916	0.893	0.882	0.878	0.877	0.877	0.875
16	0.976	0.967	0.951	0.938	0.936	0.933	0.931	0.929

Figure 7.21 compares the results given by the analytical model and by computer simulation for different window sizes. Unless very small switches are considered $(N \leq 8)$, the model overestimates the real switch capacity and its accuracy improves for larger window sizes.

Table 7.5 shows the maximum throughput of the Pipelined Three-Phase switching fabric with $N = 256$ for both packet selection algorithms (a) and (b) with a window size ranging up to 10. The two algorithms give the same switch capacity for $W \leq 2$, whereas algorithm (b) performs better than algorithm (a) for larger window sizes (the throughput increase is of the

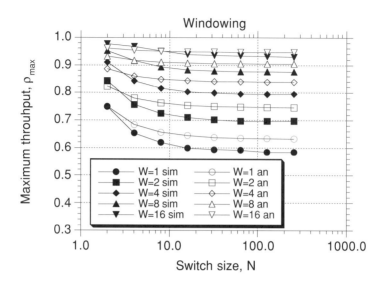

Figure 7.21. Switch maximum throughput with windowing

order of 2–3%). The almost equal performance of both algorithms for very small window sizes ($W = 2, 3$) can be explained by the very small probability of finding in the second or third position of an input queue a packet with the same address as the blocked HOL packet.

Table 7.5. Switch capacity for different windowing algorithms

W	Algorithm (a)	Algorithm (b)
1	0.587	0.587
2	0.697	0.697
3	0.751	0.757
4	0.794	0.808
5	0.821	0.838
6	0.841	0.860
7	0.860	0.874
8	0.875	0.889
9	0.878	0.900
10	0.882	0.905

7.2. Output Queueing

The queueing model of a switch with pure output queueing (OQ) is given by Figure 7.22: with respect to the more general model of Figure 7.1, it assumes $B_o > 0$, $B_i = B_s = 0$ and output speed-up $K > 1$. Now the structure is able to transfer up to $K < N$ packets from K different inlets to each output queue without blocking due to internal conflicts. Nevertheless, now there is no way of guaranteeing absence of external conflicts for the speed-up K, as N packets per slot can enter the interconnection network without any possibility for them of being stored to avoid external conflicts. So, here the packets in excess of K addressing a specific switch outlet in a slot are lost.

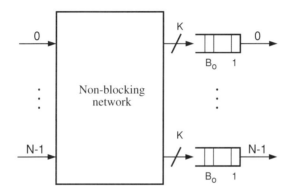

Figure 7.22. Model of non-blocking ATM switch with output queueing

7.2.1. Basic architectures

The first proposal of an OQ ATM switch is known as a *Knockout switch* [Yeh87]. Its interconnection network includes a non-blocking $N \times N$ structure followed by as many port controllers as the switch outlets (Figure 7.23), each of them feeding an output queue with capacity B_o packets. The $N \times N$ non-blocking structure is a set of N buses, each connecting one of the switch inlets to all the N output port controllers. In each port controller (Figure 7.24), each being a network with size $N \times K$, the N inlets are connected to N packet filters, one per inlet, that feed a $N \times K$ concentration network. Each packet filter drops all the packets addressing different switch outlets, whereas the concentration network interfaces through K parallel lines the output buffers and thus discards all the packets in excess of K addressing in a slot the same network outlet. The output queue capacity of B_o cells per port controller is implemented as a set of K physical queues each with capacity B_o / K interfaced to the concentrator through a shifter network.

For the $N \times K$ concentration network the structure of Figure 7.25 has been proposed (for $N = 8$, $K = 4$) which includes 2×2 contention elements and delay elements. The contention elements are very simple memoryless devices whose task is to transmit the packets to the top outlet (winner) whenever possible. So, if the element receives two packets, its state (straight/cross) does not care, while its state is straight (cross) if the packet is received on the

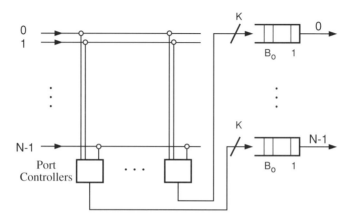

Figure 7.23. Architecture of the Knockout switch

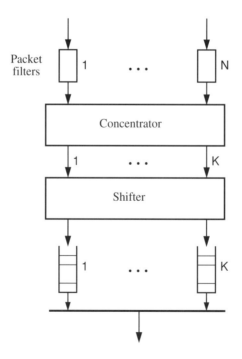

Figure 7.24. Structure of the output port controller in Knockout

top (bottom) inlet. Therefore, if two packets are received, one of them is routed to the bottom outlet (loser). By means of this structure all the packets received which are not in excess of K emerge at consecutive outlets of the concentration network starting from outlet 0. Note that we could have used a $N \times N$ banyan network with only K active outlets as the concentration network. However, such a solution has the drawback of requiring switching elements that are more complex than the contention elements. Moreover the non-blocking condition of the banyan network would require the selection of a reverse Omega or a reverse n–cube network in front of which an additional distributed device is provided to allocate consecutive output addresses to the packets received by the output concentrator slot by slot. An example of such a solution is described in Section 7.3.1 as the trap network for the Starlite switch.

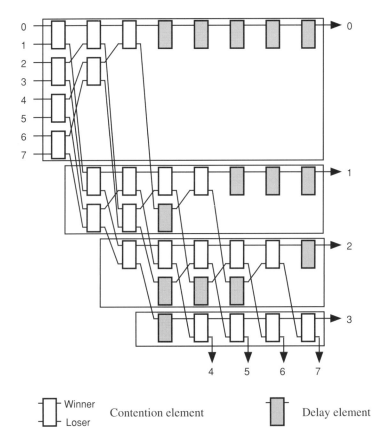

Figure 7.25. Structure of the concentrator in Knockout

The speed-up K accomplished by the Knockout switch implies that up to K packets can be received concurrently by each output queue. The distributed implementation proposed for such queue is also shown in Figure 7.24: it includes K physical queues, whose total capacity is B_0, preceded by a $K \times K$ shifter network whose task is to feed cyclically each of the K physical queues with at most one of the received packets. If the generic inlet and outlet of the shifter are denoted by i and o $(i, o = 0, ..., K-1)$, inlet i is connected at slot n to outlet $o = (i + k_n) \bmod K$ where

$$k_n = (k_{n-1} + m_{n-1}) \bmod K$$

m_n being the number of packets received by the shifter at slot n (the boundary condition $k_0 = 0$ applies). A cyclic read operation of each of the K physical queues takes place slot by slot, so that once every K slots each queue feeds the switch output line associated with the output interface. In the example of Figure 7.26 the shifter receives at slot 0 three cells, feeding queues 0–2, whereas at slot 1 it receives 7 cells entering queues 3–7, 0, 1, since $k_1 = 3$. Queue 0 holds only one cell since it transmits out a cell during slot 1. With such an implementation of the output queue, at most one read and one write operation per slot are performed in each physical queue, rather than K write operations per slot, as would be required by a standard implementation of the output queue without the shifter network. The shifter network can be implemented by an Omega or n-cube network that, owing to the very small number of permutations required (as many as the different occurrences of k_j, that is K), can be simply controlled by a state machine (packet self-routing is not required).

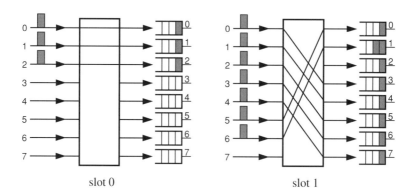

slot 0   slot 1

Figure 7.26. Implementation of the output queue

In spite of the distributed design of the output queues, structures with output queueing as Knockout show in general implementation problems related to the bus-based structure of the network and to the large complexity in terms of crosspoints of the concentrators. An alternative idea for the design of an OQ ATM switch, referred to as a *Crossbar Tree switch* and derived from [Ahm88], consists in a set of N planes each interconnecting a switch inlet to all the N output concentrators (see Figure 7.27 for $N = 16$). Each plane includes $\log_2 N$ stages of 1×2 splitters, so that each packet can be self-routed to the proper destination and packet filters in the output concentrators are not needed. The Crossbar Tree switch includes only the N left planes of the crossbar tree network (see Figure 2.7) since now several packets can be addressing each switch outlet and they are received concurrently by the concentrator.

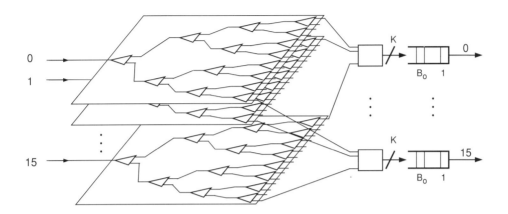

Figure 7.27. Architecture of the Crossbar Tree switch

7.2.2. Performance analysis

In a non-blocking switch with output queueing the switch capacity, that is the maximum throughput, ρ_{max} is always 1 if $K = N$ and $B_o = \infty$, independent of the switch size, since no HOL blocking takes place. Thus our interest will be focused on the packet delay and loss performance.

Owing to the random traffic assumption, that is mutual independence of packet arrival events with each packet addressing any outlet with the same probability $1/N$, then the probability a_n of n packets received in a slot that address the same "tagged" switch outlet is

$$a_n = \binom{N}{n}\left(\frac{p}{N}\right)^n\left(1 - \frac{p}{N}\right)^{N-n} \qquad n = 0, \ldots, N$$

where p is the probability of packet arrival in a slot at a switch inlet. So the probability of a packet loss event in the concentrator is given by

$$\pi_c = \frac{1}{p} \sum_{n = K+1}^{N} (n - K) a_n$$

Taking the limit as $N \to \infty$, the packet loss probability becomes

$$\pi_c = \left(1 - \frac{K}{p}\right)\left(1 - \sum_{n=0}^{K} \frac{p^n e^{-p}}{n!}\right) + \frac{p^K e^{-p}}{K!} \tag{7.7}$$

The loss performance in the concentrator of an OQ switch with $N \to \infty$ (Figure 7.28) shows that a speed-up $K = 8$ is enough to guarantee a loss in the concentrator on the order of 10^{-8} for $p = 0.5$ and 10^{-6} for $p = 1.0$. Any OQ switch with finite N always a gives better performance (Figure 7.29).

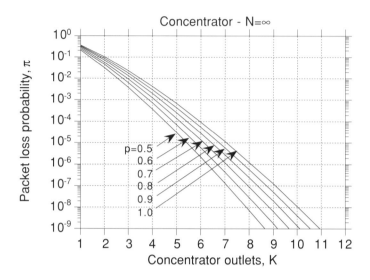

Figure 7.28. Loss performance in the concentrator for an infinitely large switch

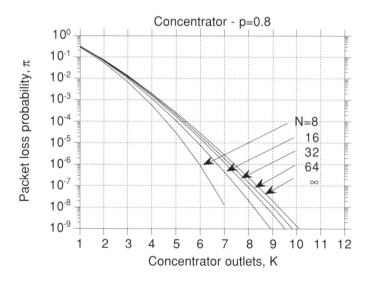

Figure 7.29. Loss performance in the concentrator for a given offered load

The output queue behaves as a $Geom(K)/D/1/B_o$ with offered load equal to

$$p_q = p(1 - \pi_c)$$

(each of the K sources is active in a slot with probability p_q/K) since up to K packets can be received in a slot. The evolution of such system is described by

$$Q_n = \min\{\max\{0, Q_{n-1} - 1\} + A_n, B_o\} \tag{7.8}$$

in which Q_n and A_n represent, respectively, the cells in the queue at the end of slot n and the new cells received by the queue at the beginning of slot n. A steady-state is assumed and the queue content distribution $q_i = Pr[Q = i]$ can be evaluated numerically according to the solution described in the Appendix for the queue $Geom(N)/D/1/B$ with internal server (the original description in [Hlu88] analyzes a queueing system with external server). The packet loss probability in the output queue is then given by

$$\pi_q = 1 - \frac{\rho}{p_q}$$

Needless to say, the output queue becomes a synchronous $M/D/1$ queue as $K \to \infty$ and $B_o \to \infty$.

The switch throughput is straightforwardly given by

$$\rho = 1 - q_0$$

and the total packet loss probability by

$$\pi = 1 - (1 - \pi_c)(1 - \pi_q)$$

The delay versus throughput performance in OQ switches is optimal since the packet delay is only determined by the congestion for the access to the same output link and thus the delay figure is the result of the statistical behavior of the output queue modelled as a $Geom(K)/D/1/B_o$ queue. Little's formula immediately gives

$$T = \frac{E[Q]}{\rho} = \frac{\sum_{i=1}^{B_o} iq_i}{1 - q_0}$$

As $B_o \to \infty$ the queue becomes a $Geom(K)/D/1$ and the delay becomes (see Appendix)

$$T = 1 + \frac{K-1}{K} \frac{p_q}{2(1 - p_q)}$$

Notice that the parameter K affects just the term representing the average packet delay as the unity represents the packet transmission time. As is shown in Figure 7.30, the offered load p must be kept below a given threshold if we like to keep small the output queue capacity ($B_o = 64$ guarantees loss figures below 10^{-7} for loads up to $p = 0.9$). The average packet delay increases with the queue capacity and its value for $K = 8$ and different queue capacities is shown in Figure 7.31.

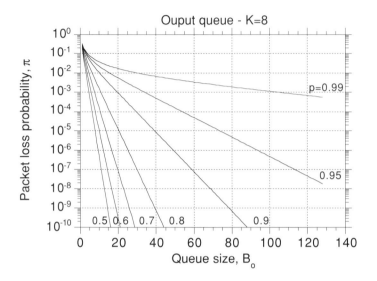

Figure 7.30. Loss performance in the output queue

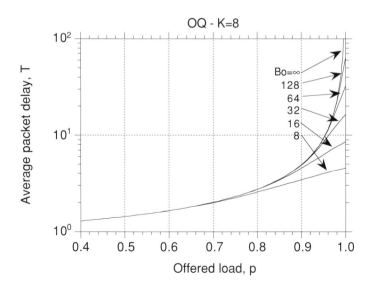

Figure 7.31. Delay performance in the output queue

7.3. Shared Queueing

The model of a non-blocking ATM switch with pure shared queueing is given by Figure 7.32; compared to the more general model of Figure 7.1, it requires $B_s > 0$, $B_i = B_o = 0$ and output speed-up $K = 1$. Thus its only queueing component is a buffer with capacity NB_s (packets) shared by all the switch inlets to contain the packets destined for all the switch outlets (B_s thus indicates the buffer capacity per switch inlet). This shared queue is said to contain N *logical queues*, one per switch outlet, each containing all the packets addressed to the associated switch outlet. The cumulative content of all the logical queues can never exceed the capacity NB_s of the physical shared queue.

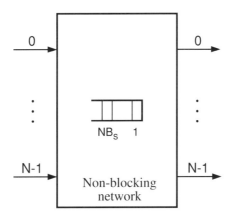

Figure 7.32. Model of non-blocking ATM switch with shared queueing

7.3.1. Basic architectures

The most simple implementation for this switch model is given by a shared memory unit that enables up to N concurrent write accesses by the N inlets and up to N concurrent read accesses by the N outlets. The actual number of concurrent write and read accesses equals the number of packets received in the slot and the number of non-empty logical queues in the shared buffer, respectively. Clearly such a structure is affected by severe implementation limits due to the memory access speed, considering that the bit rate of the external ATM links is about 150 Mb/s. The current VLSI CMOS technology seems to allow the implementation of such $N \times N$ switches up to a moderate size, say $N = 16, 32$ [Koz91].

Multistage networks still provide a good answer for the implementation of larger size ATM switches. The most important proposal is known as the *Starlite switch* [Hua84], whose block architecture is shown in Figure 7.33. As in the case of an IQ switch, its core is a non-blocking structure including a Batcher *sorting network* (SN) and a banyan *routing network* (RN) (*n*-cube or Omega network). Now, the absence of external conflicts in the packets to be routed by the banyan network and its compact and address-monotone condition, as defined in Section 3.2.2, is guaranteed by the *trap network* (TN). An additional component, the *recirculation (or delay) net-*

work (DN) of size $P \times P$, acts as the distributed shared buffer and feeds back to the routing network up to $P = NB_s$ packets that could not be switched in the preceding slot. The recirculation network is basically a set of delay units. Therefore, the sorting network, which interfaces both the N port controllers and the recirculation network, has a size $(N + P) \times (N + P)$. Note that, compared to a non-blocking sorting routing architecture with input queueing, a larger sorting network is here required as P ports must be dedicated to the transmission of recirculated cells.

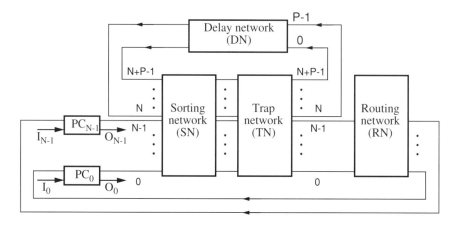

Figure 7.33. Architecture of the Starlite switch

The trap network performs the following functions:

- *Packet marking*: it marks as *winner* one packet out of those addressing the same switch outlet j (*class j packets*), $(j = 0, ..., N-1)$, and as *losers* the remaining packets.

- *Packet discarding*: it applies a suitable *selection algorithm* to choose the P packets to be recirculated (the *stored packets*) among the loser packets, if more than P packets are marked as losers among the N packet classes; the loser packets in excess of the P selected are discarded.

- *Packet routing*: it suitably arranges the winner and loser packets at the outlets of the trap network: the winner packets are offered as a compact and monotone set to the $N \times N$ banyan network and the stored packets are offered to the recirculation network.

These functions are accomplished by the trap network represented in Figure 7.34 which includes a *marker*, which marks packets as winners/losers, a *running sum adder*, which generates an increasing (decreasing) list for winner (loser) packets, and a *concentrator*. Such a trap network performs the basic selection algorithm called *fixed window* that discards first the packets addressing the switch outlets with the largest index. As with IQ, an unfairness problem affects this structure as packets addressing the lower index switch outlets have a higher probability of not being discarded. Techniques and algorithms to cope with this problem are described in [Bia96].

Cells entering the switch are given a control header that includes an activity bit AC ($AC = 0$ denotes now an active packet carrying an ATM cell), a destination address DA and a priority indicator PR (see the example in Figure 7.35 for $N = 8$, $P = 4$). Field PR codes

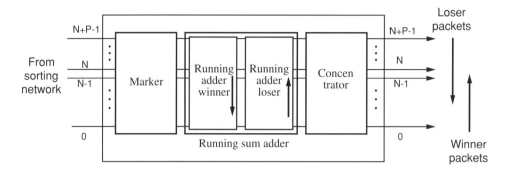

Figure 7.34. Structure of the trap network in Starlite

the packet age, that is the number of time slots spent in the switch, so that smaller PR values correspond to older packets. Field PR is updated by the recirculation network which counts the number of transits of each packet. Inactive port controllers issue an empty packet with $AC = 1$. The sorting network (SN) sorts the active packets ($AC = 0$) using field DA as primary key and field PR as secondary key, so that they emerge on adjacent outlets of network SN (nine active packets in Figure 7.35, of which six are new and three stored). The trap network performs the three above functions and offers the winner packets (four in Figure 7.35) as a compact set to the banyan network and the set of stored packets (four out of five losers in Figure 7.35) to the recirculation network.

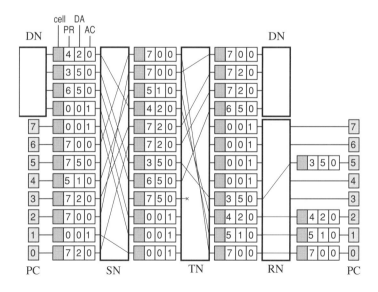

Figure 7.35. Switching example in Starlite

The operations of the trap network for the same example are shown in Figure 7.36. The marker, which receives a sorted set of active packets, is built as a single stage of 2×2 comparator elements so as to identify and mark in the marking bit MB with the first instance of each packet class as winner, and the others as loser. The active packets (winners/losers), as well as the empty packets, enter the running sum adder that adds a routing tag RT to each of them. The field RT is allocated to loser cells as descending consecutive addresses starting from $N + P - 1$, and to winner and empty cells as ascending consecutive addresses starting from 0. These $N + P$ packets are routed by the concentrator using field RT as address, so that the winner (loser) packets emerge at the outlets with smallest (largest) index, as is shown in Figure 7.36. Since in our example only $P = 4$ packets can be recirculated, one of the loser packets is discarded by the concentrator, that is one addressing outlet 5. The fields MB and RT are removed by the concentrator before transmitting the packets out.

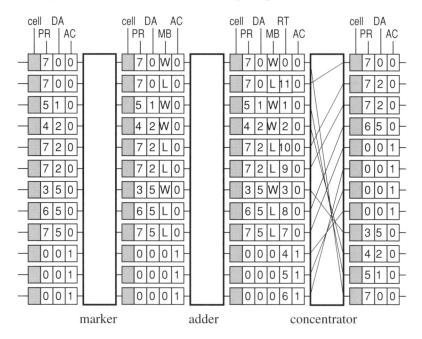

Figure 7.36. Switching example in the trap network

It can easily be shown that the concentrator can be implemented as a reverse n-cube network (see Section 2.3.1.1). In fact, if we look at the operations of the concentrator by means of a mirror (inlets becomes outlets and vice-versa, the reverse n-cube topology becomes a n-cube topology) the sequence of packets to be routed to the outlets of the mirror network is bitonic and thus an n-cube network accomplishes this task by definition of bitonic merge sorting (see Section 2.3.2.1). In the example of Figure 7.36, such a bitonic sequence of addresses is 10,7,6,4,3,0,1,2,5,8,9,11 (network inlets and outlets are numbered $N + P - 1$ through 0). Note that the suitability of a reverse n-cube network does not require $N + P$ to be a power of two. For example if $2^{n-1} < N + P < 2^n$, adding to one end of the original bitonic sequence of

$N + P$ elements another $2^n - N - P$ elements all valued $2^n - 1$ results in another bitonic sequence of 2^n elements.

7.3.2. Performance analysis

As with input queueing, the performance of an ATM switch with shared queueing (SQ) will not be referred to the Starlite architecture just described owing to its inherent unfairness. In fact its selection algorithm always discards first the packets addressing the largest-index outlets. Thus we will implicitly refer to an idealized and completely fair version of Starlite.

The performance analysis of an SQ switch is much more complex than an OQ architecture, due to the correlation of the occupancy of the different logical queues (they share the same physical buffer). Shared queueing, analogously to output queueing, gives a maximum throughput (the switch capacity) $\rho_{max} = 1$ for $B_s \to \infty$ and achieves optimal delay-throughput performance, as the only packet delay is given by the unavoidable statistical occupancy of the logical queue joined by the packet. However, pooling together the logical queues into a single physical shared buffer is expected to result in a smaller packet loss probability than with output queueing when $B_o = B_s$.

An approach to evaluate the packet loss performance of an SQ switch is based on the assumption [Hlu88] of a shared queue with infinite capacity ($NB_s \to \infty$), so that the occupancy of the shared queue is described by

$$Q_n = \max\{0, Q_{n-1} + A_n - 1\}$$

in which Q_n and A_n represent, respectively, the cells in the queue at the end of slot n and the new cells entering the queue at the beginning of slot n. Note that unlike input queueing (Equation 7.1) and output queueing (Equation 7.8) the arrivals are first summed to the queue content and then a unity is subtracted, meaning that a cell need not be stored before being switched. In fact with shared queueing in the Starlite switch, buffering occurs only if switching is not successful, whereas a buffer is always crossed with IQ and OQ. A steady state is assumed and the subscript n is removed. When N is finite, the steady-state number of packets A^i entering logical queue i ($i = 0, \ldots, N-1$) is not independent from A^j ($i \neq j$). In fact since at most N packets can be received by the queue in a slot, a larger number of packets received in a slot by logical queue i reduces the number of packets that can be received in the slot by the other logical queues. Such an observation [Eck88] implies a negative correlation in the packet arrival processes to different logical queues. So, assuming the logical queues to be independent means overestimating the content of the shared queue compared to the real buffer behavior.

Based on such independence assumption, the logical queue Q^i ($i = 0, \ldots, N-1$) is modelled as an S-ES $Geom(N)/D/1$ queue (see Appendix) statistically independent from the other logical queues Q^j. Then the average delay performance, which here represents just the waiting time in the shared queue, is

$$T = \frac{N-1}{N} \frac{p}{2(1-p)}$$

Moreover, the steady-state cumulative number of packets Q in the shared queue can be computed as the N-fold convolution of the distribution probability of the number of packets Q^i in the generic logical queue $Geom(N)/D/1$. Therefore an upper bound on the packet loss probability is obtained by

$$\pi = \Pr\left[\left(Q = \sum_{i=1}^{N} Q^i\right) \geq NB_s\right] \tag{7.9}$$

As N increases the negative correlation between logical queues diminishes. When $N \to \infty$, the queues become mutually independent and each of them can be modelled as an $M/D/1$ queue. It is intuitively clear that, given a normalized shared buffer capacity B_s, larger values of N, which increase both the switch size and the shared queue capacity, imply a better utilization of the shared queue and thus lower packet loss probabilities. Therefore a switch with $N \to \infty$ gives the smallest normalized average occupancy of the shared queue, that is $Q^i_{\min} = p^2/[2(1-p)]$. [Hlu88] shows that the normalized shared queue capacity B_s providing a given loss performance converges to $Q^i_{\min}$ as $N \to \infty$.

Unfortunately, the negative correlation between the arrival processes in different logical queues deeply affects the performance data that can be obtained by the above procedure, especially for non-large switch sizes. The result is that the loss probability evaluated for a given network configuration can overestimate the real value even by several orders of magnitude. Therefore, a few performance results, as provided through computer simulation, are now given for the Starlite switch with the DPO selection algorithm described in [Bia96], which is the one providing the best results, although it is not completely fair. The loss performance of an SQ switch is given in Figure 7.37 for $N = 32$ as a function of the normalized shared buffer size $B_s = P/N$ for different offered loads. It is seen that even a small shared queue of $NB_s = 32$ ($P/N = 1$) requires the offered load to be limited to $p = 0.6$ if loss figures below 10^{-5} are required. Note that even the implementation of such a small shared queue requires doubling of the sorting network size ($N + P = 2N$).

We expect that, given a normalized shared buffer size P/N, larger switches are more effective, as a larger number of memory positions is shared by a larger number of switch terminations. This result is confirmed in Figure 7.38 showing the switch loss probability as a function of the switch size N for a fixed offered load level, $p = 0.8$. We see that the loss performance is unacceptable for small values of the normalized shared buffer ($P/N \leq 1.5$), whereas it becomes acceptable for larger buffers only when the switch size is large enough. For example a switch with size $N = 256$ gives a reasonably low loss probability for a normalized shared buffer size $P/N = 2$.

The average packet delay, T, versus the normalized shared buffer capacity for a 32×32 Starlite switch is given in Figure 7.39. We note that the delay grows linearly with the buffer size NP as long as the shared queue is small enough to be almost always saturated (packets are very likely to enter the shared queue in its last position). As soon as the shared queue size is larger than its average occupancy, which occurs when cell loss occurs with a smallest probability, the delay remains constant in spite of any increase in the number of recirculation ports P. The load value $p = 0.9$ is such that the queue is almost always saturated even for $P = 3N$ (see also the corresponding loss curve in Figure 7.37).

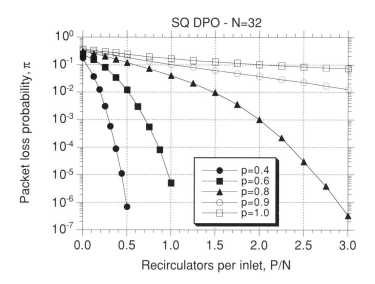

Figure 7.37. Loss performance in Starlite

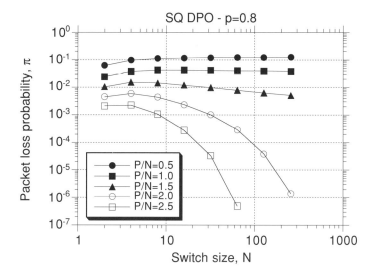

Figure 7.38. Loss performance in Starlite

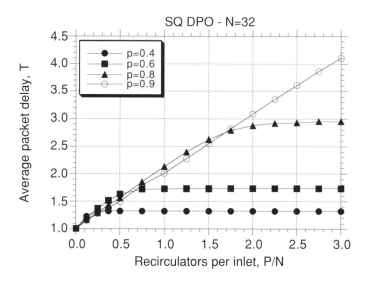

Figure 7.39. Delay performance in Starlite

7.4. Performance Comparison of Different Queueings

Adoption of input queueing in an ATM switch architecture implies the acceptance of the inherent throughput limit characterizing the technique according to which no more than 58% of the switch theoretical capacity can be exploited, independently of how large the single queues are. Such performance impairment is however compensated by the relative simplicity of the hardware architecture where the internal operations require at most a minimal internal overhead (and hence an internal bit rate a little higher than the external). The head-of-line blocking causing this throughput degradation can be relieved with the two techniques of channel grouping and windowing, both of which require a significant amount of additional hardware to be built. In some cases these techniques can also cause loss of cell sequence on a given virtual call. HOL blocking does not occur with output and shared queueing which therefore are characterized by the maximum switch capacity of 100%. However, more hardware problems can arise in these cases due to the multiple writing operations required in the output or shared queue.

The packet delay performance of the different queueing strategies is basically governed by the switch capacity of the architecture: in fact the switch capacity determines the input load at which the average cell delay T goes to infinity. If an infinitely large queue capacity is assumed for the different strategies ($B_i = B_o = B_s = \infty$), the average cell delay T is shown in Figure 7.40: the delay curve for IQ is obtained by means of Equation 7.3, whereas the delay performance for OQ and SQ are clearly given by the $M/D/1$ queue (the minimum value $T = 1$ represents the transmission time since there is no waiting time). Therefore under random traffic the offered load p must be limited to, say, 0.5 (IQ) and 0.9 (OQ, SQ) if an average delay value less than $T = 10$ slots is required.

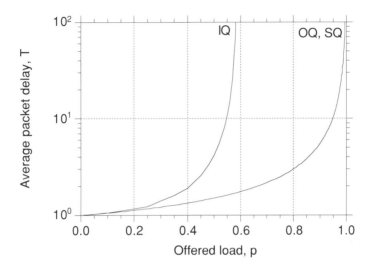

Figure 7.40. Delay performance with IQ, OQ and SQ

In order to compare the loss performance of the three queueing strategies, the two load values $p = 0.5, 0.8$ have been selected so that at least for the smallest of them all the three strategies can be compared to each other (the switch capacity is higher than 0.5 for all of them). The packet loss probability with IQ has been obtained by the analytical model described in Section 8.1.2.2 for a switch with combined input–output queueing and here used with no output queueing ($B_o = 1$) and no output speed-up ($K = 1$). The loss performance for OQ is straightforwardly given by a $Geom(N)/D/1/B_o$ queue (for the sake of a fair comparison the concentrators have been disregarded). Finally Equation 7.9 has been used to evaluate the loss performance with SQ. The corresponding results for the three queueing strategies are plotted in Figure 7.41. For OQ and SQ the two switch sizes $N = 16, 32$ have been considered: the loss probability is about the same with output queueing for the two switch sizes, whereas a very small difference is noticed for shared queueing (a little less buffer capacity is required in the larger switch to provide the same loss performance). With offered load $p = 0.5$ the buffer capacity $B_i = 27$ needed with IQ to give a loss probability $\pi < 10^{-9}$ reduces to $B_o = 16$ and $B_s = 1$ with the other queueing modes. With $p = 0.8$ the same loss performance is given by $B_o = 44$ with output queueing and by $B_s = 8$ with shared queueing. These example clearly show how the loss performance significantly improves for a given buffer size passing from IQ to OQ and from OQ to SQ.

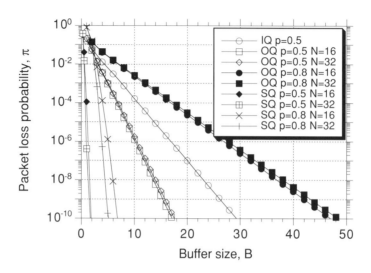

Figure 7.41. Loss performance with IQ, OQ and SQ

7.5. Additional Remarks

The concept of channel grouping has been widely investigated in non-blocking ATM switches either with input queueing [Lau92] or with output queueing [Lin93]. The effect of adopting channel grouping on the input side of the switch has also been studied [Li91]. A different type of input queueing in which each input buffer is split into separate queues, each associated with an output, is described in [Del93] where the analysis of the pure input queueing under finite switch size is also given.

As already mentioned, the traffic performance under random traffic carried out in this chapter for non-blocking ATM switches with single queueing gives optimistic results compared to those provided under different offered traffic patterns. Performance analyses with correlated traffic patterns have been reported in [Li92], [Cao95] for pure input queueing and in [Hou89], [Che91] for pure output queueing; priority classes have also been taken into account with queues on inputs in [Li94], [Jac95]. A simulation study of either input or output queueing under correlated traffic is reported in [Lie94]. Non-uniformity in traffic patterns has been studied in [Li90] with input queueing, in [Yoo89] with output queueing.

Proposals on how to improve the basic switch architectures with input and output queueing are described in [Kar92] (IQ), which is based on output port scheduling techniques, and in [Che93] (OQ), which avoids using a fully interconnected topology by means of a sorting network and switching modules. The performance analysis of this latter architecture under non-uniform traffic is given in [Che94]. Another kind of non-blocking architecture in which the buffers are placed at the crosspoints of a crossbar network is described and analyzed in [Gup91]. Adding input queues to this structure has been studied in [Gup93].

7.6. References

[Ahm88] H. Ahmadi, W.E. Denzel, C.A. Murphy, E. Port, "A high-performance switch fabric for integrated circuit and packet switching", *Proc. of INFOCOM 88*, New Orleans, LA, Mar. 1988, pp. 9-18.

[Ara90] N. Arakawa, A. Noiri, H. Inoue, "ATM switch for multimedia switching systems", *Proc. of Int. Switching Symp.*, Stockholm, May-June 1990, Vol. V, pp. 9-14.

[Bas76] F. Baskett, A.J. Smith, "Interference in Multiprocessor computer systems with interleaved memory", *Commun. of the ACM*, Vol. 19, No. 6, June 1976, pp. 327-334.

[Bha75] D.P. Bhandarkar, "Analysis of memory interference in multiprocessors", *IEEE Trans. on Comput.*, Vol. C-24, No. 9, Sep. 1975, pp. 897-908.

[Bia96] G. Bianchi, A. Pattavina, "Architecture and performance of non-blocking ATM switches with shared internal queueing", *Computer Networks and ISDN Systems*, Vol. 28, 1996, pp. 835-853.

[Bin88a] B. Bingham, H.E. Bussey, "Reservation-based contention resolution mechanism for Batcher-banyan packet switches", *Electronics Letters*, Vol. 24, No. 13, June 1988, pp. 722-723.

[Bin88b] B.L. Bingham, C.M. Day, L.S. Smooth, "Enhanced efficiency Batcher-banyan packet switch", *U.S. Patent 4,761,780*, Aug. 2, 1988.

[Bus89] H.E. Bussey, J.N. Giacopelli, W.S. Marcus, "Non-blocking, self-routing packet switch", *US Patent 4,797,880*, Jan. 10, 1989.

[Cao95] X.-R. Cao, "The maximum throughput of a nonblocking space-division packet switch with correlated destinations", *IEEE Trans. on Commun.*, Vol. 43, No. 5, May 1995, pp. 1898-1901.

[Che91] D.X. Chen, J.W. Mark, "Performance analysis of output buffered fast packet switches with bursty traffic loading", *Proc. of GLOBECOM 91*, Phoenix, AZ, Dec. 1991, pp. 455-459.

[Che93] D.X. Chen, J.W. Mark, "SCOQ: a fast packet switch with shared concentration and output queueing", *IEEE/ACM Trans. on Networking*, Vol. 1, No. 1, Feb. 1993, pp. 142-151.

[Che94] D.X. Chen, J.W. Mark, "A buffer management scheme for the SCOQ switch under non-uniform traffic loading", *IEEE Trans. on Commun.*, Vol. 42, No. 10, Oct. 1994, pp. 2899-2907.

[Del93] E. Del Re, R. Fantacci, "Performance evaluation of input and output queueing techniques in ATM switching systems", *IEEE Trans. on Commun.*, Vol. 41, No. 10, Oct. 1993, pp. 1565-1575.

[Eck88] A.E. Eckberg, T.-C. Hou. "Effects of output buffer sharing on buffer requirements in an ATDM packet switch", *Proc. of INFOCOM 88*, New Orleans, LA, Mar. 1988, pp. 459-466.

[Gup91] A.K. Gupta, L.O. Barbosa, N.D. Georganas, "16x16 limited intermediate buffer switch module for ATM networks", *Proc. of GLOBECOM 91*, Phoenix, AZ, Dec. 1991, pp. 939-943.

[Gup93] A.K. Gupta, L.O. Barbosa, N.D. Georganas, "Switching modules for ATM switching systems and their interconnection networks", *Comput. Networks and ISDN Systems*, Vol. 26, 1993, pp. 443-445.

[Hlu88] M.G. Hluchyj, K.J. Karol, "Queueing in high-performance packet switching", *IEEE J. on Selected Areas in Commun.*, Vol. 6, No. 9, Dec. 1988, pp. 1587-1597.

[Hou89] T.-C. Hou, "Buffer sizing for synchronous self-routing broadband packet switches with bursty traffic", *Int. J. of Digital and Analog Commun. Systems*, Vol. 2, Oct.-Dec. 1989, pp. 253-260.

[Hua84] A. Huang, S. Knauer, "Starlite: a wideband digital switch", *Proc. of GLOBECOM 84*, Atlanta, GA, Nov. 1984, pp. 121-125.

[Hui87] J.Y. Hui, E. Arthurs, "A broadband packet switch for integrated transport", *IEEE J. on Selected Areas in Commun.*, Vol. SAC-5, No. 8, Oct. 1987, pp. 1264-1273.

[Hui90] J.Y. Hui, *Switching and Traffic Theory for Integrated Broadband Networks*, Kluwer Academic Publisher, 1990.

[Jac95] L. Jacob, A. Kumar, "Saturation throughput analysis of an input queueing ATM switch with multiclass bursty traffic", *IEEE Trans. on Commun.*, Vol. 43, No. 6, June 1995, pp. 2149-2156.

[Kar87] M.J. Karol, M.G. Hluchyj, S.P. Morgan, "Input versus output queueing on a space-division packet switch", *IEEE Trans. on Commun.*, Vol. COM-35, No. 12, Dec. 1987, pp. 1347-1356.

[Kar92] M.J. Karol, K.Y. Eng, H. Obara, "Improving the performance of input-queued ATM packet switches", *Proc. of INFOCOM 92*, Florence, Italy, May 1992, pp. 110-115.

[Koz91] T. Kozaki, Y. Sakurai, O. Matsubara, M. Mizukami, M. Uchida, Y. Sato, K. Asano, "32 X 32 Shared buffer type ATM switch VLSIs for BISDN", *Proc. of ICC 91*, Denver, CO, June 1991, pp. 711-715.

[Lau92] P.S.Y. Lau, A. Leon-Garcia, "Design and analysis of a multilink access subsystem based on the Batcher-banyan network architecture", *IEEE Trans. on Commun.*, Vol. 40, No. 11, Nov. 1992, pp. 1757-1766.

[Li90] S.-Q Li, "Nonuniform traffic analysis of a nonblocking space-division packet switch", *IEEE Trans. on Commun.*, Vol. 38, No. 7, July 1990, pp. 1085-1096.

[Li91] S.-Q. Li, "Performance of trunk grouping in packet switch design", *Performance Evaluation*, Vol. 12, 1991, pp. 207-218.

[Li92] S.-Q. Li, "Performance of a nonblocking space-division packet switch with correlated input traffic", *IEEE Trans. on Commun.*, Vol. 40, No. 1, Jan. 1992, pp. 97-108.

[Li94] L. Li, P. Liu, "Maximum throughput of an input queueing packet switch with two priority classes", *IEEE Trans. on Commun.*, Vol. 42, No. 2-4, Feb.-Apr. 1994, pp. 757-761.

[Lie94] S.C. Liew, "Performance of various input-buffered and output-buffered ATM switch design principles under bursty traffic: simulation study", *IEEE Trans. on Commun.*, Vol. 42, No. 2-4, Feb.-Apr. 1994, pp. 1371-1379.

[Lin93] A.Y.-M. Lin, J.A. Silvester, "On the performance of an ATM switch with multichannel transmission groups", *IEEE Trans. on Commun.*, Vol. 41, No. 5, May 1993, pp. 760-769.

[Pat88] A. Pattavina, "Multichannel bandwidth allocation in a broadband packet switch", *IEEE J. on Selected Areas in Commun.*, Vol. 6, No. 9, Dec. 1988, pp. 1489-1499.

[Pat89] A. Pattavina, "Improving efficiency in a Batcher-banyan packet switch", *Proc. of GLOBECOM 89*, Dallas, TX, Nov. 1989, pp. 1483-1487.

[Pat90]   A. Pattavina, "A multiservice high-performance packet switch for broadband networks", *IEEE Trans. on Commun.*, Vol. 38, No. 9, Sep. 1990, pp. 1607-1615.

[Pat91]   A. Pattavina, "Performance evaluation of a Batcher-banyan interconnection network with output pooling", *IEEE J. on Selected Areas in Commun.*, Vol. 9, No. 1, Jan. 1991, pp. 95-103.

[Yeh87]   Y.S. Yeh, M.G. Hluchyj, A.S. Acampora, "The knockout switch: a simple, modular architecture for high-performance packet switching", *IEEE J. on Selected Areas in Commun.*, Vol. SAC-5, No. 8, Oct. 1987, pp. 1274-1283.

[Yoo89]   H. Yoon, M.T. Liu, K.Y. Lee, Y.M. Kim "The knockout switch under nonuniform traffic", *IEEE Trans. on Commun.*, Vol. 43, No. 6, June 1995, pp. 2149-2156.

7.7. Problems

7.1   Compute numerically the maximum throughput of an ATM switch with input queueing and finite size $N = 16$ using Equation 7.3 and compare it with the value given by computer simulation (see Table 7.2).

7.2   Plot Equation 7.4 as a function of the offered load p for $B_i = 1, 2, 4, 8, 16, 32$ and evaluate the accuracy of the bound using the simulation data in Figure 7.10.

7.3   Compute the switch capacity of a non-blocking ATM switch of infinite size with input queueing and windowing for a window size $W = 2$ using the approach followed in Section 7.1.3.2.

7.4   Provide the expression of the switch throughput as a function of the offered load only, p, for a non-blocking ATM switch of infinite size with input queueing and windowing with a window size $W = 3$ using the approach followed in Section 7.1.3.2.

7.5   Plot the average delay as a function of the offered load p of an ATM switch with input queueing, FIFO service in the virtual queue, and finite size $N = 32$ using Equation 7.3. Use the analysis of a $Geom(N)/D/1/B$ queue reported in the Appendix to compute the first two moments of the waiting time in the virtual queue. Compare these results with those given by computer simulation and justify the difference.

7.6   Repeat Problem 7.5 for a random order service in the virtual queue.

7.7   Repeat Problems 7.5 and 7.6 for an ATM switch with channel grouping and group size $R = 8$ using the appropriate queueing model reported in the Appendix for the virtual queue.

7.8   Explain why the two priority schemes, local and global, for selecting the winner packets of the output contention in an ATM switch with input queueing and channel grouping give the same asymptotic throughput, as is shown in Figure 7.16.

7.9   Derive Equation 7.7.

7.10   Draw the concentration network for the Knock-out switch with $N = 8$, $K = 4$ adopting the same technique as in Figure 7.25.

Chapter 8 *ATM Switching with Non-Blocking Multiple-Queueing Networks*

We have seen in the previous chapter how a non-blocking switch based on a single queueing strategy (input, output, or shared queueing) can be implemented and what traffic performance can be expected. Here we would like to investigate how two of the three different queueing strategies can be combined in the design of a non-blocking ATM switch.

The general structure of a non-blocking switch with size $N \times M$ is represented in Figure 8.1. Each input port controller (IPC) and output port controller (OPC) are provided with a FIFO buffer of size B_i and B_o cells, respectively. A FIFO shared buffer with capacity NB_s cells is also associated with the non-blocking interconnection network (IN). Therefore B_i, B_o and B_s represent the input, output and shared capacity per input–output port in a squared switch. Apparently having $B_x = 0$ for $x = i, o, s$ corresponds to absence of input, output and shared queueing, respectively.

Usually, unless required by other considerations, IPC and OPC with the same index are implemented as a single port controller (PC) interfacing an input and an output channel, so that the switch becomes squared. Unless stated otherwise, a $N \times N$ squared switch is considered in the following and its size N is a power of 2.

Output queueing is adopted when the interconnection network is able to transfer more than one cell to each OPC, since only one cell per slot can leave the OPC. Then the switch is said to have an *(output) speed-up* K meaning that up to K packets per slot can be received by each OPC. Note that the condition $K \leq \min[B_o, N]$ always applies. While the second bound is determined by obvious physical considerations, the former bound is readily explained considering that it would make no sense to feed the output queue in a slot with a number of packets larger than the queue capacity. Therefore even the minimum speed-up $K = 2$ requires an output queueing capability $B_o \geq 2$. The switch can also be engineered to accomplish an *input speed-up* K_i in that up to K_i packets per slot can be transmitted by each IPC. When an input speed-up is performed ($K_i \geq 2$), the output speed-up factor K will be indicated by K_o. Accordingly in the general scheme of Figure 8.1 the interconnection network is

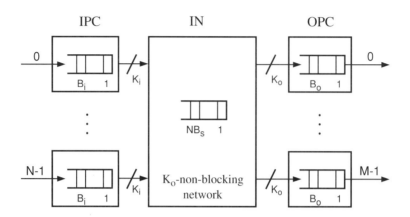

Figure 8.1. Model of K_o-non-blocking ATM switch

labelled as a K_o-non-blocking network. Nevertheless, to be more precise, the network needs only be K_o-rearrangeable due to the ATM switching environment. In fact, all the I/O connections through the interconnection network are set up and cleared down at the same time (the slot boundaries).

In general if the switch operates an input speed-up K_i, then the output speed-up must take a value $K_o \geq K_i$. In fact it would make no sense for the OPCs to be able to accept a number of packets NK_o smaller than the NK_i packets that can be transmitted by the IPCs in a slot. Therefore, output speed-up is in general much cheaper to implement than input speed-up, since the former does not set any constraint on K_i which can be set equal to one (no input speed-up). For this reason most of this section will be devoted to networks adopting output speed-up only.

The availability of different queueing capabilities makes it possible to engineer the switch in such a way that packets are transferred from an upstream queue to a downstream queue, e.g. from an input queue to an output queue, only when the downstream queue does not overflow. Therefore we will study two different internal operations in the switch:

- *Backpressure* (BP): signals are exchanged between upstream and downstream queues so that the former queues transmit downstream packets only within the queueing capability of the latter queues.

- *Queue loss* (QL): there is no exchange of signalling information within the network, so that packets are always transmitted by an upstream queue independent of the current buffer status of the destination downstream queue. Packet storage in any downstream queue takes place as long as there are enough idle buffer positions, whereas packets are lost when the buffer is full.

Thus, owing to the non-blocking feature of the interconnection network, packets can be lost for overflow at the upstream queues only in the BP mode and also at the downstream queues in the QL mode. In the analytical models we will assume that the selection of packets to be backpressured in the upstream queues (BP) or to be lost (QL) in case of buffer saturation is always random among all the packets competing for the access to the same buffer.

Our aim here is to investigate non-blocking ATM switching architectures combining different queueing strategies, that is:

- *combined input–output queueing* (IOQ), in which cells received by the switch are first stored in an input queue; after their switching in a non-blocking network provided with an output speed-up, cells enter an output queue;
- *combined shared-output queueing* (SOQ), in which cells are not stored at the switch inlets and are directly switched to the output queues; an additional shared storage capability is available to hold those cells addressing the same switch outlet in excess of the output speed-up or not acceptable in the output queues due to queue saturation when backpressure is applied;
- *combined input-shared queueing* (ISQ), in which cells received by the switch are first stored in an input queue; an additional queueing capability shared by all switch inputs and outputs is available for all the cells that cannot be switched immediately to the desired switch outlet.

If a self-routing multistage interconnection network is adopted, which will occur in the case of IOQ and SOQ structures, the general model of Figure 8.1 becomes the structure shown in Figure 8.2 where only output speed-up, with a factor K, is accomplished. Following one of the approaches described in Section 3.2.3, the K-non-blocking interconnection network is implemented as a two-block network: an $N \times N$ sorting network followed by K banyan networks, each with size $N \times N$. The way of interconnecting the two blocks is represented as a set of N splitters $1 \times N$ in Figure 8.2, so that the overall structure is K-non-blocking. The other implementations of the non-blocking network described in Section 3.2.3 could be used as well.

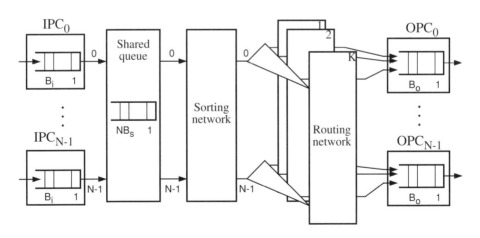

Figure 8.2. Model of K-non-blocking self-routing multistage ATM switch

In the following, Section 8.1 describes architectures and performance of ATM switches with combined input–output queueing. Section 8.2 and Section 8.3 do the same with combined shared-output and input-shared, respectively. The switch capacities of different non-blocking switches, either with single queueing or with multiple queueing, are summarized and compared with each other in Section 8.4. Some additional remarks concerning the class of ATM switches with multiple queueing are given in Section 8.5.

8.1. Combined Input–Output Queueing

By referring to the general switch model of Figure 8.1, an ATM switch with combined input–output queueing (IOQ) is characterized by $B_i \geq 1$, $B_o \geq 1$ and $B_s = 0$. However, since the switch accomplishes an output speed-up $K_o \geq 2$, the minimum value of the output queues is $B_o = 2$ (see the above discussion on the relation between K_o and B_o).

A rather detailed description will be first given of basic IOQ architectures without input speed-up operating with both the internal protocols BP and QL. These architectures adopt the K-non-blocking self-routing multistage structure of Figure 8.2 where the shared queue is removed. A thorough performance analysis will be developed and the results discussed for this structure by preliminarily studying the case of a switch with minimum output queue size $B_o = K$. Then a mention will be given to those architectures adopting a set of parallel non-blocking switch planes with and without input speed-up.

8.1.1. Basic architectures

The basic switch implementations that we are going to describe are based on the self-routing multistage implementation of a K-non-blocking network described in Section 3.2.3, under the name of K-rearrangeable network. Therefore the interconnection network is built of a sorting Batcher network and a banyan routing network with output multiplexers, which are implemented as output queues located in the OPCs in this ATM packet switching environment. As with the interconnection network with pure input queueing, the IPC must run a contention resolution algorithm to guarantee now that at most K of them transmit a packet to each OPC. Additionally here the algorithm must also avoid the overflow of the output queues when backpressure is adopted.

The switch architectures with the queue loss (QL) internal protocol and without input speed-up ($K_i = 1$) will be described first, by upgrading the structures already described with pure input queueing (Section 7.1.1). Then a possible implementation of one of these architectures with internal backpressure between input and output queueing will be studied.

8.1.1.1. Internal queue loss

The two basic architectures described with pure input queueing, that is the Three-Phase switch and the Ring-Reservation switch can be adapted to operate with an output speed-up K. In both cases no input speed-up is assumed ($K_i = 1$).

The architecture of the Three-Phase switch with combined input–output queueing is shown in Figure 8.3: it includes a K-non-blocking network, composed of a sorting network (SN), a routing banyan network (RN), and an allocation network (AN). Such structure differs

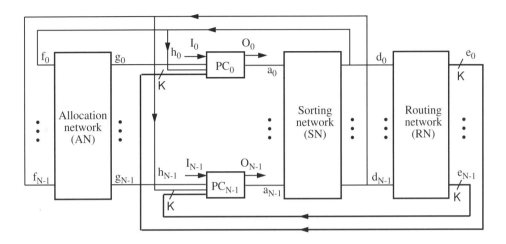

**Figure 8.3. Architecture of the K-non-blocking Three-Phase switch
with internal queue loss**

from that in Figure 7.3 only in the routing network, which has a size $N \times NK$ and is able to transfer up to K packets to each output queue. The coordination among port controllers so as to guarantee that at most K IPCs transmit their HOL cell to an OPC is achieved by means of a slightly modified version of the three-phase algorithm described for the Three-Phase switch with channel grouping (Section 7.1.3.1). As is usual, three types of packets are used[1]:

- *request packet* (REQ), including the fields
  — *AC* (activity): identifier of a packet carrying a request $(AC = 1)$ or of an idle request packet $(AC = 0)$;
  — *DA* (destination address): requested switch outlet;
  — *SA* (source address): address of the IPC issuing the request packet;
- *acknowledgment packet* (ACK), which includes the fields
  — *SA* (source address): address of the IPC issuing the request packet;
  — *GR* (grant): indication of a granted request $(GR < K)$, or of a denied request $(GR \geq K)$;
- *data packet* (DATA), including the fields
  — *AC* (activity): identifier of a packet carrying a cell $(AC = 1)$ or of an idle data packet $(AC = 0)$;
  — *routing tag*: address of the routing network outlet feeding the addressed output queue; this field always includes the physical address *DA* (destination address) of the switch outlet used in the request phase; depending on the implementation of the routing network it can also include a routing index *RI* identifying one of the K links entering the addressed output queue;
  — *cell*: payload of the data packet.

---

1. The use of a request priority field as in a multichannel IQ switch is omitted here for simplicity, but its adoption is straightforward.

In the *request phase*, or Phase I, (see the example in Figure 8.4 for $N = 8$, $K = 2$) all IPCs issue a packet REQ that is an idle request packet ($AC = 0$) if the input queue is empty, or requests the outlet DA addressed by its HOL packet ($AC = 1$). The packet also carries the address SA of the transmitting IPC. These request packets are sorted by network SN so that requests for the same switch outlet emerge on adjacent outlets of the sorting network. This kind of arrangement enables network AN to compute the content of the field GR of each packet in such a way that the first K requests for the same address are given the numbers $0, ..., K-1$, whereas $GR \geq K$ for other eventual requests.

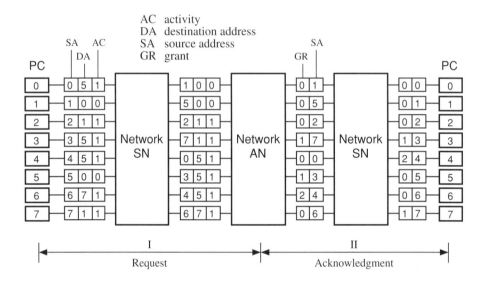

Figure 8.4. Example of packet switching (Phases I and II) in the QL IOQ Three-Phase switch

In the *acknowledgment (ack) phase*, or Phase II (see the example in Figure 8.4), packets ACK are generated by each port controller that receives field SA directly from network SN and field GR from network AN. Notice that the interconnection between networks guarantees that these two fields refer to the same request packet. Packets ACK are sorted by network SN. Since N packets ACK are received by the sorting network, all with a different SA address in the interval $[0, N-1]$ (each PC has issued one request packet), the packet ACK with $SA = i$ is transmitted on output d_i ($i = 0, ..., N-1$) and is thus received by the source port controller PC_i. A PC receiving $GR \leq K-1$ realizes that its request is granted; the request is not accepted if $GR \geq K$. Since all different values less than K are allocated by network AN to the field GR of different request packets with the same DA, at most K port controllers addressing the same switch outlet have their request granted.

In the *data phase*, or Phase III, the port controllers whose requests have been granted transmit their HOL cell in a packet DATA through the sorting and banyan networks, with the packet header including an activity bit and a self-routing tag. The structure of this last field depends on the type of routing network adopted. With reference to the different solutions for implementing output multiplexing described in Section 3.2.3, the example in Figure 8.5 shows the data phase for the two solutions b and c'. The routing network is implemented as an 8×16 banyan network in the former case, thus requiring the use of a one-bit field RI following field DA, and as a set of two 8×8 banyan networks in the latter case with EGS interconnection from the sorting network (no field RI is here required). In the example the PCs issue six requests, of which one is not granted owing to a number of requests for outlet 5 larger than the output speed-up $K = 2$.

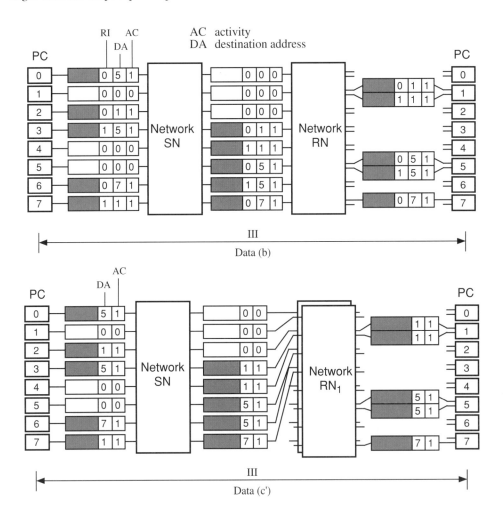

Figure 8.5. Example of packet switching (Phase III) in the QL IOQ Three-Phase switch

About the hardware required by this IOQ Three-Phase switch $N \times N$ with QL protocol, the K-non-blocking network is assumed as the minimum-cost implementation described in Section 3.2.3 with a Batcher sorting network $N \times N$ cascaded through an EGS pattern to K $N \times N$ banyan networks. The allocation network is implemented basically as the running sum adder network of Figure 7.14, which now includes $k' = \lceil \log_2 K \rceil$ stages of adders. Some minor changes have to be applied to take into account that now an activity bit distinguishes idle packets from true request packets so that the running sum in network AN is started in correspondence of the first non-idle packet. With such a structure different values for the field GR in the range $[0, 2^{k'} - 1]$ will be associated with the first $2^{k'}$ requests for the same switch outlet, received on adjacent inputs.

The Ring-Reservation switch with pure input queueing can more easily be adapted to operate an output speed-up K. In this case the interconnection network is a K-non-blocking network, the same above adopted in the IOQ Three-Phase switch. The reservation process for the NK outlets of the routing network (K per switch outlet) simply requires that a reservation frame of NK fields makes a round along the ring crossing all the PCs. As in the case of pure input queueing, each PC reserves the first idle field of the K associated to the desired switch outlet. If all the K fields are already reserved (they have been booked by upstream PCs), the PC will attempt a new reservation in the following slot. Unlike an IQ ring reservation switch, the use of the routing network cannot be avoided here, since now a total of NK lines enter the OPCs but at most N packets are transmitted by the IPCs.

By implicitly assuming the same hypotheses and following the same procedure as for the IQ multichannel Three-Phase switch described in Section 7.1.3.1 (MULTIPAC switch), the switching overhead $\eta = T_{I-II}/T_{III}$ required to perform the three-phase algorithm in the IOQ QL switch is now computed. The duration of Phase I is given by the latency $n(n+1)/2$ in the Batcher network and the transmission time $1+n$ of the first two fields in packet REQ (the transmission time of the other field in packet REQ is summed up in Phase II). The duration of Phase II includes the latency $n(n+1)/2$ in the Batcher network and the transmission time $n+1+k'$ of packet ACK since the time to cross network AN need not be summed (see the analogous discussion for the multichannel IQ switch in Section 7.1.3.1). Hence, the duration of the first two phases for an IOQ switch is given by

$$T_{I-II} = n(n+3) + k' + 2 = \log_2 N (\log_2 N + 3) + \lceil \log_2 K \rceil + 2$$

whereas $T_{I-II} = \log_2 N (\log_2 N + 4) + 1$ in the basic IQ Three-Phase switch. Therefore adding the output speed-up to the basic IQ architecture does not increase the channel internal channel rate $C(1+\eta)$, owing to the use of idle request packets that make it useless for packets ACK to cross network RN. In the ring reservation IOQ switch the minimum bit rate on the ring is clearly K times the bit rate computed for the basic IQ switch, since now the reservation frame is K times longer. Therefore, the minimum bit rate is equal to $KNC/(53 \cdot 8)$.

8.1.1.2. Internal backpressure

Basic architecture. An architecture of IOQ switch with internal backpressure (BP) that prevents packet loss in the output queues [Pat91] is obtained starting from the IOQ Three-Phase switch with QL protocol described in the previous section. The architecture of an $N \times N$ BP IOQ switch is represented in Figure 8.6: it includes an $N \times N$ sorting network (SN), an $N \times NK$ routing network (RN), and three networks with size $N \times 2N$, that is a merge network (MN), an allocation network (AN) and a concentration network (CN).

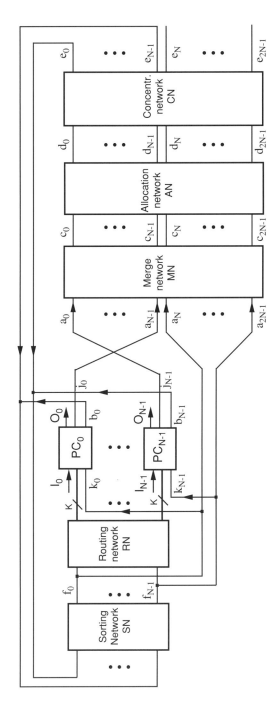

**Figure 8.6. Architecture of the K-non-blocking Three-Phase switch
with internal backpressure**

Since now the information about the content of each output queue is needed for the reservation process, a new packet type, the queue status packet, is used in combination with the three other types of packet. Therefore, the packet types are (as in the QL switch the use of a request priority field is omitted for simplicity):

- *request packet* (REQ), including the fields
  - — *AC* (activity): indicator of a packet REQ carrying a request $(AC = 1)$ or of an idle packet REQ $(AC = 0)$;
  - — *DA* (destination address): requested switch outlet;
  - — *PI* (packet indicator): identifier of the packet type, always set to 1 in a packet REQ;
  - — *CI* (concentration index): field initially set to 0 that will be filled by the sorting network;
  - — *SA* (source address): indicates the address of the IPC issuing the request packet;
  - — *GR* (grant): information used to signal back to each requesting IPC if its request is granted; it is initially set to 0;
- *queue status packet* (QUE), including the fields
  - — *AC* (activity): always set to 1 in a packet QUE;
  - — *DA* (destination address): output queue address;
  - — *PI* (packet indicator): identifier the packet type, always set to 0 in a packet QUE;
  - — *IF* (idle field): field used to synchronize packets REQ and QUE;
  - — *QS* (queue status): indication of the empty positions in the output queue;
- *acknowledgment packet* (ACK), which is generated by means of the last two fields of a request packet, and thus includes the fields
  - — *SA* (source address): address of the IPC issuing the request packet;
  - — *GR* (grant): indication of a granted request if $GR < K$, or of a denied request if $GR \geq K$;
- *data packet* (DATA), including the fields
  - — *AC* (activity): identifier of a packet carrying a cell $(AC = 1)$ or of an idle data packet $(AC = 0)$;
  - — *DA* (destination address): switch outlet addressed by the cell;
  - — *cell*: payload of the data packet.

In the *request phase* (see the example of Figure 8.7) each PC issues a packet REQ, either idle or containing the request for a switch outlet, and a packet QUE, indicating the status of its output queue. The field QS of the packet QUE transmitted by PC_i is equal to $\max[0, K - q_i]$ where q_i is the number of empty cell positions in the output queue of the port controller after the eventual transmission in the current slot. Since one packet per slot is always transmitted by the OPC on the output channel O_t, then $q_i \geq 1$, that is at least one packet per slot can be received by any OPC. Packets REQ, which are issued first, are sorted by network SN and offered to merge network MN synchronously with packets QUE. MN merges the two sets of packets so that packets REQ requesting a given outlet and the packet QUE carrying the corresponding output queue status are adjacent to each other.

The packets REQ requesting a specific switch outlet, whose number is k, emerge grouped on the adjacent outlets $c_{i+1} - c_{i+k}$, the outlet c_i carrying the packet QUE associated with that outlet. This configuration enables the allocation network AN to assign a different field GR to each packet REQ whose request is granted. Let $x(y_i)$ denote the content of the field x

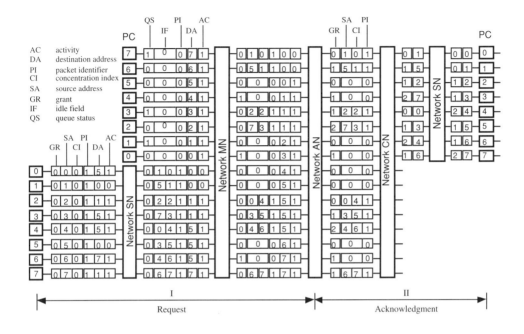

Figure 8.7. Example of packet switching (Phases I and II) in the BP IOQ Three-Phase switch

transmitted on lead y_i and $C(i, k) = \{c_{i+1}, ..., c_{i+k}\}$ represent the complete set of MN outlets carrying the k packets REQ with the same address *DA*. The field *GR* is computed by network AN so that

$$GR(d_{i+j}) = QS(c_i) + j - 1 \qquad (1 \le j \le k, c_i \in C(i, k))$$

A request is granted if its associated field *GR* is less than *K*. Since the value of *QS* is set equal to 0 if the queue has at least *K* empty positions, or to the speed-up factor decreased by the number of empty positions otherwise, then such an operation of network AN guarantees that

1. no more than *K* requests for a given switch outlet are granted;

2. the number of granted request for the switch outlet O_i in a slot never exceeds the output queueing capability of PC$_i$ in the slot.

The first two fields of packets REQ and QUE are dropped by network AN which then starts the *acknowledgment phase* (see again the example of Figure 8.7). Field *PI* acts now as activity bit in the concentration network CN, which separates packets REQ from packets QUE using field *CI* as a routing index within CN. Note that the field *CI* has been filled at the output of network SN using the physical address of the SN outlet engaged by the packet REQ. Therefore CN routes the *N* "active" packets to the top *N* outlets and drops the fields *PI* and *CI* of packets REQ. So the remaining two fields *SA* and *GR*, which represent now a packet ACK, enter network SN to be now sorted based on field *SA*. In such a way each PC receives the outcome of the contention phase carried by field GR: $GR < K$ means a granted request, whereas $GR \geq K$ means a denied request.

In the *data phase* the PCs that either lose contention in the request phase, or issue an idle packet REQ, transmit an idle packet DATA, whereas the contention winners transmit a packet DATA carrying their HOL cell (see the example in Figure 8.8 referred to in the solution c' for the implementation of the routing network described in Section 3.2.3). The sorting–routing structure is K-non-blocking so that all the non-idle packets DATA reach the addressed OPC.

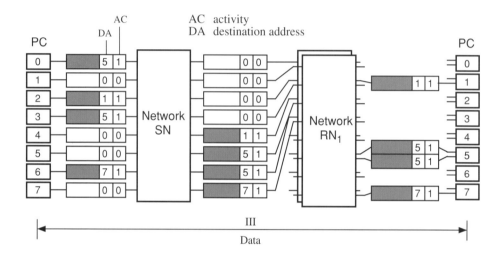

Figure 8.8. Example of packet switching (Phase III) in the BP IOQ Three-Phase switch

In the example of Figures 8.7–8.8 for $N = 8$, $K = 2$, the PCs issue six requests, of which two are denied: one addressing outlet 1 owing to queue saturation $(q_1 = 1)$ and one addressing outlet 5 owing to a number of requests larger than K. The same packet configuration of this example, applied to an IOQ switch without backpressure, has been shown in Figures 8.4–8.5: in that case five packets have been successfully switched rather than four in the BP switch due to absence of queue saturation control.

Hardware structure. In the above IOQ Three-Phase switch with BP protocol and size $N \times N$, the K-non-blocking network is the minimum-cost implementation described in Section 3.2.3 with a Batcher sorting network $N \times N$ cascaded through an EGS pattern onto K banyan networks $N \times N$. As already discussed, the only self-routing tag DA suffices in such a K-non-blocking structure to have the packet self-routing.

The merge network MN receives a packet bitonic sequence, that is the juxtaposition of an address-ascending sequence of packets QUE on inlets $a_0 - a_{N-1}$ and an address-descending sequence of packets REQ on inlets $a_N - a_{2N-1}$. Therefore MN can be implemented as an Omega (or n-cube) network with $\log_2(2N) = n + 1$ stages of 2×2 sorting elements.

The allocation network AN is a variation of the running sum adder network already described for the channel grouping application in an IQ switch (see Section 7.1.3.1) that sums on partitioned sets of its inlets. Here the network size is doubled, since also the queue status information is used in the running sum. Now the allocation network has size $2N \times 2N$ and includes $k' = \lceil \log_2 K \rceil$ stages of adders. Its structure for $N = 8$ and $5 \le K \le 8$ is shown in Figure 8.9. The component with inlets c_{i-1} and c_i, outlets $reset_i$ and $info_i$, which is referred to as an *i-component*, has the function of identifying the partition edges, that is the inlets with smallest and largest address carrying packets REQ with the same address DA. The network AN drops the fields AC and DA, while transferring transparently all the other fields of packets REQ and QUE except their last fields GR and QS, respectively. As long as this transparent transfer takes place, the adders are disabled, so that they start counting when they receive the first bit of GR and QS. If $(0)_b$ and $(1)_b$ denote the binary numbers 0 and 1, each with $k' + 1$ bits, the outlets of the *i-component* $(1 \le i \le N - 1)$ assume the following status:

$$reset_i = \begin{cases} \text{low} & \text{if } DA(c_i) \ne DA(c_{i-1}) \text{ or } AC(c_{i-1}) = 0 \\ \text{high} & \text{if } DA(c_i) = DA(c_{i-1}) \text{ and } AC(c_{i-1}) = 0 \end{cases}$$

$$info_i = \begin{cases} QS(c_i) & \text{if } PI(c_i) = 0 \\ (0)_b & \text{if } PI(c_i) = 1, PI(c_{i-1}) = 0 \\ (1)_b & \text{if } PI(c_i) = 1, PI(c_{i-1}) = 1 \end{cases}$$

The 0-component always transfers transparently on $info_0$ the field QS or GR received on c_0. Figure 8.9 also shows the output of the *i-components* and adders at each stage, when the running sum is actually performed with the same pattern of packets of Figure 8.7.

The $2N \times 2N$ concentration network CN is required to route N packets from N inlets out of the $2N$ inlets $d_0 - d_{2N-1}$ to the top N outlets $e_0 - e_{N-1}$. The outlets requested by the packets REQ represent a monotonic sequence of increasing addresses, as is guaranteed by the pattern of indices written in field CI by SN and by the topological mapping between adjacent networks. Therefore it can easily be shown that CN can be implemented as a reverse *n-cube* network with $\log_2 N + 1$ stages. In fact, if we look at the operations of the concentrator by means of a mirror (inlets becomes outlet and viceversa, the reverse *n-cube* network becomes an *n-cube* network) the sequence of packets to be routed to the outlets of the mirror network is compact and monotone (CM sequence) and thus an *n-cube* network, which is functionally equivalent to an Omega network, accomplishes this task (see Section 3.2.2).

As in all the other applications of the three-phase algorithm, an internal channel rate $C(1 + \eta)$ bit/s higher than the external rate C bit/s must be used since the request, acknowledgment and data phases share the same hardware. The switching overhead factor η is computed assuming the same hypotheses (e.g., each sorting/switching/adding stage accounts for a 1-bit latency) and following the same procedure as for the IQ switch (Section 7.1.1.1). Therefore, we would come up with the latencies $n(n+1)/2$ in SN, $n+1$ in MN and k' in AN for the request phase as well as $n+1$ in CN and $n(n+1)/2$ in SN for the acknowledgment phase. Since the length of the six fields in a packet REQ sums up to $1 + n + 1 + n + n + k' + 1$ bit, then the total duration of the first two phases is

$$T_{I-II} = n(n+6) + 2k' + 5 = \log_2 N(\log_2 N + 6) + 2\lceil \log_2 K \rceil + 5$$

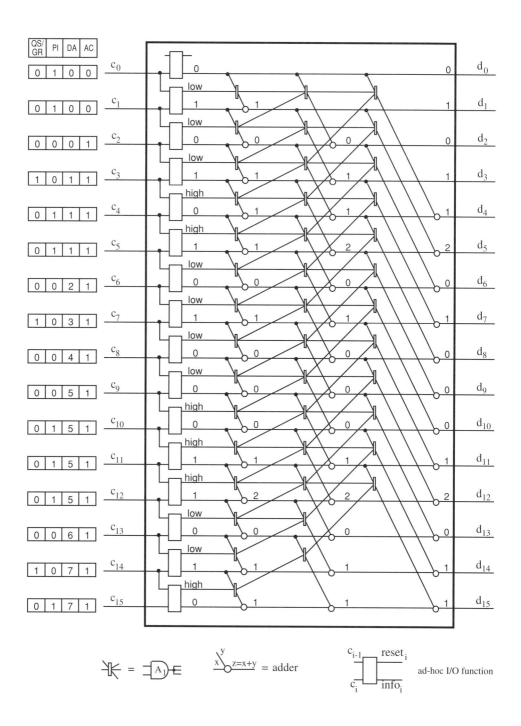

Figure 8.9. Hardware structure of the allocation network

For $N = 1024$ and $K = 4$, the overhead factor is $\eta = T_{I-II}/T_{III} = 0.399$ (the data phase lasts $53 \cdot 8$ bit times). Such an overhead, which is higher than in a multichannel switch (see Section 7.1.3.1), can be reduced by pipelining as much as possible the transmission of the different packets through the interconnection network. According to the procedure explained in [Pat90], a reduced duration of the first two phases is obtained, $T_{I-II} = n(n+11)/2 + 2k' + 3$, which gives an overhead $\eta = 0.264$ for the same network with $N = 1024$ and $K = 4$.

8.1.2. Performance analysis

The analysis of a non-blocking $N \times N$ switch with input and output queueing (IOQ) is now performed under the random traffic assumption with average value of the offered load p $(0 < p \le 1)$. The analysis relies as in the case of pure input queueing on the concept of *virtual queue* VQ_i $(i = 0, \ldots, N-1)$ defined as the set of HOL positions in the different input queues holding a cell addressed to outlet i. A cell with outlet address i entering the HOL position also enters the virtual queue VQ_i. So, the capacity of each virtual queue is N (cells). The virtual queue VQ_i feeds the output queue OQ_i, whose server is the switch output O_i. A graphical representation of the interaction among these queues is given in Figure 8.10.

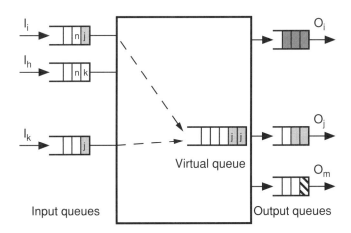

Figure 8.10. Queue model for the non-blocking IOQ switch

The analysis always assumes a FIFO service in the input, virtual and output queue and, unless stated otherwise, an infinite switch size is considered $(N = \infty)$. Thus the number of cells entering the virtual queues in a slot approaches infinity and the queue joined by each such cell is independently and randomly selected. Furthermore, since the arrival process from individual inlets to a virtual queue is asymptotically negligible, the interarrival time from an input queue to a virtual queue becomes sufficiently long. Therefore, the virtual queues, the output queues, as well as the input queues, are mutually independent discrete-time systems. Owing to the random traffic assumption, the analysis will be referred to the behavior of a generic "tagged" input, virtual and output queue, as representative of any other queue of the same type. Let p_i, p_v and p_o denote the probability that a packet is offered to the tagged input queue, that a packet leaves the tagged input queue and that a packet leaves the tagged output queue, respectively. If the external offered load is p, then $p_i = p$.

As usual, the three main performance parameters will be computed, that is the switch capacity ρ_{max} ($\rho_{max} \leq 1$), the average packet delay T ($T \geq 1$) and the packet loss probability π ($0 \leq \pi \leq 1$). These two latter measures will be computed as

$$T = E[\eta_i] + E[\eta_v] + E[\delta_o] \qquad (8.1)$$

$$\pi = 1 - \frac{p_o}{p_i} = 1 - (1 - \pi_i)(1 - \pi_o) = \pi_i + \pi_o - \pi_i \pi_o \qquad (8.2)$$

where η denotes a waiting time, δ a queueing time (waiting plus service) and the subscripts i, v and o indicate the input, virtual and output queue, respectively. Note that η_i denotes the time it takes just to enter the HOL position in the input queue, as the waiting time before winning the output contention is represented by η_v.

The analysis will be first developed for the case of an output queue size equal to the speed-up factor and infinite input queues and then extended to the case of arbitrary capacities for input and output queues.

8.1.2.1. Constrained output queue capacity

Our aim here is to analyze the IOQ switch with backpressure, output queue capacity equal to the switch speed-up ($B_o = K$) and very large input queue size ($B_i = \infty$), so that $p_v = p_i$ [Ili90]. Note that the selected output queue capacity is the minimum possible value, since it would make no sense to enable the switching of K packets and to have an output queue capacity less than K. Rather than directly providing the more general analysis, this model in which the output queues has the minimum capacity compatible with the switch speed-up has the advantage of emphasizing the impact of the only speed-up factor on the performance improvements obtained over an IQ switch.

Let us study first the tagged input queue, which is fed by a Bernoulli process. Therefore it can be modelled as a $Geom/G/1$ queue with arrival rate $p_i = p$ and service time θ_i given by the waiting time η_v it takes for the tagged packet to begin service in the tagged virtual queue plus the packet transmission time, that is $\theta_i = \eta_v + 1$. Using a well-known result [Mei58] about this queue, we get

$$E[\eta_i] = \frac{p_i E[\theta_i(\theta_i - 1)]}{2(1 - p_i E[\theta_i])} = \frac{p_i(E[\eta_v^2] + E[\eta_v])}{2(1 - p_i E[\eta_v + 1])} \qquad (8.3)$$

in which the first and second moment of the waiting time in the virtual queue are obtained from the study of the tagged virtual and output queue.

Let us introduce the following notations:

A_m number of packets entering the tagged virtual queue at the beginning of slot m;

R_m number of packets in the tagged virtual queue after the packet switch taking place in slot m;

V_m number of packets switched from the tagged virtual queue to the tagged output queue in slot m;

Q_m number of packets in the tagged output queue at the end of slot m, that is after the packet switch taking place in slot m and after the eventual packet transmission to the corresponding switch outlet.

Based on the operations of the switch in the BP mode, the state equations for the system can be easily written (an example is shown in Figure 8.11 for $R_{m-1} > 0$, $Q_{m-1} = 3$, $B_o = 3$):

$$R_m = R_{m-1} + A_m - V_m$$

$$V_m = \min\{B_o - max\{0, Q_{m-1} - 1\}, R_{m-1} + A_m\} \tag{8.4}$$

$$Q_m = \max\{0, Q_{m-1} - 1\} + V_m$$

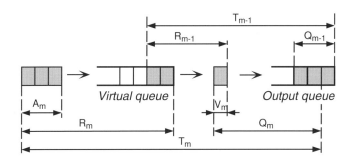

Figure 8.11. Example of tagged VQ-OQ switching for $B_o = K$

These equations take into account that the packet currently transmitted by the output queue to the switch outlet occupies a position of the output queue until the end of its transmission. Then, the cumulative number T_m of packets in the tagged virtual queue and tagged output queue is given by

$$T_m = R_m + Q_m = R_{m-1} + \max\{0, Q_{m-1} - 1\} + A_m \tag{8.5}$$

Note that according to this equation, if one packet enters an empty tagged virtual queue and the output queue is also empty, then $T_m = 1$. In fact, the packet is immediately switched from the tagged virtual queue to the tagged output queue, but the packet spends one slot in the tagged output queue (that is the packet transmission time).

Since $Q_m = 0$ implies $R_m = 0$, Equation 8.5 can be written as

$$T_m = \max\{0, R_{m-1} + Q_{m-1} - 1\} + A_m = \max\{0, T_{m-1} - 1\} + A_m \tag{8.6}$$

As with pure input queueing, the procedure described in [Kar87] enables us to prove that the random variable A_m, also representing the number of packets becoming HOL in their input queue and addressing the tagged output queue, has a Poisson distribution as $N \to \infty$. Since Equation 8.6 describes the evolution of a single-server queue with deterministic server (see Equation A.1), the system composed by the cascade of the virtual queue and tagged output queue behaves as an $M/D/1$ queue with an infinite waiting list. Note that T_m represents here the total number of users in the system including the user currently being served.

In order to compute the cumulative delay spent in the tagged virtual queue and tagged output queue, we can say that a packet experiences two kinds of delay: the delay for transmitting the packets still in the virtual queue arrived before the tagged packet, whose number is R_{m-1}, and the time it takes for the tagged packet to be chosen in the set of packets with the same age, i.e. arriving in the same slot. By still relying on the approach developed in [Kar87], it is possible to show that the cumulative average delay in the tagged virtual queue and tagged output queue is given by the average delay (waiting time plus service time) of an $M/D/1$ queue with arrival rate p, that is

$$E[\eta_v] + E[\delta_o] = \frac{p_v}{2(1-p_v)} + 1 \tag{8.7}$$

However, a more detailed description of the virtual queue operation is needed since the average waiting time in the input queue is a function of the first two moments of the waiting time in the virtual queue (see Equation 8.3). Since it has been proved [Ili90] that

$$E[\eta_v] = \frac{E[R]}{p}$$

$$E[\eta_v^2] = \frac{E[R^2]}{p}$$

we simply have to compute the first two moments of the virtual queue content. The assumption $B_o = K$ greatly simplifies the computation of these two moments, since the occurrence of at least one empty position in the tagged output queue implies that the tagged virtual queue is empty. So, the only state variable T_m fully describes the content of the two queues, which can be seen as a single queue with the first B_o positions representing the output queue and the other N positions modelling the virtual queue (Figure 8.12). Therefore

$$E[R] = \sum_{i=1}^{\infty} i \cdot Pr[T = B_o + i] = \sum_{i=B_o+1}^{\infty} (i - B_o) t_i$$

$$E[R^2] = \sum_{i=1}^{\infty} i^2 \cdot Pr[T = B_o + i] = \sum_{i=B_o+1}^{\infty} (i - B_o)^2 t_i$$

(8.8)

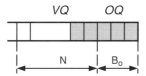

Figure 8.12. Combined representation of tagged VQ and OQ for $B_o=K$

which after some algebraic manipulations become

$$E[R] = E[T] - B_o(1-t_0) + \sum_{i=1}^{B_o-1} (B_o - i)\, t_i$$

$$(8.9)$$

$$E[R^2] = E[T^2] - 2B_o E[T] - B_o^2(1-t_0) - \sum_{i=1}^{B_o-1} (B_o - i)^2 t_i$$

The numerical values of the first and second moment of the variable T, as well its distribution t_i, are those of an $M/D/1$ queue with internal server (see Appendix).

The asymptotic throughput ρ_{max} of the switch, i.e. its capacity, is obtained by computing the limiting load conditions for the input queue, that is the root of the function $F(p_i) = 1 - p_i - p_i E[\eta_v]$ (see Equation 8.3). The average packet delay is obtained as the sum of the waiting times in the three queues increased by one (the service time, see Equation 8.1). No cell loss takes place as the backpressure prevents loss in the output queues, in spite of their finite capacity, and infinite input queues are lossless by definition.

The switch capacity ρ_{max} is shown in Figure 8.13 for increasing values of the speed-up K. It increases starting from the value $\rho_{max} = 0.58$ for $K = 1$, which means pure input queueing (output queues are not needed as each output port controller receives at most $K = 1$ packet per slot). The increase of the maximum throughput is not as fast as we would have expected, since a speed-up $K = 19$ is needed to have $\rho_{max} > 0.9$.

The average packet delay T, given by Equations 8.1, 8.3 and 8.7, is plotted in Figure 8.14 as a function of the offered load p. Again, the curve with $K = 1$ refers to the case of pure input queueing, whereas the delay approaches the behavior of an $M/D/1$ queue as $K \to \infty$. In fact if the output queue has an infinite capacity the tagged virtual queue is always empty and A_m, whose distribution is Poisson, describes directly the arrival process to the output queue. In the following section we will see that a much better performance can be obtained, even with a very small speed-up value, by simply providing a large output queueing capability.

8.1.2.2. Arbitrary input and output queue capacities

Now the more general analysis of the IOQ switch with arbitrary size B_i of the input queue and B_o of the output queue is provided [Pat93]. The study of the tagged input queue is first developed and then the behavior of the tagged virtual and output queues is analyzed.

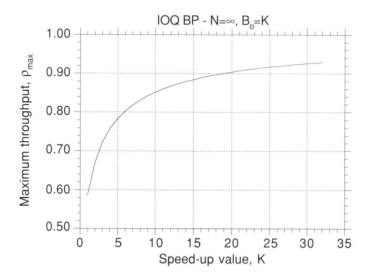

Figure 8.13. Switch capacity of a BP IOQ switch when $B_o=K$

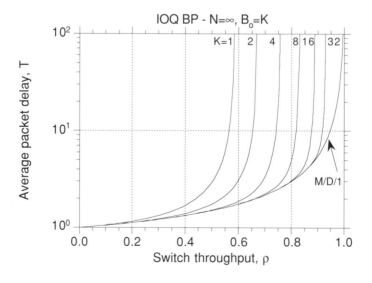

Figure 8.14. Delay performance of a BP IOQ switch when $B_o=K$

In the case of infinite input queue, the previous analysis of a $Geom/G/1$ queue applies here too, so that the average waiting time is given by Equation 8.3, in which the first two moments of the waiting time in the virtual queue are given by definition as

$$E[\eta_v] = \sum_{n=0}^{\infty} n\eta_{v,n}$$

$$E[\eta_v^2] = \sum_{n=0}^{\infty} n^2 \eta_{v,n} \tag{8.10}$$

Unlike the previous case of a constrained output queue size, now the distribution function $\eta_{v,n} \equiv \Pr[\eta_v = n]$ is needed explicitly, since the condition $K < B_o$ makes it possible that the tagged output queue is not full and the tagged virtual queue is not empty in the same slot. The distribution of the random variable η_v is computed later in this section.

When the input queue has a finite capacity, the tagged input queue behaves as a $Geom/G/1/B_i$ queue with a probability p_i of a packet arrival in a slot. Also in this case the service time distribution is general and the tagged virtual queue receives a load that can still be considered Poisson with rate $p_v = p_i(1 - \pi_i)$, where π_i is the cell loss probability at the input queue. An iterative approach is used to solve the $Geom/G/1/B_i$ queue, where B_i represents the total number of users admitted in the queue (including the user being served), which starting from an initial admissible value for p_v, consists in

- evaluating the distribution function $\theta_{i,n}$ of the service time in the input queue, which is given by the distribution function $\eta_{v,n-1}$ computed in the following section as a function of the current p_v value;
- finding the cell loss probability π_i in the input queue according to the procedure described in the Appendix;
- computing the new value $p_v = p_i(1 - \pi_i)$.

These steps are iterated as long as the new value of p_v differs from the preceding value for less than a predefined threshold. The interaction among different input queues, which occurs when different HOL packets address the same output queue, is taken into account in the evaluation of the distribution function of the waiting time in the virtual queue, as will be shown later.

The packet loss probability π_i is found by observing in each slot the process (n, j) where n is the number of packets in the input queue and j is the number of time slots needed to complete the service of the HOL packet starting from the current slot. Note that the service time for a new packet entering the input queue is given by the above-mentioned distribution $\theta_{i,n}$. The Markov Chain analysis of such a $Geom/G/1/B_i$ system is given in the Appendix which evaluates the measures of interest to us, that is the queueing delay $E[\eta_i + \theta_i] = E[\delta_i]$ and the packet loss probability π_i.

The analysis of virtual queue and output queue will be developed separately for the two internal protocols BP and QL.

The BP protocol. The operations of the switch model with backpressure is illustrated by the example in Figure 8.15 for $R_{m-1} > K$, $Q_{m-1} = 2$, $K \geq 2$, $B_o = 3$. Equations 8.4 describing the system evolution still apply for the description of variables R_m and Q_m, whereas the one describing V_m must be modified to take into account that the speed-up too can limit the packet transfer from the virtual queue to the output queue, that is

$$V_m = \min \{ K, B_o - \max \{ 0, Q_{m-1} - 1 \}, R_{m-1} + A_m \}$$

The cumulative number T_m of packets in the tagged virtual queue and tagged output queue and the total average delay spent in the two queues are still expressed by Equations 8.6 and 8.7.

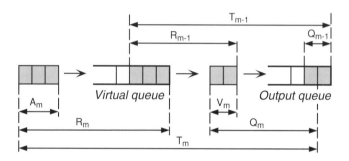

Figure 8.15. Example of tagged VQ-OQ switching for BP operation

As previously discussed, the evaluation of the delay and loss performance in the input queue requires the knowledge of the distribution function of the service time θ_i of the HOL cell in the input queue, which is obtained through the distribution function of the waiting time η_v in the virtual queue.

Let the couple (i, j) denote the generic state of the system $R_m = i$, $Q_m = j$, that is the tagged virtual queue and the tagged output queue hold at the end of the m-th slot i and j packets, respectively, and let $p_{i,j}$ indicate the probability of the state (i, j). In order to compute the probability distribution $\eta_{v,n} \equiv \Pr[\eta_v = n]$ that the tagged packet waits n slots before being served in the tagged virtual queue, a deterministic function $F_{n,j}$ is introduced: it represents the number of cells that can be switched from the tagged virtual queue to the tagged output queue in n time slots starting from an initial state of j cells in the output queue. If (i, j) is the system state found by the tagged packet entering the HOL position in the tagged input queue and given that it is selected as k-th among those packets entering the same virtual queue in the same slot, then

$$\eta_{v,n} = \Pr[F_{n,j} < i + k \leq F_{n+1,j}] \tag{8.11}$$

In fact, $i + k - 1$ packets must be switched before the tagged packet.

The evaluation of the number $F_{n,j}$ of packets that can be switched in n slots from the tagged virtual queue to the tagged output queue can be evaluated by computing the number of packets $f_{l,j}$ that can be switched in the generic slot l of these n slots $(l \leq n)$. It is simple to show that $f_{l,j}$ is given by the minimum between the speed-up K and the number of idle positions in the output queue, that is

$$f_{l,j} = \min\left\{ K, B_o - \sum_{m=1}^{l-1} f_{m,j} + l - 1 - \max\{0, j-1\} \right\} \qquad (j \geq 0; \; 1 \leq l \leq n)$$

Since by definition

$$F_{n,j} \equiv \sum_{l=1}^{n} f_{l,j} \qquad (n \geq 1, j \geq 0)$$

$$F_{0,j} = 0 \qquad (j \geq 0)$$

through a simple substitution we obtain the recursive relations for $F_{n,j}$:

$$F_{n,j} = \sum_{k=1}^{n} \min\{ K, B_o - F_{k-1,j} + k - 1 - \max\{0, j-1\} \} \qquad (n \geq 1, j \geq 0)$$

The probability distribution $\eta_{v,n}$ is then given by Equation 8.11 considering the range of variation of k and saturating on all the possible i and j values that give the state (i, j). That is

$$\eta_{v,n} = \sum_{j=0}^{B_o} \sum_{m=F_{n,j}+1}^{F_{n+1,j}} \sum_{k=1}^{m} b_k p_{m-k,j} \qquad (n \geq 0) \tag{8.12}$$

in which b_k is the probability that the tagged packet is the k-th among the set of packets entering the virtual queue in the same slot. The factor b_k is obtained as the probability that the tagged packet arrives in a set of i packets times the probability $1/i$ of being selected as k-th in the set of i packets. The former probability is again found in [Kar87] to be ia_i/p_v, in which a_i is the probability of i arrivals in one slot given by the Poisson distribution (we are assuming the switch size $N = \infty$). So

$$b_k = \sum_{i=k}^{\infty} \frac{1}{i} \frac{ia_i}{p_v} = \frac{1}{p_v} \sum_{i=k}^{\infty} a_i \qquad (k \geq 1)$$

$$a_i = p_v^i \frac{e^{-i}}{i!}$$

In order to determine the probability $p_{i,j}$, we can write the equations expressing the Markov chain of the system (virtual queue, output queue). Four cases are distinguished:

1. **Empty tagged virtual queue and non-saturated output queue** $(i = 0, \; 0 \leq j \leq B_o - 1)$. Here $p_{i,j}$ is evaluated by considering all the possible state transitions leading to state (i, j). This occurs when the system state is $(k, j - m + 1)$ and $m - k$ packets enter the tagged virtual queue for each k from 0 to m. Since the tagged virtual queue will be empty at the end of the slot, all the k packets already in the tagged virtual queue together with the new

packets are switched to the output queue, whose final state is $(j - m + 1) + k + (m - k) - 1 = j$ (recall that one packet per slot is transmitted to the switch outlet). Thus we obtain

$$p_{i,j} = \sum_{m=0}^{\min\{K,j\}} \sum_{k=0}^{m} p_{k, j-m+1} a_{m-k} + u_{-1}(K-j) p_{0,0} a_j \qquad (8.13)$$

in which

$$u_{-1}(x) = \begin{cases} 1 & \text{if } x \geq 0 \\ 0 & \text{if } x < 0 \end{cases}$$

Note that the upper bound of the first sum comes from the condition that the index m must be less than the speed-up factor in order for the virtual queue to be empty at the end of the slot. The second term takes into account that a transition from state $(0,0)$ to state $(0,j)$ is possible only when $j \leq K$, by also considering that no packet is transmitted by the empty output queue.

2. Non-empty tagged virtual queue and not more than $K–1$ packets in the tagged output queue $(i > 0, \ 0 \leq j \leq K-1)$. In this case we simply have

$$p_{i,j} = 0 \qquad (8.14)$$

since the tagged virtual queue is non-empty only if the output queue holds at least K packets.

3. Non-empty tagged virtual queue and more than $K–1$ packets in a non-saturated tagged output queue $(i > 0, \ K \leq j \leq B_o - 1)$. Analogously to Case 1, we get

$$p_{i,j} = \sum_{k=0}^{i+K} p_{k, j-K+1} a_{i+K-k} + u_{-1}(K-j) p_{0,0} a_{i+K} \qquad (8.15)$$

4. Saturated tagged output queue $(i \geq 0, j = B_o)$. Through considerations similar to those in Case 1, the state probability is given by

$$p_{i,j} = \sum_{m=i+1}^{i+K} \sum_{k=0}^{m} p_{k, B_o-(m-i)+1} a_{m-k} + u_{-1}(K-B_o) p_{0,0} a_{i+K} \qquad (8.16)$$

In fact, the total number of packets in the virtual queue after the new arrivals becomes $k + (m - k) = m$. Only $m - i$ of these packets can be switched to occupy the residual idle capacity of the output queue, whereas the other i packets remain stored in the virtual queue. The last term gives a non-null contribution only for architectures with a speed-up equal to the output queue capacity (remember that our assumption is $K \leq B_o$).

Some performance results given by the analytical model just described are now provided, whose accuracy is also shown by providing as well the results obtained through computer simulation of the switch architecture. A switch size 256×256 has been selected in the simulation model, so that the mismatch between the finite size of the simulated switch and the infinite switch size of the model can be neglected. In fact, it is well known (see, e.g., [Kar87], [Pat90]) that the maximum throughput of a non-blocking switching structure with input (or input and output) queueing converges quite rapidly to the asymptotic throughput as the switch size increases. Unless specified otherwise, solid lines in the graphics represent analytical results, whereas plots are simulation data.

The switch capacity ρ_{max} is obtained as in the case of constrained output queue capacity, that is as the limiting load of an infinite capacity $Geom/G/1$ input queue. Therefore the maximum throughput is again given by the root of the function $F(p_i) = 1 - p_i - p_i E[\eta_v]$ (see Equation 8.3). Note that now a completely different procedure has been applied to compute the moments of the waiting time η_v in the virtual queue starting from its probability distribution function (Equation 8.12). Figure 8.16 shows the asymptotic throughput performance of the switch in the BP mode for increasing output buffer sizes starting with $B_0 = K$ and different speed-up values. The throughput ρ_{max} increases with the output buffer size starting from a minimum value of 0.671 ($B_0 = 2$) up to an asymptotic value 0.885 for $K = 2$ and from 0.723 ($B_0 = 3$) up to 0.973 for $K = 3$. Compared to this latter curve, assuming a speed-up $K = 4$ provides almost the same performance for small buffer sizes ($B_0 < 64$), but the asymptotic throughput is now 0.996. Thus a speed-up $K = 4$ gives an asymptotic throughput very close to the theoretical maximum $\rho_{max} = 1$.

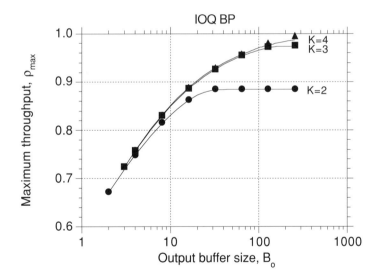

Figure 8.16. Switch capacity of a BP IOQ switch

The average packet delay is given by Equation 8.1: the input queue component is provided by Equation 8.3 with the moments of the waiting time computed by the iterative procedure explained above, whereas the virtual and output queue component is given by Equation 8.7. The performance figure T is plotted in Figure 8.17 as a function of the offered load $p = p_i$ for $K = 2$ with $B_i = \infty$. With the BP mode the average delay decreases for a given offered load as the output buffer size is increased. This is clearly due to the increase of the asymptotic throughput with the output buffer size that implies a better performance for an increasing "idleness factor" $\rho_{max}(B_o) - p$ of the input queue server. Note that in this case the offered load level of a curve coincides with the switch throughput (the input queues have infinite capacity). All the delay curves move to the right with higher speed-up values K.

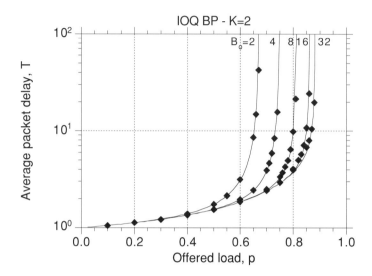

Figure 8.17. Delay performance of a BP IOQ switch

The cell loss performance of the switch, given by the packet loss probability in the input queue $Geom/G/1/B_i$ (see the above iterative procedure), is shown in Figure 8.18 for different output queue values as a function of the input buffer size for $K = 2$ and $p = 0.8$. Apparently, smaller cell loss figures are obtained as the input buffer size is increased. A very good matching between analytical and simulation results has been obtained for throughput and delay figures, as well as for the cell loss figure with a small input buffer size, say up to $B_i = 8$. However, the model is accurate for large input buffer sizes too, as the analytical loss probability values, although very small and hard to evaluate through simulation, lie within the 95% confidence interval given by the simulation model.

The QL protocol. Without backpressure under queue loss operation cell losses can occur also at the output queues, so that we have to distinguish between the cell flow outgoing from the virtual queue and the cell flow entering the output queue. So, we introduce the two random variables

V_m' number of packets leaving the tagged virtual queue in slot m;

V_m'' number of packets entering the tagged output queue in slot m.

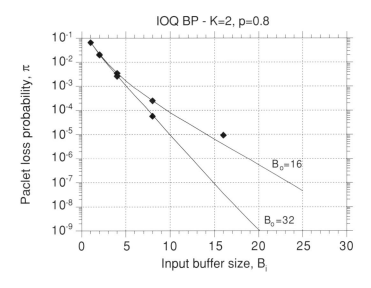

Figure 8.18. Loss performance of a BP IOQ switch

An example of packet switching taking place in slot m is shown in Figure 8.19 for $R_{m-1} > K$, $Q_{m-1} = 2$, $K = 3$, $B_o = 3$. The state equations are now given by

$$R_m = R_{m-1} + A_m - V_m'$$

$$V_m' = \min\{K, R_{m-1} + A_m\}$$

$$V_m'' = \min\{V_m', B_o - \max\{0, Q_{m-1} - 1\}\} \qquad (8.17)$$

$$Q_m = \max\{0, Q_{m-1} - 1\} + V_m''$$

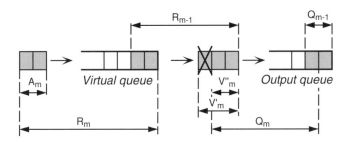

Figure 8.19. Example of tagged VQ-OQ switching for QL operation

Unlike the BP protocol, in this case the cumulative time spent by the tagged packet in the virtual queue and output queue cannot be expressed through an $M/D/1$ system, because of the difference between $V_m{}'$ and $V_m{}''$. However, now the number of packets in the virtual queue, expressed as

$$R_m = R_{m-1} + A_m - \min\{K, R_{m-1} + A_m\} \tag{8.18}$$

may be computed separately from Q_m. Thus, from Equation 8.18 we obtain the equations for the computation of the probability r_i of i packets in the virtual queue, that is

$$r_0 = \sum_{m=0}^{K} \sum_{k=0}^{m} r_k a_{m-k}$$

$$r_i = \sum_{m=0}^{K+i} r_m a_{i+K-m} \qquad (i > 0)$$

The distribution function of the waiting time in the virtual queue is then computed similarly to Equation 8.12 considering that in the QL mode $F_{n,j} = nK$, that is

$$\eta_{v,n} = \sum_{j=0}^{B_o} \sum_{m=nK+1}^{(n+1)K} \sum_{k=1}^{m} b_k p_{m-k,j} = \sum_{m=nK+1}^{(n+1)K} \sum_{k=1}^{m} b_k r_{m-k} \tag{8.19}$$

Note that the distribution $p_{i,j}$ is not needed to compute $\eta_{v,n}$.

In order to obtain the distribution function of the waiting time in the output queue, the joint statistics of the virtual queue and output queue has to be evaluated also in this case. In fact, even if there is no backpressure mechanism, the output queue state Q_m is still dependent on the virtual queue state R_{m-1} in the preceding time slot. The equations expressing the joint probability of the state (i, j) are the same as in the BP mode for a non-saturated output queue (cases 1–3) as the backpressure mechanism will have been operating only when $Q_m = B_o$. The equation of Case 4 $(j = B_o)$ of the BP mode (Equation 8.16) must be replaced by

$$p_{0,B_o} = \sum_{l=B_o-K+1}^{B_o} \sum_{m=B_o-l+1}^{K} \sum_{k=0}^{m} p_{k,l} a_{m-k} + u_{-1}(K-B_o) p_{0,0} a_{B_o}$$

$$p_{i,B_o} = \sum_{l=B_o-K+1}^{B_o} \sum_{k=0}^{i+K} p_{k,l} a_{i+K-k} + u_{-1}(K-B_o) p_{0,0} a_{i+K} \qquad (i > 0) \tag{8.20}$$

The distribution function of the number of packets in the output queue is easily obtained as the marginal distribution of $p_{i,j}$

$$q_j = \sum_{i=0}^{\infty} p_{i,j}$$

and by applying Little's formula the average waiting time in the output queue is obtained, that is

$$E[\delta_o] = \frac{\displaystyle\sum_{i=0}^{B_o} jq_j}{P_o} = \frac{\displaystyle\sum_{i=0}^{B_o} jq_j}{1 - q_0} \tag{8.21}$$

The packet loss probability π_o in the output queue in the QL mode can be found considering the probability $\pi_{o,l}$ of l cells lost in a generic time slot and then computing the ratio between lost cells and offered cells in the output queue. Thus,

$$\pi_{o,l} = \sum_{m=l+1}^{\infty} \sum_{i=0}^{m} p_{i, B_o - \min\{m, K\} + l + 1} a_{m-i} \qquad (1 \le l \le K - 1)$$

$$\pi_o = \frac{\displaystyle\sum_{l=1}^{K-1} l\pi_{o,l}}{p_v} = 1 - \frac{p_o}{p_v} \tag{8.22}$$

In fact, for $m < K$, exactly l packets out of the $(m - i) + i = m$ packets in the virtual queue are lost if the residual output queue capacity is $m - l - 1$ packets (remember that one packet is always transmitted by the non-empty output queue). For $m \ge K$, the residual queue capacity must be $K - l - 1$, owing to the limiting speed-up factor. Thus, the output queue must hold $B_o - \min\{m, K\} + l + 1$ packets in order for exactly l packets to be lost.

The Markov chain representing the state (i, j) can be solved iteratively by setting a suitable threshold for the state probability value assumed in two consecutive iterations. However, in the BP mode the bidimensional Markov chain can be also solved without iterations by: (a) determining the distribution of the total number of cells in the tagged virtual and output queues by means of a monodimensional Markov chain; (b) using this distribution and the Equations 8.13–8.16 in such a way that each unknown probability can be expressed as a function of already computed probabilities. In the QL mode a similar approach does not yield the same simplifying procedure. Thus, the solution of the bidimensional Markov chain becomes critical in the QL mode as the state space increases, that is when the product $K \cdot B_o$ becomes large. To cope with these situations an approximate analysis of the QL mode can be carried out in which the output queue state is assumed to be independent of the virtual queue state. By using Equations 8.13–8.15, 8.20, replacing $p_{i,j}$ by $r_i q_j$ and summing over the virtual queue state index i, we obtain the distribution function of the packets in the output queue:

$$q_j = \sum_{m=\max\{1, j-K+1\}}^{j+1} q_m v'_{j-m+1} + u_{-1}(K-j) q_0 v'_j \qquad (0 \le j \le B_o - 1)$$

$$q_{B_o} = \sum_{m=\max\{1, B_o-K+1\}}^{B_o} q_m \sum_{i=B_o-m+1}^{K} v'_i + u_{-1}(K-B_o) q_0 v'_{B_o}$$

in which

$$v_k' = \sum_{i=0}^{k} r_i a_{k-i} \qquad (k < K)$$

$$v_K' = \sum_{m=K}^{\infty} \sum_{i=0}^{m} r_i a_{m-i}$$

Note that v_k' is the distribution function of the number of packet arrivals at the output queue (see Equations 8.17).

The switch capacity $\rho_{\max}$ is obtained as in the cases of constrained output queue capacity and BP protocol, that is as the limiting load of an infinite capacity $Geom/G/1$ input queue. Therefore the maximum throughput is again given by the root of the function $F(p_i) = 1 - p_i - p_i E[\eta_v]$ (see Equation 8.3). Now the moments of the waiting time η_v in the virtual queue are computed by means of Equations 8.10 and 8.19. Figure 8.20 shows the asymptotic throughput of the switch in the QL mode for output buffer sizes ranging from $B_o = K$ to $B_o = 256$. The results of the exact model, which owing to the mentioned computation limits are limited up to $B_o = 64$, are plotted by a solid line, whereas a dashed line gives the approximate model (as usual, plots represent simulation results). By comparing these results with those under backpressure operation (Figure 8.16), the QL mode provides better performance than the BP mode. The reason is that the cumulative number of packets lost by all the output queues is random redistributed among all the virtual queues in the former case (in saturation conditions the total number of packets in the virtual queues is constant). Such a random redistribution of the packet destinations reduces the HOL blocking that occurs in the BP mode where the HOL packets are held as long as they are not switched. According to our expectations, the asymptotic throughput of the QL and BP mode for very large B_o values are the same, since a very large output buffer makes useless the backpressure mechanism between output queue and virtual queue. A very good matching is found between analytical and simulation results for the exact model of QL, whereas the approximate model overestimates the maximum throughput for small output queue sizes. The approximation lies in independence of the arrival process to the output queue from the output queue status in the model, so that a smoother traffic is offered to the output queue (in the approximate model the output queue can well be empty even if the virtual queue is not empty).

The average packet delay T for the QL mode is given by Equation 8.1: the output queue component is given by Equation 8.21, the virtual queue component by Equation 8.19, whose expression is used in the iterative computation of the input queue component as provided by Equation 8.3. The parameter T is plotted in Figure 8.21 as a function of the offered load $p = p_i$ for $K = 2$ with $B_i = \infty$. Unlike the BP mode, with the QL operation all the delay curves have the same asymptotic load value independently of the actual B_o value. In fact the packet delay approaches infinity when the input queues saturate and the saturation conditions are functions only of the speed-up and are independent of the output buffer size. Such asymptotic offered load, $p_{\max}(K)$, coincides with the asymptotic switch throughput given by an infinite output buffer size. The relatively small increase of packet delay for increasing B_o values

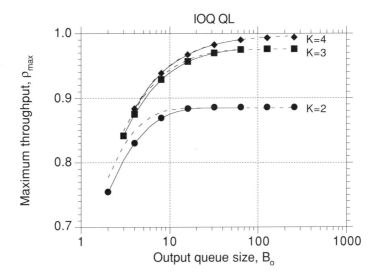

Figure 8.20. Switch capacity of a QL IOQ switch

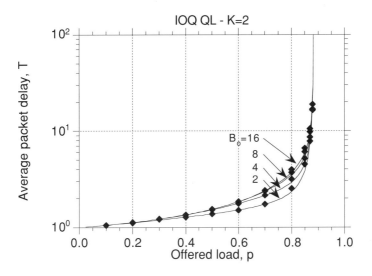

Figure 8.21. Delay performance of a QL IOQ switch

is only due to the larger amount of time spent in the output queue. All the delay curves move to the right with higher speed-up values K.

In the QL mode cells can be lost both at input queues and at output queues. The former loss probability, π_i, is evaluated by means of the analysis of the tagged input queue $Geom/G/1/B_i$, whereas the latter loss probability is computed by Equation 8.22. These two loss performance figures are given in Figure 8.22 and in Figure 8.23, respectively for speed-up $K = 2$ and different output queue sizes under a variable offered load. In general we note that a very low loss value under a given offered load requires larger buffers at outputs than at inputs. The model accuracy in evaluating delay and loss performance figures give very good results, as shown in Figures 8.21–8.23 (plots are again simulation results). If the speed-up is increased, much smaller input queues are required. For example, $B_i = 2$ gives a loss probability below 10^{-8} with $p = 0.1$ for $K = 2$ and with $p = 0.4$ for $K = 3$.

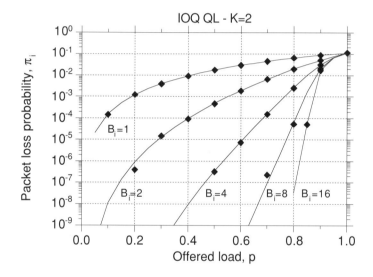

Figure 8.22. Input loss performance of a QL IOQ switch

As for the comparison of the total cell loss between the BP and QL modes as a function of the output buffer size, for different input queue sizes, Figure 8.24 shows the cell loss probability given by the only analytical model for $K = 4$ and $p = 0.8$. The loss curves for the two modes approach the same asymptotic level for large output buffer sizes, as the absence of output buffer saturation makes the backpressure mechanism ineffective so that the two modes operate in the same way. Apparently, smaller cell loss figures are obtained as the input buffer size is increased. Below a certain size of the output buffer the two modes provide a different cell loss performance that is a function of the offered load and input buffer size. For smaller input queues the QL mode provides lower loss figures, whereas for higher values of B_i the BP mode performs better. Such behavior, already observed above, occurs for loads up to a certain value, since for load values close to the maximum throughput the QL mode always gives the best loss results. This phenomenon can be explained by considering that the backpressure

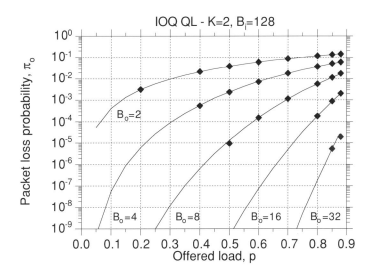

Figure 8.23. Output loss performance of a QL IOQ switch

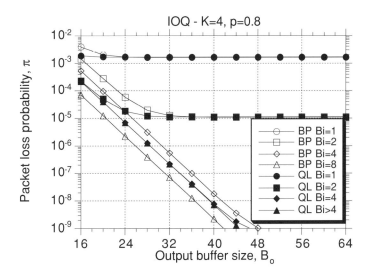

Figure 8.24. Comparison of cell loss performance between BP and QL

mechanism makes effective, especially at low loads and with greater than minimal input queue sizes, a kind of sharing of input and output queueing resources in order to control cell loss. However, if the load increases, this effect becomes negligible compared to the larger maximum throughput shown by the QL mode when the output queues saturate, thus explaining a better performance for the QL mode.

In order to identify the buffer configuration providing the best cell loss performance given a total queueing budget, Figure 8.25 shows for the same network parameters the loss probability as a function of the input queue size for a total buffer $B_t = B_i + B_o$ (here the plots have also been used to distinguish between BP and QL, even if all the data in the figures represent analytical results). As one might expect, an optimum value B_i^* exists for each total buffer value B_t. When backpressure is applied, the service time of the input queue grows with B_i for a given B_t as the corresponding reduction of B_o emphasizes the backpressure. However, the cell loss decreases for increasing values of $B_i < B_i^*$ since the growth of the input queue size overcompensates the negative effect of the service time increase. Apparently, beyond B_i^* such an advantage does not apply any more. Without backpressure, the existence of an optimum value B_i^* is explained considering that for smaller input buffers the total cell loss is mainly due to input queues, whereas most of the loss takes place at output queues for larger input buffers (recall that the total buffer B_t is constant). The figure shows that the relation $B_i^* \leq B_o^*$ applies for QL and $B_i^* \geq B_o^*$ applies for BP, consistently with the intuitive explanation that the QL mode requires in general larger output queues to guarantee a certain loss performance. The results shown here for the loss probability, given a total buffer budget, are analogous to those obtained for a switch with the same queueing strategy (input and output) with an unbuffered blocking interconnection network (see Figure 6.45).

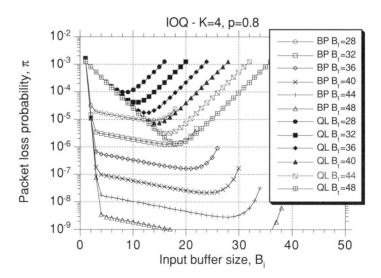

Figure 8.25. BP-QL loss performance comparison for a total buffer budget

8.1.3. Architectures with parallel switching planes

Rather than using an ATM network that is K-non-blocking, K switching planes $(K > 1)$ can be used to build a non-blocking ATM switch with combined input–output queueing [Oie90], in which each plane is 1-non-blocking. Therefore, the output speed-up K_o of the switch must also be equal to K. An $N \times N$ architecture with K parallel switching planes and *shared input queues* is represented in Figure 8.26 showing how a device (a splitter) per input port is needed that routes each packet from the shared input queue to a specific plane. Depending on the splitter operation, the network performance can be substantially different. Two operation modes of the splitters are examined, that is with or without speed-up at the input queues.

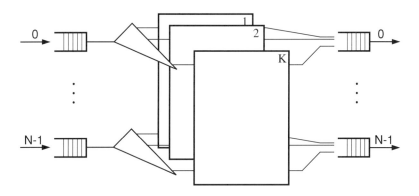

Figure 8.26. IOQ switch with multiple parallel planes and shared input queues

When input speed-up is not accomplished $(K_i = 1)$, each queue transmits at most one packet per slot and the plane where to transmit the HOL packet has been selected randomly with even probability for all the switch outlets. Since each plane is 1-non-blocking, if two or more HOL packets select the same plane and address the same switch outlet, all but one of them are blocked in their respective input queues (one of the algorithms described in Section 7.1.1 is assumed to be applied here for resolving the conflicts for the same switch outlets). The asymptotic throughput performance of this structure is obtained quite easily. In fact the operation of random selection of a switching plane, coupled with the usual assumption of a random selection of the switch outlet, corresponds to having a non-squared interconnection network of size $N \times M$ with $M = NK$ without output speed-up. The analysis of such a structure, reported in Section 7.1.2.1, shows that the switch capacity is

$$\rho_{max} = K + 1 - \sqrt{K^2 + 1}$$

The numerical values of the maximum throughput, given in Table 7.1, show that $\rho_{max} = 0.764, 0.877$ for $K = 2, 4$, respectively. It is interesting to note that the analogous asymptotic throughput values of a non-blocking IOQ switch with only output speed-up $K_o = K$, computed in Section 8.1.2.2, are higher: $\rho_{max} = 0.885$ for $K = 2$ and $\rho_{max} = 0.996$ for $K = 4$. Such difference in the throughput performance could look not correct at a first glance, since both of them have the same input $(K_i = 1)$ and output

$(K_o = K)$ speed-up. Nevertheless, note that the operation of selecting the switch plane followed by the output contention resolution in each 1-non-blocking plane is worse than operating the contention resolution first and then finding the path in a K-non-blocking switching plane. For example, if $K = 2$ and only two packets address a given switch outlet, one of the packets is blocked in the former case if both packets select the same plane, whereas both packets are always switched in the latter case.

A different operation of the splitter is given when an input speed-up $K_i = K$ is accomplished. An implementation of this operation is proposed in [Hui87] where $K = 2$ planes switch packets in half slot from the input to the output queues. Each plane is a Three-Phase switch and their operations are staggered by half slot. In such a way, the HOL packets that are blocked due to an output conflict on switching plane 1 in the first half slot can run a second contention resolution in the same slot for their transfer through switching plane 2. Since a single plane with dedicated input queues has a maximum throughput $\rho_{max} = 0.58$, it follows that the above structure with $K = 2$ planes is capable of carrying the maximum offered load $\rho_{max} = p = 1.0$ and is thus free from throughput limitations.

A different architecture of non-blocking IOQ switch with multiple switching planes and *dedicated input queues* is shown in Figure 8.27. Now packet queueing follows the splitters, so that each input termination is now served by K physical queues, each dedicated to a specific switch plane. Again each splitter selects randomly with even probability the dedicated input queue to which to send the packet, so that each plane receives a load p/K (p is the load offered to each switch inlet). This operation corresponds to accomplishing an input speed-up $K_i = K$ since up to K_i packets received on a switch inlet can be switched in a slot (one per plane). As in the previous case of shared input queueing with input speed-up, each plane is capable of carrying a throughput $\rho_{max} = 0.58$. Since each plane receives a load p/K, it follows that $K = 2$ planes are enough to remove any throughput limitation to the switch architecture $(\rho_{max} = 1.0)$. It is worth noting that using shared input queues guarantees the packet sequence integrity in each virtual connection supported by the switch. Packet sequencing cannot be guaranteed when dedicated input queues are used, each serving a single plane inlet.

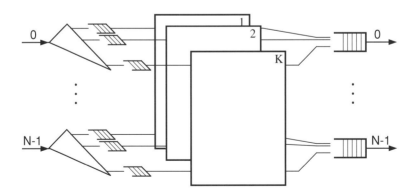

Figure 8.27. IOQ switch with multiple parallel planes and dedicated input queues

The windowing technique, described in Section 7.1.3.2 and aimed at reducing the throughput limitation due to the HOL blocking, can be applied in principle also when multiple switching planes are equipped. Apparently we refer here to the only switch configuration where the switch capacity is smaller than unity, that is the multiplane structure with shared input queues and plane preassignment for each packet. When the windowing technique is applied with window size W, it means that the k-th cell $(1 < k \leq W)$ in an input queue is transmitted to its addressed outlet if it is the winner of its outlet contention and the $k-1$ cells ahead in the queue have lost their respective contention. The implementation of the windowing technique can be realized by the techniques already described in Section 7.1.3.2 for a pure input queueing switch. Figure 8.28 shows the switch capacity obtained by computer simulation for a 256×256 switch when windowing is applied for different values of the number of planes K. The curve $K = 1$ refers to the limiting case of one single plane, so that output queueing is not needed: its capacity values are the same given in Table 7.5. The results show that windowing is more effective for smaller values of the switch speed-up K and that most of the capacity gain is accomplished with the very first cycles of the windowing technique.

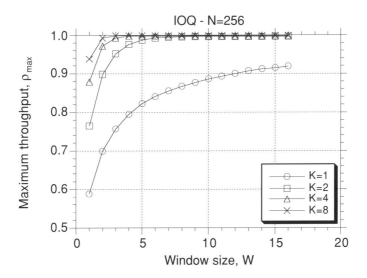

Figure 8.28. Switch capacity for a multiplane IOQ switch with dedicated input queues

8.2. Combined Shared-Output Queueing

The queueing model of a switch with combined shared-output queueing (SOQ) is given by Figure 8.1 assuming $B_i = 0$, $B_o \geq 1$, $B_s > 0$ and output speed-up $K \geq 2$. As in the case of an IOQ switch, $K \geq 2$ implies $B_o \geq 2$. The structure is able to transfer up to $K < N$ packets to each output queue without blocking due to internal conflicts. Now the packets in excess of K addressing a specific outlet remain stored in the shared queue, whose capacity is NB_s cells, to be switched in the next slot. Therefore in each slot the packets contending for the N outlets (at most K can be switched to each of them) are the new packets that just entered the switch together with those held in the shared queue as losers of the previous outlet contentions.

The basic proposal of a SOQ switch architecture is based on the self-routing multistage structure whose general model is represented in Figure 8.2 without the input queues. As with pure shared queueing, the shared queue in a SOQ switch is implemented as a set of recirculating lines.

8.2.1. Basic architecture

The basic proposal of an ATM switch with combined shared-output queueing is an extension of the Starlite switch, a pure shared queueing architecture described in Section 7.3.1. This architecture is called a *Sunshine switch* [Gia91] and is shown in Figure 8.29 for a switch with size $N \times N$. It includes a *sorting network* (SN), a *trap network* (TN), both of size $(N + P) \times (N + P)$, a *routing network* of size $N \times NK$ and a *recirculation (or delay) network* (DN) of size $P \times P$. This last network acts as the distributed shared buffer and feeds back to the routing network up to $P = NB_s$ packets that could not be switched in the preceding slot. Input and output links I_i and O_i are terminated on port controller PC_i $(0 \le i \le N-1)$, which also contains the i-th output queue.

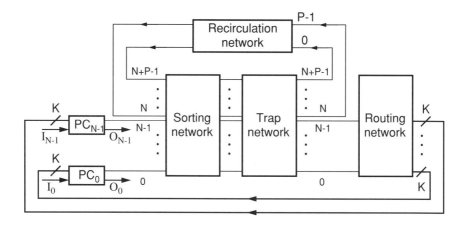

Figure 8.29. Architecture of the Sunshine switch

The internal packet format is assumed to be the same as in Starlite. A packet carrying an ATM cell is preceded by a control header that includes an activity bit AC set to 0, the cell destination address DA and a priority field PR (a lower content of this field denotes a higher priority cell). If no cell is received in the slot on input I_i, PC_i issues an empty packet indicated by an activity bit $AC = 1$.

The sorting network (SN) receives up to N new packets from the N port controllers, as well as up to P packets recirculated in the interconnection network as losers of the contention in the previous slot. The trap network (TN) performs the following functions:

- *Packet marking*: it marks as *winner* K packets out of those packets addressing the same switch outlet j $(j = 0, ..., N-1)$ *(class j packets)*, and as *losers* the remaining packets.
- *Packet discarding*: it applies a suitable *selection algorithm* to choose the P packets to be recirculated (the *stored packets*) out of the loser packets, if more than P packets are marked as losers among the N packet classes; the loser packets in excess of the P selected are discarded.
- *Packet routing*: it suitably arranges the winner and loser packets at the outlets of the trap network: the winner packets are offered as a compact and monotone set to the $N \times NK$ banyan network and the stored packets are offered to the recirculation network.

These functions are accomplished by the same trap network used in the Starlite switch for the fixed window selection algorithm (see Figure 7.34). This network includes a *marker*, a *running sum adder*, which generates an increasing (decreasing) list for winner (loser) packets, and a *concentrator*. The only difference between the two trap networks is that now the marker selects as winners up to K packets in each packet class j, whereas in Starlite only the first occurrence in each packet class is marked as winner. In order to identify the sets of the first K adjacent cells addressing the same network outlet, the marker compares the field DA of the packets received on inlets i and $i - K$ $(i \geq K)$. If they are different, the packet received on inlet i is marked as winner as it belongs to the set of the first K packets in its class of the sorted list generated by the sorting network. If the two addresses are equal, the packet on inlet i is marked as loser, since at least K other packets in the same class received on lower index inlets have already been marked as winners. Details about the hardware implementation of the Sunshine switch using the CMOS VLSI technology can be found in [Hic90].

The same switching example illustrated for Starlite with $N = 8$, $P = 4$ in Section 7.3.1 is here discussed assuming a Sunshine switch with speed-up $K = 2$, as shown in Figure 8.30. A set of nine active packets $(AC = 0)$ and three empty packets $(AC = 1)$ are sorted by SN and offered to the trap network. The marker marks as winners (W) up to two packets for each packet class (seven winner packets in total in the example) and as losers (L) the other active packets. The adder generates the addresses needed for the packet routing through the concentrator starting from 0 for winners and from $N + P - 1$ for losers (empty cells are given the remaining addresses in between). These $N + P$ packets are routed by the concentrator, so that the winner (loser) packets emerge at the outlets with smallest (largest) index. In this case there is no packet discarding as there are only three losers and four recirculation lines.

The overall switching example in Sunshine, which the trap network operations of Figure 8.30 refer to, are shown in Figure 8.31. The basic difference with the analogous example for Starlite is that here seven packets are successfully switched by the routing network to four port controllers (three of them receive $K = 2$ packets). Note that here we have assumed that the $N \times NK$ routing network is implemented as a set of two banyan networks $N \times N$ interconnected to the upstream trap network by an EGS pattern. As shown in Section 3.2.3, this kind of arrangement, denoted as solution c', is the minimum cost (it does not require the use of splitters between TN and RN) and ensures the non-blocking condition in each banyan network.

It is worth noting that Sunshine is implicitly an architecture of the queue loss (QL) type, since packets can be lost both at the shared queue and at the output queues. In fact there is no guarantee that the packets switched by the interconnection network can be stored into the

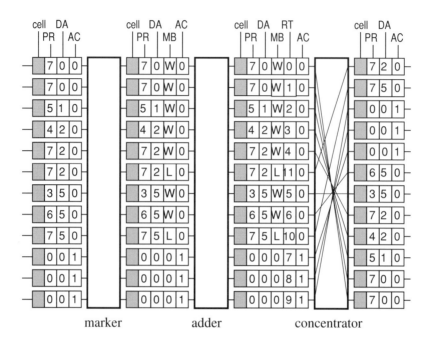

Figure 8.30. Switching example in the trap network

addressed output queue: up to K packets can require output queue storage in a slot, whereas only one packet per slot leaves the queue (it is sent downstream by the port controller). Therefore if queue saturation occurs, cells can be lost in the output queue.

A different architecture can be derived from Sunshine that adds backpressure (BP) between the shared queue and the output queues. In this structure, shown in Figure 8.32, a concentration network of size $(NK' + P) \times P$ is added, whose function is to enable the feedback into the shared queue of those packets that, although successfully switched to the output queues, cannot be stored there due to queue saturation. For this purpose each port controller is provided with K' output links feeding the concentrator. Since each port controller (PC_i) always transmits downstream one packet on its output link (O_i) and receives up to K packets from SN, the condition to have full backpressure between shared queue and output queues is $K' = K - 1$. Since the concentrator size is basically given by the product NK', this solution is feasible only for very small values of K.

8.2.2. Performance analysis

The traffic performance of a non-blocking switch with mixed shared and output queueing (SOQ) is now evaluated under the random traffic assumption by means of computer simulation. In fact developing an accurate analytical model for this kind of switch is a rather complex task.

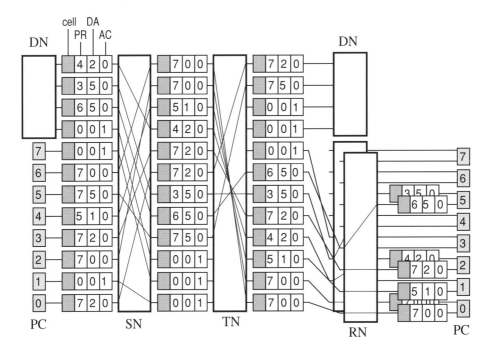

Figure 8.31. Switching example in Sunshine

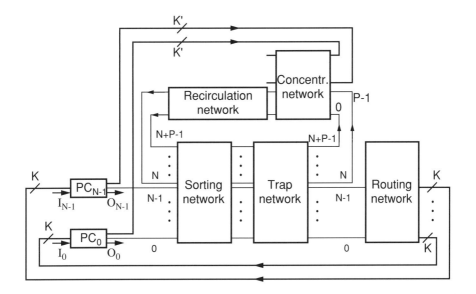

Figure 8.32. Architecture of a BP SOQ switch

The theoretical switch capacity is unity in this kind of switch that is implicitly free from the HOL blocking phenomenon. Nevertheless this requires a very large capacity (theoretically infinite) in the shared and the output queues. The packet loss probability π $(0 \leq \pi \leq 1)$, due to the finite size of these queues, can be computed as

$$\pi = 1 - (1 - \pi_s)(1 - \pi_o) = \pi_s + \pi_o - \pi_s \pi_o \qquad (8.23)$$

where π_s and π_o are the loss probabilities in the shared queue and in the output queue, respectively. It is rather clear that the bottleneck of the system is the capacity of the former queue, which must be implemented as a set of recirculation lines whose number is proportional to the switch size. An output queue is simply implemented in each output port controller of the switch.

We examine first the effect of the normalized shared queue size P/N on the switch capacity, expressed now as the packet loss probability $\pi_s = 1 - \rho_s$ by only looking at the shared queue behavior (ρ_s denotes its capacity). Recall that computing the switch capacity implies an offered load $p = 1.0$. This result applies to both QL and BP modes as it assumes an output queue size large enough to prevent saturation. The performance parameter π_s is shown in Figure 8.22 for a squared switch 32×32 with a speed-up value K ranging up to 4. Note that the case $K = 1$ corresponds to the Starlite switch, a pure shared queueing architecture. The result of adding speed-up and output queueing to Starlite is a remarkable reduction of the number of recirculation lines to obtain a given switch capacity. In particular a switch capacity very close to unity, say $1 - 10^{-5}$, is easily obtained with a very small shared buffer size, $P = 0.5N$ and $P = 0.1N$ for the speed-up values $K = 2$ and $K = 4$, respectively.

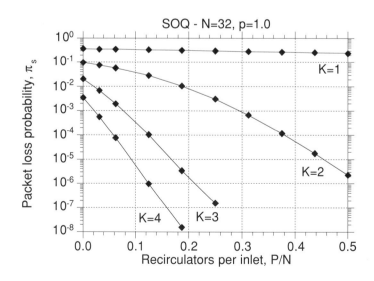

Figure 8.33. Recirculation loss performance of an SOQ switch

The overall switch capacity ρ_{max}, obtained by taking into account also the finite size of the output queue, is shown in Figure 8.20 for the output queue sizes $B_o = 4, 16, 64$. As we might expect, the switch capacity grows with the output buffer size. Note however that for large shared queue capacities the backpressure operation outperforms the QL mode for small values of B_o. The reason is clearly that the (large) shared queue can be effectively exploited by the backpressure mechanism as an additional common buffer space for holding those cells that cannot be stored in their addressed (small) output queues. Moreover the switch capacity becomes the same for both speed-up values $K = 2, 3$, above a certain value of the output buffer size, e.g. $B_o = 16$. Interestingly enough, under QL operations the switch capacity with speed-up $K = 2$ is larger than with $K = 3$, when the shared queue is large and the output queue small $(B_o = 4)$. Apparently the larger number of packets that can be switched in the latter case worsens the throughput performance as most of these packets are discarded due to the continuous saturation of the small output queues.

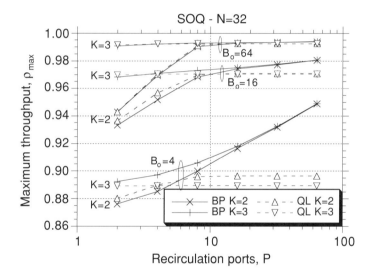

Figure 8.34. Switch capacity of a SOQ switch

The better performance provided in an SOQ switch when backpressure is accomplished is better described by looking at the overall packet loss probability, given by Equation 8.23, as shown in Figure 8.23, which assumes $N = 128$, $P = 16$. When the output buffer size is very small $(B_o = 4)$, the BP mode provides a loss performance remarkably better than the QL mode (several orders of magnitude in the region of low loss probabilities) for both speed-up values considered. As B_o is increased $(B_o = 16)$, the backpressure still gives a performance improvement when the number of packets capable of leaving the shared queue is enough: this is the case of the speed-up $K = 4$. Apparently the performance advantages provided by the backpressure between shared and output queue reduce as B_o grows (output queue saturations occur less often). Again as previously observed, increasing the speed-up with a very small output buffer $(B_o = 4)$ degrades the performance in absence of backpressure.

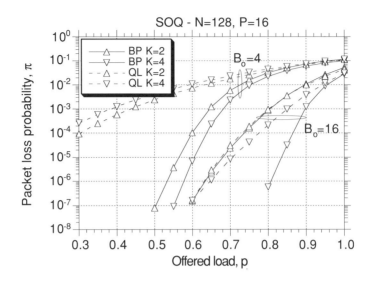

Figure 8.35. Overall loss performance of an SOQ switch

The average packet delay for the SOQ switch is plotted in Figure 8.21 for $N = 32$, $P = 16$. The results are rather straightforward: higher delays are given under back-pressure as cells spend in general more time in the shared queue due to backpressure (some of them would be lost in the QL mode). This additional time is higher for small output buffers, e.g. for $B_o = 4$, and shrinks as the output queue size increases (backpressure operates less often). The average delay value has a minimum value $T = 1$ for low offered loads: it is given basically by the transmission time. The waiting time in the two queues grows with the load level and becomes apparently higher as the buffer capacity enlarges: this is the case of the output queue in Figure 8.21 as the shared queue capacity is fixed $(P = 16)$. The speed-up value seems not to affect significantly the delay performance.

8.3. Combined Input-Shared Queueing

The model of a non-blocking ATM switch with combined input-shared queueing is given by Figure 8.1 assuming $B_i \geq 1$, $B_o = 0$ and $B_s > 0$. Now we have neither input speed-up, as each input queue transmits one packet per slot, nor output speed-up, as there is no output queueing; hence $K_i = K_o = 1$. However, analogously to the operation of the downstream output queues in an IOQ switch, now the downstream shared queue can receive up to $K_s = \min[NB_s, N]$ packets from the input queues. Thus we say that the ISQ architecture is characterized by an *internal speed-up* K_s.

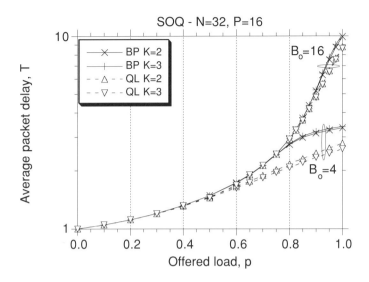

Figure 8.36. Delay performance of an SOQ switch

8.3.1. Basic architectures

The mixed input-shared queueing arrangement has been proposed in the backpressure (BP) configuration [Den92]. This proposal is based on the adoption of a shared memory operated at a very fast speed that receives packets from N upstream input queues. Apparently, the current VLSI CMOS technology sets up a bound on the maximum switch size, e.g. $N = 16$, even if these elements can be used as the basic building blocks to design larger switches.

The scheme of a generic switching element with shared queueing has already been described in Section 6.2.3, even if with reference to the minimum size 2×2. The technology limit inherent to that scheme, if referred to a size $N \times N$, is that the access to the shared queue must be serialized to enable N read and N write accesses in a bit time of the external links. To relieve this limit, techniques implementing bit parallelism would be exploited. Serial to parallel converters would be placed on the input port controllers, so that W bits in parallel can be stored in the shared queue (W planes of shared memory must be used). Due to the serial access to each memory plane by all the switch inlets and outlets, the maximum digital speed that can be handled on the external links is dependent not only on the memory access time, but also on the switch size. For example, assume that the memory access time is 20 ns and the maximum degree of parallelism is accomplished, that is $W = 424$ (the whole ATM cell length). Then the maximum bit rate on the external links for a 16×16 switch (that requires 16 read and 16 write accesses on each memory plane in a slot time) would be 662.5 Mbit/s. This bit rate could be small for very fast switches and its increase can be obtained only by reducing the memory access time, which is technology dependent.

The switch proposal described in [Den92] overcomes this limit by proposing an architecture with combined input-shared queueing where the switch size does not limit the external link speed. The architecture of the $N \times N$ switch is shown in Figure 8.37: it includes a data section and a control section. The data section contains the shared queue of NB_s cells accessed by N splitters $1 \times NB_s$, each associated with a network inlet. The input queues are not shown in the figure (they are located into the input port controllers). A set of N combiners $NB_s \times 1$ enables the full access of the N output ports to all the positions of the shared queue. The setting of splitters and combiners slot by slot is provided by the control section, which includes a queue (EQ) containing the addresses of the currently empty positions in the shared memory. This section also contains one control queue per outlet (OQ) to hold the list of pointers in the shared queue of the packets addressing the outlet.

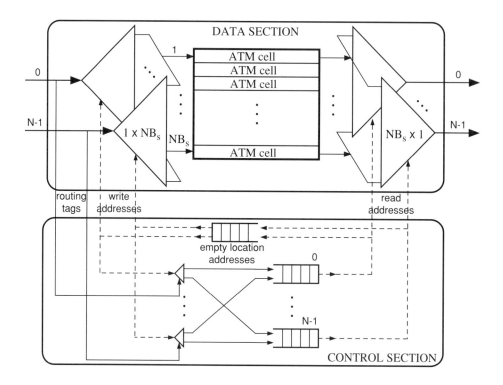

Figure 8.37. Architecture of a BP ISQ switch

The self-routing tag of the ATM cell received on each input is transferred to the control section to be combined with the shared queue address given to the cell by the queue EQ and stored through a $1 \times N$ splitter in the OQ associated with the addressed switch outlet. The queue EQ provides the same address to the splitter in the data section receiving the cell, so as to set its path to the selected empty memory position. On the outlet side, each control queue OQ gives slot per slot the address of the cell to be read and transmitted downstream to its associated combiner and frees the cell by storing the same address in the queue EQ.

Note that the maximum speed of the external links in this case is not dependent on the switch size, N, as each switch inlet (outlet) writes (reads) a different memory cell by virtue of the conflict resolution operated by the control section. Each ATM cell position in the shared queue can be seen as a shift register with the first position written by the incoming data flow. It is apparent that a limited degree of internal parallelism must be accomplished, for example byte wide, in order to limit the internal clock requirements.

8.3.2. Performance analysis

The function of the downstream queue in this architecture (the shared queue) is different from the analogous queue (the output queue) in the two previous cases IOQ and SOQ. Here the downstream shared queue need not be crossed necessarily by packets being switched to the output ports. In fact access to the output ports is enabled to the shared queue and to the input queues for the transmission of their HOL packet. A packet entering the HOL position of its input queue is transmitted immediately to the addressed output port if (i) the shared queue holds no other packet for the same output port and (ii) that HOL packet is the contention winner among the HOL packets addressing the same output port. If either of the two conditions does not hold, the input queue offers the HOL packet to the shared queue which accepts it unless queue saturation occurs. In this case backpressure is operated and the HOL packet is held in the HOL position in the input queue.

Analogously to the model developed for an IOQ switch, we say that the set of HOL packets in the input queues with the same address represent a virtual queue with capacity N cells (one per input) (see Figure 8.38). Therefore a packet entering the HOL position in the input queue enters at the same time the virtual queue. Since both the shared queue and the HOL positions in the input queues have access to the output ports, we can see the union of the virtual and shared queue, referred to as *VSQ queue*, as the total downstream capacity available to access the output port. The total capacity B_t of this VSQ queue is thus $N(B_s + 1)$, which also means a normalized capacity of $B_s + 1$ cells per output port.

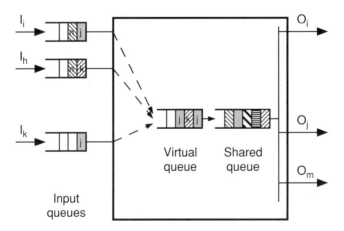

Figure 8.38. Queue model for the non-blocking ISQ switch

The analysis always assumes a FIFO service in the input, virtual and shared queue as well as an infinitely large switch $(N = \infty)$. Thus the number of cells entering the virtual queue and addressing a specific switch outlet in a slot approaches infinity with the outlet being independently and randomly selected by each cell. However, since the arrival process from individual inlets to the virtual queue is asymptotically negligible, the interarrival time of cells with a given address from an input queue to the virtual queue becomes sufficiently long. Therefore, the input queues are mutually independent discrete-time system with identical probability distributions. Owing to the random traffic assumption, the analysis will be referred to the behavior of a generic "tagged" input queue, as representative of any other input queue, whose offered traffic values is $p_i = p$. The average offered load values to the virtual and shared queue are denoted as p_v and p_s, respectively.

We evaluate here the throughput and delay switch performance by computing the maximum throughput ρ_{max} $(\rho_{max} \leq 1)$ and the average packet delay T $(T \geq 1)$. This latter parameter is given by

$$T = E[\eta_i] + E[\eta_v] + E[\eta_s] + 1 \tag{8.24}$$

where η_i is the waiting time in the input queue, i.e. before entering the HOL position, η_v is the waiting time in the virtual queue (that is in the HOL position of the input queue) before its transmission to the output port or to the shared queue, and η_s is the waiting time in the shared queue before its service.

By following the analysis in [Ili91], we can compute the limiting load conditions for the shared queue. Let *VSQ logical queue* and *SQ logical queue* denote the portion of the VSQ and SQ, respectively, containing cells with a given address. Reference will be made to a tagged VSQ logical queue and to a tagged SQ logical queue. By means of arguments similar to those developed in the analysis of an IOQ switch (see Section 8.1.2), the tagged VSQ logical queue can be regarded as fed by a Poisson process, so that it behaves as an M/D/1 queue. Since the input queue HOL packet finding an empty tagged SQ logical queue is directly transmitted to the output line, it follows that the tagged VSQ logical queue behaves as an M/D/1 queue with external server (see Appendix). The saturation load ρ_{smax} for each SQ logical queue is then given when its content in the VSQ logical queue equals the normalized shared queue capacity B_s, that is

$$\frac{\rho_{smax}^2}{2(1 - \rho_{smax})} = B_s \tag{8.25}$$

which gives

$$\rho_{smax} = -B_s + \sqrt{B_s^2 + 2B_s}$$

It follows that for loads below ρ_{smax} the waiting time in the virtual queue is null (the HOL packet is either transmitted directly to the output port, or enters the non-saturated shared queue). For loads above ρ_{smax} this waiting time can be computed by Little's formula once we know the average number of packets with the same address waiting for service in the virtual queue, that is $E[W_i] - B_s$ where W_i is the number of cells waiting for service in the tagged VSQ logical queue[Ili91]. So we obtain

$$E[\eta_v] = \begin{cases} 0 & (\rho \le \rho_{smax}) \\ \dfrac{E[W_i] - B_s}{p_i} = \dfrac{1}{p_i}\left(\dfrac{p_i^2}{2(1-p_i)} - B_s\right) & (\rho > \rho_{smax}) \end{cases} \tag{8.26}$$

It is worth noting that ρ_{smax} does not represent the switch capacity ρ_{max} since not all packets cross the shared queue to be switched. As with the IOQ switch, the switch capacity is given by the limiting load ρ_{imax} of the input queue. Owing to the random traffic assumption, the arrival process to the tagged input queue is Bernoulli. Therefore, the tagged input queue can be modelled as $Geom/G/1$ whose average waiting time can be expressed as a function of the first two moments of average waiting time in the virtual queue by means of Equation 8.3. Therefore the saturation load at input queues ρ_{imax} is computed as the root of the function $F(p_i) = 1 - p_i - pE[\eta_v]$ with $p_i = p$. Since $F(p_i) > 0$ for each $p < \rho_{imax}$, using Equations 8.25–8.26 we obtain

$$F(\rho_{smax}) = 1 - \rho_{smax} - \rho_{smax}\frac{E[W_i] - B_s}{\rho_{smax}} = 1 - \rho_{smax} > 0$$

It follows that, according to our expectation, $\rho_{imax} > \rho_{smax}$. The switch capacity ρ_{max}, which equals the saturation load of the tagged input queue ρ_{imax}, is then computed as the root of

$$F(\rho_{imax}) = 1 - \rho_{imax} - (E[W_i] - B_s)$$

which by means of Equation 8.25 is given by

$$\rho_{imax} = B_s + 2 - \sqrt{B_s^2 + 2B_s + 2}$$

Table 8.1 compares the two saturation loads ρ_{imax} and ρ_{smax} in the input and shared queue, respectively, for some values of the normalized shared queue capacity. The limiting case of $B_s = 0$ corresponds to pure input queueing and hence the well-known value $\rho_{max} = 0.586$ is obtained. When the normalized shared queue has size $B_s = 1$, the shared queue gives a capacity of 0.732 with the HOL positions of the input queues providing a throughput increase of about 5%. As the shared queue capacity is increased, the throughput gain given by the direct transmission of the virtual queue to the output ports becomes negligible.

It is worth noting that B_s need not be integer, as it simply represents the ratio between the total shared queue size NB_s and the switch size N. The switch capacity ρ_{max} for the normalized switch capacity ranging in the interval [0–32] is shown in Figure 8.39. It is observed that most of the capacity increase compared to pure input queueing is obtained for small values of B_s. Nevertheless, this does not imply that the total shared queue is small, as it is proportional to the switch size N.

We are now able to evaluate the average packet delay, T, given by Equation 8.24. As already discussed, we have to distinguish between two load regions, since the shared queue saturation completely modifies the switch operating conditions. In underload conditions for the shared

Table 8.1. Saturation loads for different shared queue capacities

B_s	ρ_{smax}	ρ_{imax}
0	0.000	0.586
1	0.732	0.764
2	0.828	0.838
4	0.899	0.901
8	0.944	0.945
16	0.971	0.971
∞	1.000	1.000

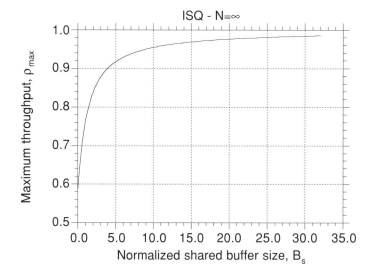

Figure 8.39. Switch capacity of a BP ISQ switch

queue $(p \leq \rho_{smax})$, the waiting time in the virtual queue is always null and so is the waiting time in the input queue due to the Bernoulli arrival process. Therefore the only waiting time is experienced in the tagged VSQ logical queue, which behaves as an $M/D/1$ queue. Therefore Equation 8.24 becomes

$$T = E[\eta_v] + E[\eta_s] + 1 = \frac{p}{2(1-p)} + 1 \qquad (\rho \leq \rho_{smax}) \qquad (8.27)$$

In the overload region for the shared queue $(p > \rho_{smax})$, packets can be backpressured into the HOL position if the addressed output port is busy and the shared buffer is full. Therefore delay is also experienced in the input queue in addition to the delay experienced in the tagged VSQ logical queue. Thus, by using Equation 8.3 the average delay is given by

$$T = E[\eta_i] + E[\eta_v] + E[\eta_s] + 1 = \frac{p(E[\eta_v^2] + E[\eta_v])}{2(1 - p(E[\eta_v] + 1))} + \frac{p}{2(1 - p)} + 1 \qquad (8.28)$$

$$(\rho > \rho_{smax})$$

where the first moment of the waiting time in the virtual queue is given by Equation 8.26. Since the second moment is rather difficult to evaluate, an approximation is used now that assumes $E[\eta_v^2] = E^2[\eta_v]$. The resulting average delay as a function of the offered load p, as given by Equations 8.27–8.28, is shown in Figure 8.40 for some values of the normalized shared queue capacity. It is seen that as long as the shared queue does not saturate the delay performance is optimal (it is that of an $M/D/1$ queue), whereas it worsens sharply when saturation occurs. The delay curve for loads above ρ_s is steeper as B_s increases: in fact ρ_s becomes closer to ρ_i as the shared queue capacity grows.

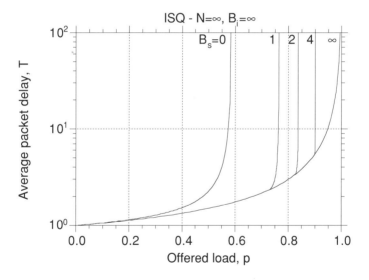

Figure 8.40. Delay performance of a BP ISQ switch

8.4. Comparison of Switch Capacities in Non-blocking Switches

We would like to report here some comparative evaluations about the capacity of a non-blocking switch adopting input queueing, either alone or coupled with another queueing strategy. As already discussed, all these architectures have a capacity $\rho_{max} < 1$, which is due to the head-of-line blocking occurring in the input queues. For these architectures Table 8.2 provides the switch capacity values for different queueing arrangements and internal operations, that is pure input queueing (IQ), combined input–output queueing (IOQ) or combined input-shared queueing (ISQ). All the values contained therein assume $B_i = \infty$ and, unless stated otherwise, $N = \infty$. Note that column A only contains the numerical values of the independent variable for which the switch capacity is evaluated and its meaning is specified in each of the following columns. All the values of columns C, F, G, H, I for a switch with finite size are given by computer simulation, all the others by the analytical models developed in the relevant section.

Table 8.2. Switch capacity for various non-blocking configurations

A	B	C	D	E	F	G	H	I	J	K
		IQ				IOQ				ISQ
	expan.	wind.	group.			BP		QL		BP
	$M>N$			$K = B_C$	$K = 2$	$K = 4$	$K = 2$	$K = 4$	$B_O = \infty$	
	$\rho_m\left(\dfrac{M}{N}\right)$	$\rho_m(W)$	$\rho_m(R)$	$\rho_m(B_o)$	$\rho_m(B_o)$	$\rho_m(B_o)$	$\rho_m(B_o)$	$\rho_m(B_o)$	$\rho_m(K)$	$\rho_m(B_i)$
1	0.586	0.587	0.586	0.586	0.586	0.586	0.623	0.633	0.586	0.586
2	0.764	0.697	0.686	0.671	0.671	0.671	0.754	0.785	0.885	0.764
4	0.877	0.794	0.768	0.758	0.748	0.758	0.830	0.883	0.996	0.877
8	0.938	0.875	0.831	0.832	0.814	0.831	0.869	0.938	1.000	0.938
16	0.969	0.929	0.878	0.889	0.862	0.889	0.882	0.967		0.969
32	0.984	0.953	0.912	0.930	0.883	0.929	0.885	0.982		0.984
64	0.992	0.957	0.937		0.885	0.957	0.885	0.990		0.992
128			0.955			0.976		0.994		
256			0.968			0.987		0.996		
∞	1.000	1.000	1.000	1.000	0.885	0.996	0.885	0.996	1.000	1.000

The starting point of this throughput comparison is that the basic input queueing switch with infinite size is characterized by $\rho_{max} = 0.58$. We have seen in Section 7.1 that the HOL blocking can be reduced in different ways: adopting a switch *expansion ratio*, a *windowing* technique or a *channel grouping* technique. In the former case the number of switch outlets M is larger than the number of switch inlets N, so as to reduce the number of conflicts for each switch outlet, as shown in Column B. An expansion ratio $M/N = 8$ is enough to increase the switch capacity above 0.9. Windowing relieves the HOL blocking by allowing also non-HOL cells to contend for the switch outlets. The throughput results for window depths ranging up to 64 are given in Column C for a switch with $N = 256$. A window $W = 4$ gives a switch capacity close to 0.8. With channel grouping the switch outlets are subdivided into groups of size R each and packets now address groups, not single links of a group. A switch capacity above 0.9 is provided by a group size $R = 32$ as is shown in column D. Channel grouping is in general simpler to implement than windowing for the same numerical value $R = W$ (implementing a group size $R = 32$ is relatively simple, whereas a window $W = 4$ represents a significantly difficult task). Apparently a switch capacity $\rho_{max} = 1$ is given for the examined technique when the expansion ratio M/N, or the window size W, or the channel group size R becomes infinite.

The switch capacity of an IOQ switch with backpressure and speed-up equal to the output buffer size $(K = B_o)$ increases as K grows (see Column E), since the HOL blocking phenomenon is now relieved: for example $\rho_{max} = 0.93$ for $K = B_o = 32$. It is interesting to note how this performance is still significantly limited by the limited output buffer capacity available. Columns F-I give the switch capacity as a function of the output queue size B_o for the two speed-up values $K = 2, 4$ and for the two internal protocols BP and QL. Even with such small values of the speed-up, the switch capacity grows remarkably. The BP protocol gives in general a smaller switch capacity than the QL, for the same reason for which an IQ switch has a smaller capacity than a crossbar switch. Apparently the two sets of values converge when the output queues do not saturate any more, which occurs for an output queue size $B_o = 64\,(256)$ with speed-up $K = 2\,(4)$. These limiting values are given in Column J: the switch capacity grows very fast with the speed-up value K for an infinitely large output queue, since $\rho_{max} = 0.996$ for $K = 4$. Note that in general it is cheaper to provide larger output queues than increasing the switch speed-up, unless the switching modules are provided in VLSI chips that would prevent memory expansions. Thus, these results suggest that a prescribed switch capacity can be obtained acting on the output buffer size, given a speed-up compatible with the selected architecture and the current technology.

Having a shared buffer of size NB_s rather than separate output queues with capacity $B_o = B_t$ (recall that the total downstream normalized buffer capacity in an ISQ switch is $B_t = B_s + 1$) provides a higher switch capacity as shown by the comparison of Columns E and K. Note that in both cases the (internal) speed-up $(K_i)\,K$ fully exploits the available buffer capacity $(NB_s)\,B_o$. In both configurations $B_o = B_s = 0$ means pure input queueing and $B_o = B_s = \infty$ corresponds to a configuration free from HOL blocking. Interestingly enough, the capacity of an ISQ switch with normalized buffer size per output port B_t is numerically equal to that provided by an $N \times M$ IQ switch with $M/N = B_t$.

8.5. Additional Remarks

The traffic performance of non-blocking ATM switching architectures has been evaluated by relying on the random and uniform traffic assumption More realistic evaluation can be obtained by taking into account the features of correlation and non-uniformity in the offered traffic pattern. Performance analyses of non-blocking switches with combined input–output queueing have been reported in [Li92] for correlated traffic patterns and in [Che91], [Li91] for non-uniform traffic patterns.

8.6. References

[Che91]   J.S.-C. Chen, T.E. Stern, "Throughput analysis, optimal buffer allocation, and traffic imbal-
          ance study of a generic nonblocking packet switch", *IEEE J. on Selected Areas in Commun.*,
          Vol. 9, No. 3, Apr. 1991, pp. 439-449.

[Den92]   W.E. Denzel, A.P.J. Engbersen, I. Iliadis, G. Karlsson, "A highly modular packet switch for
          Gb/s rates", *Proc. of Int. Switching Symp.*, Yokohama, Japan, Oct. 1992, pp. 236-240.

[Gia91]   J.N. Giacopelli, J.J. Hickey, W.S. Marcus, W.D. Sincoskie, M. Littlewood, "Sunshine: a high
          performance self-routing broadband packet switch architecture", *IEEE J. on Selected Areas in
          Commun.*, Vol. 9, No. 8, Oct. 1991, pp. 1289-1298.

[Hic90]   J.J. Hickey, W.S. Marcus, "The implementation of a high speed ATM packet switch using
          CMOS VLSI", *Proc. of Int. Switching Symp.*, Stockholm, May-June 1990, Vol. I, pp. 75-83.

[Hui87]   J. Hui, E. Arthurs, "A broadband packet switch for integrated transport", *IEEE J. on Selected
          Areas in Commun.*, Vol. SAC-5, No. 8, Oct. 1987, pp. 1264-1273.

[Ili90]   I. Iliadis, W.E. Denzel, "Performance of packet switches with input and output queueing",
          Proc. of ICC 90, Atlanta, GA, Apr. 1990, pp. 747-753.

[Ili91]   I. Iliadis, "Performance of packet switch with shared buffer and input queueing", *Proc. of
          13th Int. Teletraffic Congress*, Copenhagen, Denmark, June 1991, pp. 911-916.

[Kar87]   M.J. Karol, M.G. Hluchyj, S.P. Morgan, "Input versus output queueing on a space-division
          packet switch", *IEEE Trans. on Commun.*, Vol. COM-35, No. 12, Dec. 1987, pp. 1347-
          1356.

[Li91]   S.-Q. Li, "Performance of trunk grouping in packet switch design", *Performance Evaluation*,
          Vol. 12, 1991, pp. 207-218.

[Li92]   S.-Q. Li, "Performance of a nonblocking space-division packet switch with correlated input
          traffic", *IEEE Trans. on Commun.*, Vol. 40, No. 1, Jan. 1992, pp. 97-108.

[Mei58]   T. Meisling, "Discrete-time queueing theory", *Operations Research*, Vol. 6, Jan.-Feb. 1958,
          pp. 96-105.

[Oie90]   Y. Oie, T. Suda, M. Murata, D. Kolson, H. Miyahara, "Survey of switching techniques in
          high-speed networks and their performance", *Proc. of INFOCOM 90*, San Francisco, CA,
          June 1990, pp. 1242-1251.

[Pat90]   A. Pattavina, "Design and performance evaluation of a packet for broadband central
          offices", *Proc. of INFOCOM 90*, San Francisco, CA, June 1990, pp. 1252-1259.

[Pat91]   A. Pattavina, "An ATM switch architecture for provision of integrated broadband services",
          IEEE J. on Selected Areas in Commun., Vol. 9, No. 9, Dec. 1991, pp. 1537-1548.

[Pat93]   A. Pattavina, G. Bruzzi, "Analysis of input and output queueing for non-blocking ATM
          switches", *IEEE/ACM Trans. on Networking*, Vol. 1, No. 3, June 1993, pp. 314-328.

8.7. Problems

8.1   Derive Equations 8.9 from Equations 8.8.

8.2   Provide the expression of the capacity for a non-blocking $N \times N$ switch with combined input–output queueing, infinite buffer sizes, speed-up factor K assuming that the packets losing the contention for the addressed outlets are discarded.

8.3   Plot the capacity value found in Problem 8.2 as a function of N for $K = 2, 3, 4$ with values up to $N = 1024$ (N is a power of 2) and compare these values with those given by computer simulation in which losing packets are not discarded; justify the difference between the two curves.

8.4   Repeat Problem 8.2 for $N = \infty$. Compare the switch capacity for $K = 2, 3, 4$ so obtained with the values given in Table 8.2, justifying the differences.

8.5   Express the average packet delay of a non-blocking switch with combined input–output queueing, infinite switch size and input buffer size, speed-up factor K assuming that the tagged input queue and the tagged output queue are mutually independent. Use the appropriate queueing models and corresponding results described in Appendix by assuming also $E[\eta_v^2] = E^2[\eta_v]$.

8.6   Plot the average delay expressed in Problem 8.5 as a function of the offered load p for the output buffer sizes $B_o = 2, 4, 8, 16$ and compare these results with those given in Figure 8.21, justifying the difference.

8.7   Express the range of capacity values of a switch with combined input–output queueing without internal backpressure with $B_o = 1$ and increasing values of the speed-up factor K. Justify the values relevant to the two extreme values $K = 1$ and $K = \infty$.

8.8   Assuming that an infinite capacity is available for the shared queue of a switch with combined shared-output queueing and parameters K, B_o, find the queueing model that fully describes the behavior of the switch.

8.9   Explain how the switch capacity changes in a non-blocking architecture with combined input-shared queueing if the packets must always cross the shared queue independent of its content.

Chapter 9 *ATM Switching with Arbitrary-Depth Blocking Networks*

We have seen in Chapter 6 how an ATM switch can be built using an interconnection network with "minimum" depth, in which all packets cross the minimum number of self-routing stages that guarantees the network full accessibility, so as to reach the addresses switch outlet. It has been shown that queueing, suitable placed inside or outside the interconnection network, allows the traffic performance typical of an ATM switch. The class of ATM switching fabric described in this Chapter is based on the use of very simple unbuffered switching elements (SEs) in a network configuration conceptually different from the previous one related to the use of banyan networks. The basic idea behind this new class of switching fabrics is that packet loss events that would occur owing to multiple packets requiring the same interstage links are avoided by deflecting packets onto unrequested output links of the switching element. Therefore, the packet loss performance is controlled by providing several paths between any inlet and outlet of the switch, which is generally accomplished by arranging a given number of self-routing stages cascaded one to other. Therefore here the interconnection network is said to have an "arbitrary" depth since the number of stages crossed by packets is variable and depends on the deflections occurred to the packets. Now the interconnection network is able to switch more than one packet per slot to a given switch output interface, so that queueing is mandatory on the switch outputs, since at most one packet per slot can be transmitted to each switch outlet.

As in the previous chapters, a switch architecture of size $N \times N$ will be considered with the notation $n = \log_2 N$. Nevertheless, unlike architectures based on banyan networks, now n no longer represents the number of network stages, which will be represented by the symbol K.

The basic switch architectures adopting the concept of deflection routing are described in Section 9.1, whereas structures using simpler SEs are discussed in Section 9.2. Additional functionalities of the interconnection network that enhance the overall architectures are presented in Section 9.3. The traffic performance of the interconnection network for all these structures is studied in Section 9.4 by developing analytical modes whenever possible. The network performances of the different switches are also compared and the overall switch performance is discussed. The use of multiple switching planes is examined in Section 9.5. Additional remarks concerning deflection routing switches are finally given in Section 9.6.

9.1. Switch Architectures Based on Deflection Routing

The general model of an ATM switch based on deflection routing is first described and then such a model is mapped onto the specific switch architectures. These architectures basically differ in the structure of the interconnection routing and in its routing algorithm.

All the switch architectures that will be considered here share several common features: the interconnection network is internally unbuffered and provides several paths between any inlet and outlet of the switch. Each switch outlet interface is equipped with a queue that is able to receive multiple packets per slot, whereas only one packet per slot is transmitted by the queue. The general model of an $N \times N$ switch architecture, shown in Figure 9.1, includes two basic subsystems: the *interconnection network* and the set of the N *output queues*. In the interconnection network K_b switching blocks of size $N \times 2N$ are cascaded one to the other by means of proper interblock connection patterns. The expansion factor in each block is due to the direct connection of each block to each of the N output queue interfaces. Therefore, unlike all other ATM switch classes, cells cross a variable number of blocks before entering the addressed output queue. The N outlets of each switching block to the output queues are referred to as *local outlets*, whereas the other N are called *interstage outlets*.

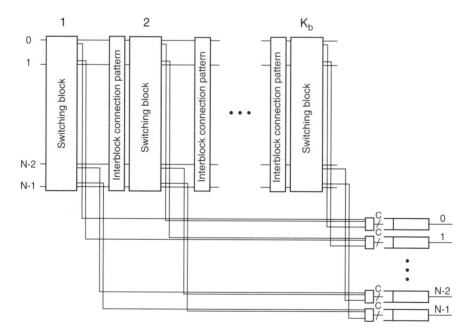

Figure 9.1. General model of ATM switch architecture based on deflection routing

The configuration of the switching block, which is a memoryless structure composed of very simple switching elements arranged into one or more stages, and of the interblock pattern depends on the specific structure of the ATM switch. In general we say that the interconnection network includes K switching stages with $K = n_s K_b$, n_s being the number of SE stages per switching block. Also the routing strategy operated by the switching block depends on the specific architecture. However, the general switching rule of this kind of architectures is to route as early as possible the packets onto the local outlet addressed by the cell. Apparently, those cells that do not reach this outlet at the last switching block are lost. As for single–path banyan networks, interconnection networks based on deflection routing can be referred to as self-routing, since each cell carries all the information needed for its switching in a *destination tag* that precedes the ATM cell. However now, depending on the specific network architecture, the packet self-routing requires the processing of more than one bit; in some cases the whole cell address must be processed in order to determine the path through the network for the cell.

Each output queue, which operates on a FIFO basis, is fed by K_b lines, one from each block, so that up to K_b packets can be concurrently received in each slot. Since K_b can range up to some tens depending on the network parameter and performance target, it can be necessary to limit the maximum number of packets entering the queue in the same slot. Therefore a *concentrator* with size $K_b \times C$ is generally equipped in each output queue interface so that up to C packets can enter the queue concurrently. The number of outputs C from the concentrator and the output queue size B (cells) will be properly engineered so as to provide a given traffic performance target.

The model of a deflection network depicted in Figure 9.1 is just a generalization of the basic functionalities performed by ATM switches based on deflection routing. Nevertheless, other schemes could be devised as well. For example the wiring between all the switch blocks and the output queue could be removed by having interstage blocks of size $N \times N$ operating at a speed that increases with the block index so that the last block is capable of transmitting K_b packets in a slot time to each output queue. This particular solution with internal speed-up is just a specific implementation that is likely to be much more expensive than the solution based on earlier exits from the interconnection network adopted here.

9.1.1. The Shuffleout switch

The *Shuffleout switch* will be described here in its Open-Loop architecture [Dec91a], which fits in the general model of Figure 9.1. A switching block in the Shuffleout switch is just a switching stage including $N/2$ switching elements of size 2×4 and the interblock connection pattern is just an interstage connection pattern. Therefore the general scheme of Figure 9.1 simplifies into the scheme of deflection routing architecture of Figure 9.2.

The network thus includes K stages of SEs arranged in $N/2$ rows of SEs, numbered 0 through $N/2 - 1$, each including K SEs. An SE is connected to the previous stage by its two *inlets* and to the next stage by its two *interstage outlets*; all the SEs in row i $0 \le i \le N/2 - 1$ have access to the output queues interfacing the network outlets $2i$ and $2i + 1$, by means of the *local outlets*. The destination tag in the Shuffleout switch is just the network output address. More specifically, the interstage connection pattern is the shuffle pattern for all the stages, so that the interconnection network becomes a continuous interleaving of switching stages and

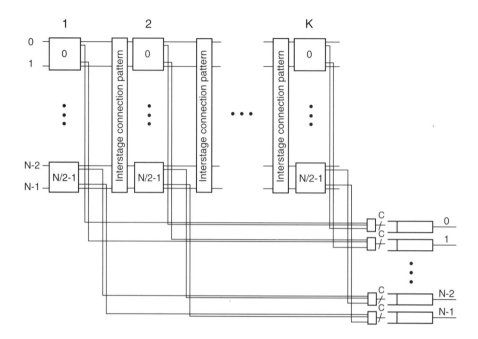

Figure 9.2. Model of deflection routing ATM switch with single-stage switching block

shuffle patterns, that is a Shuffle-exchange network. An example of 16×16 interconnection network for the Shuffleout switch with $K = 8$ stages is shown in Figure 9.3, where the local outlets have been omitted for the sake of readability.

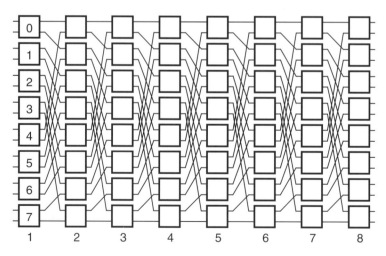

Figure 9.3. Shuffleout interconnection network

The distributed routing algorithm adopted in the interconnection network is jointly based on the *shortest path* and *deflection routing* principles. Therefore an SE attempts to route the received cells along its outlets belonging to the minimum I/O path length to the required destination. The *output distance d* of a cell from the switching element it is crossing to the required outlet is defined as the minimum number of downstream stages to be crossed by the cell in order to enter an SE interfacing the addressed output queue. After reading the cell output address, the SE can compute very easily the cell output distance whose value ranges from 0 to $\log_2 N - 1$ owing to the shuffle interstage pattern. A cell requires a local outlet if $d = 0$, whereas it is said to require a *remote outlet* if $d > 0$.

In fact consider an $N \times N$ network with inlets and outlets numbered 0 through $N - 1$ and SEs numbered 0 through $N/2 - 1$ in each stage (see Figure 9.3 for $N = 16$). The inlets and outlets of a generic switching element of stage k with index $x_{n-1}x_{n-2}...x_1$ ($n = \log_2 N$) have addresses $x_{n-1}x_{n-2}...x_1 0$ and $x_{n-1}x_{n-2}...x_1 1$. Owing to the interstage shuffle connection pattern, outlet $x_{n-1}x_{n-2}...x_1 x_0$ of stage k is connected to inlet $x_{n-2}x_{n-3}...x_0 x_{n-1}$ in stage $k + 1$, which also means that SE $x_{n-1}x_{n-2}...x_1$ of stage k is connected to SEs $x_{n-2}x_{n-3}...x_1 0$ and $x_{n-2}x_{n-3}...x_1 1$ in stage $k + 1$. Thus a cell received on inlet $x_{n-1}x_{n-2}...x_1 y$ ($y = 0, 1$) is at output distance $d = 1$ from the network outlets $x_{n-2}x_{n-3}...x_1 yz$ ($z = 0, 1$). It follows that the SE determines the cell output distance to be $d = k$, if k cyclic left-rotations of its own address are necessary to obtain an equality between the $n - k - 1$ most significant bits of the rotated address and the $n - k - 1$ most significant bits of the cell address. In order to route the cell along its shortest path to the addressed network outlet, which requires to cross k more stages, the SE selects for the cell its interstage outlet whose address $x_{n-1}x_{n-2}...x_1 y$ ($y = 0, 1$) after k cyclic left-rotations has the $n - k$ most significant bits equal to the same bits of the cell network outlet. Therefore, the whole output address of the cell must be processed in the Shuffleout switch to determine the routing stage by stage.

When two cells require the same SE outlet (either local or interstage), only one can be correctly switched, while the other must be transmitted to a non-requested interstage outlet, due to the memoryless structure of the SE. Conflicts are thus resolved by the SE applying the deflection routing principle: if the conflicting cells have different output distances, the closest one is routed to its required outlet, while the other is deflected to the other interstage link. If the cells have the same output distance, a random choice is carried out. If the conflict occurs for a local outlet, the loser packet is deflected onto an interstage outlet that is randomly selected.

An example of packet routing is shown in Figure 9.4 for $N = 8$. In the first stage the SEs 2 and 3 receive two cells requiring the remote switch outlets 0 and 2, so that a conflict occurs in the latter SE for the its top interstage link. The two cells in SE 2 are routed without conflict so that they can enter the addressed output queue at stage 2. The two contending cells in SE 3 have the same distance $d = 2$ and the random winner selection results in the deflection of the cell received on the bottom inlet, which restarts its routing from stage 2. Therefore this cell enters the output queue at stage 4, whereas the winner cell enters the queue at stage 3.

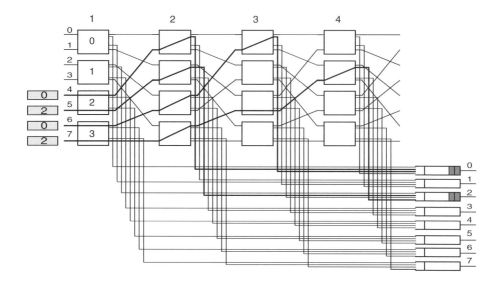

Figure 9.4. Routing example in Shuffleout

9.1.2. The Shuffle Self-Routing switch

The interconnection network of the *Shuffle Self-Routing switch* [Zar93a] is topologically identical to that of the Shuffleout switch; therefore the switching block is a stage of $N/2$ switching elements with size 2×4 and the interblock pattern is the shuffle pattern. Therefore, as in the previous switch, the network includes K stages of SEs arranged in $N/2$ rows of SEs, so that the same network model of Figure 9.2 applies here with an example of 16×16 interconnection network with $K = 8$ also given by Figure 9.3 with the local outlets omitted.

The difference between Shuffle Self-Routing and Shuffleout switch lies in the different mode of operating the packet self-routing within the SE. In the former switch the cell routing is not based on the previous concept of minimum distance, rather on the classical bit-by-bit self-routing typical of banyan networks. However, unlike a single-path Omega network including only $\log_2 N$ stages of SEs with interstage shuffle patterns, now the cells can be deflected from their requested path owing to conflicts. For example, a cell entering the network reaches the requested row, i.e. the row interfacing the addressed output queue, at stage $n = \log_2 N$ in absence of conflicts. Should a conflict occur at stage i ($i < n$) and the tagged packet is the loser, the routing of this packet can start again from stage $i + 1$ and last until stage $i + n$ since any set of n adjacent stages represent a banyan network with Omega topology (see Section 2.3.1.1). Apparently what is needed now in the cell is an indication of the distance of a cell from the addressed network outlet, that is the remaining number of stages to be crossed by the tagged cell from its current location. Unlike Shuffleout where the output distance (ranging up to $n - 1$) indicating the downstream stages to be crossed is computed by the SE, now the output distance d is carried by the cell.

The destination tag of the cell thus includes two fields: the addressed network outlet $o = o_{n-1}o_{n-2}\cdots o_0$ and the output distance d $(0 \leq d \leq n-1)$. The initial value of the output distance of a cell entering the network is $n-1$. If the cell can be switched without conflicts, the cell is routed. If the distance is $d > 0$, the SE routes the cell onto the top SE interstage outlet if $o_d = 0$, onto the bottom SE interstage outlet if $o_d = 1$ by also decreasing the distance by one unit. If $d = 0$, the SE routes the cell onto the local top (bottom) outlet if $o_0 = 0$ $(o_0 = 1)$. Note that this routing rule is exactly the same that would be applied in an Omega network, which includes n stages each preceded by a shuffle pattern (see Section 2.3.1.2). In fact crossing n adjacent stages of the Shuffle Self-Routing switch without deflections is equivalent to crossing an Omega network, if we disregard the shuffle pattern preceding the first stage in this latter network. Nevertheless, removing this initial shuffle permutation in the Omega network does not affect its routing rule, as it simply corresponds to offering the set of cells in a different order to the SEs of the first stage.

The rules to be applied for selecting the loser cell in case of a conflict for the same interstage or local SE outlet are the same as in the Shuffleout switch. The cell distance of the deflected cell is reset to $n-1$, so that it starts again the switching through n stages. It follows that the interconnection network can be simplified compared to the architecture shown in Figure 9.2, since the local outlets are not needed in the first $n-1$ stages, as the cells must cross at least n stages. The advantage is not in the simpler 2×2 SEs that could be used in the first $n-1$ stages, rather in the smaller number of links entering each output queue interface, that is $K - (n-1)$.

With this architecture only one bit of the outlet address needs to be processed in addition to the distance field. Nevertheless, owing to the occurrence of deflections, it is not possible to foresee a technique for routing the packet by delaying the cell until the first bit of the outlet address is received by the SE. In fact the one-bit address rotation that would make the address bit to be processed the first to be received does not work in presence of a deflection that requires a restoration of the original address configuration.

The routing example in a 8×8 network already discussed for Shuffleout is reported in Figure 9.5 for the Shuffle Self-Routing switch, where the SEs are equipped with local outlets only starting from stage $n = \log_2 N = 3$. Now only one cell addressing outlet 0 and outlet 2 can reach the output queue, since at least three stages must now be crossed by all the cells.

9.1.3. The Rerouting switch

In the Rerouting switch [Uru91] the switching block is again given by a column of 2×4 switching elements, so that the general $N \times N$ switch architecture of Figure 9.2 applies here too. Nevertheless, unlike the previous switches, the interstage pattern here varies according to the stage index. In particular the interstage patterns are such that the subnetwork including n adjacent stages $(n = \log_2 N)$ starting from stage $1 + k(n-1)$ (k integer) has the topology of a banyan network. If the network includes exactly $K = 1 + k(n-1)$ stages, the whole interconnection network looks like the cascading of k reverse SW-banyan networks (see Section 2.3.1.1) with the last stage and first stage of adjacent networks merged together. Since in general the network can include an arbitrary number of stages $1 + k(n-1) + x$ $(0 < x < n-1)$ the subnetwork including the last $x + 1$ switching stages has the topology of

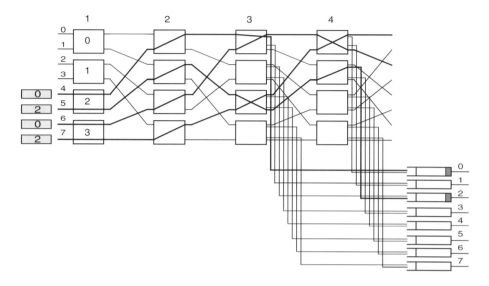

Figure 9.5. Routing example in Shuffle Self-Routing

the first $x + 1$ stages of a reverse SW-banyan network. An example of 16×16 interconnection network with $K = 8$ stages is shown in Figure 9.6.

As for the Shuffle Self-Routing switch, the destination tag must include two fields: the addressed network outlet $o = o_{n-1}o_{n-2}\cdots o_0$ and the output distance d $(0 \le d \le n-1)$. The distance is defined again as the number of stages to be crossed before entering the addressed local SE outlet. Unlike the previous switches with only shuffle interstage patterns, the routing rule is now a little more complex since the topology of subnetwork including n stages varies according to the index of the first stage. The cell distance is initially set to $n-1$ as the cell enters the network and is decreased by one after each routing operation without deflection. If the received distance value is $d > 0$, the SE at stage k routes the cell onto the top (bottom) outlet if $o_j = 0$ $(o_j = 1)$ with $j = n-1-(k-1) \bmod (n-1)$. If $d = 0$, the SE routes the cell onto the top or bottom local outlet if $o_0 = 0$ or $o_0 = 1$, respectively. Upon a deflection, the cell distance is reset to the value $n-1$, so that at least n more stages must be crossed before entering the addressed output queue. Again the cell routing requires the processing of only one bit of the destination address, in addition to the distance field. However, as in the previous architecture, the whole destination address needs to be received by the SE before determining the cell routing.

As with the Shuffle Self-Routing switch, also in this case the number of links entering each output queue interface reduces from the original value of K of the general model of Figure 9.2 to $K - (n-1)$ since all cells must cross at least n stages.

An interesting observation arises concerning the particular network topology of the rerouting switch compared to the shuffle-based switch. When a packet is deflected in Shuffleout, which is based on the shortest-path distance routing, it gets a new distance from the addressed outlet which depends on the SE address and on the specific link on which deflection takes

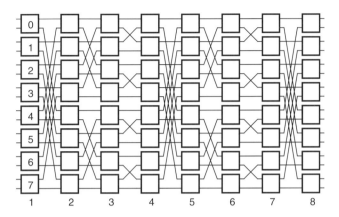

Figure 9.6. Rerouting interconnection network

place. Consider for example the case of a packet switched by the SE 0 $(N/2-1)$ that is deflected onto the top (bottom) outlet: in this case it keeps the old distance. In any other case the distance increases up to the value n. Even if these observations apply also to the Shuffle Self-Routing switch, the routing based on a single bit status always requires resetting of the distance to n. The network topology of the basic banyan network of the Rerouting switch is such that the two outlets of the first stage SEs each accesses a different $N/2 \times N/2$ network, the two outlets of the second stage SEs each accesses a different $N/4 \times N/4$ network and so on. It follows that a deflected cell always finds itself at a distance n to the addressed network outlet (this is true also if the deflection occurs at a stage different from the first of each basic banyan n-stage topology). Therefore the new path of a deflected cell always coincides with the shortest path to the addressed destination only with the Rerouting switch.

The same switching example examined for the two previous architectures is shown in Figure 9.7 for the Rerouting switch. It is to be observed that both cells losing the contention at stage 1 restart their routing at stage 2. Unlike the previous cases where the routing restarts from the most significant bit, if no other contentions occur (this is the case of the cell entering the switch on inlet 7) the bit o_1, o_2, and o_0 determine the routing at stage 2, 3 and 4, respectively.

9.1.4. The Dual Shuffle switch

The switch architecture model that describes the Dual Shuffle switch [Lie94] is the most general one shown in Figure 9.1. However, in order to describe its specific architecture, we need to describe first its building blocks, by initially disregarding the presence of the local outlets. An $N \times N$ Dual Shuffle switch includes two networks, each with K switching stages: an $N \times N$ *shuffle network* (SN) and an $N \times N$ *unshuffle network* (USN): the USN differs from the SN in that a shuffle (unshuffle) pattern always precedes (follows) a switching stage in SN (USN). Therefore each set of $n = \log_2 N$ adjacent stages in SN (USN) including the permutation that precedes (follows) the first (last) stage can be seen as an Omega (reverse Omega) network in which the routing rules described in Section 2.3.1.2 can be applied. The two net-

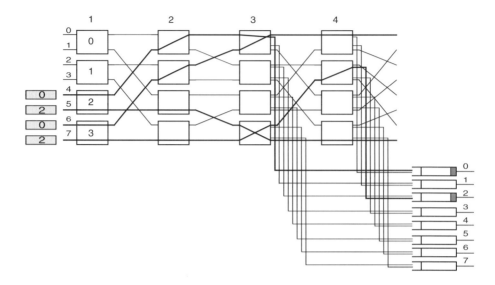

Figure 9.7. Routing example in Rerouting

works are arranged side-by-side so that packets to be switched by an SE of a network can be transmitted also to the outlets of the homologous switching element of the other network. The general structure of this network for $N = 16$ and $K = 8$ is shown in Figure 9.8, where the local outlets, as well as the last unshuffle pattern in the USN, are omitted for simplicity. It will be shown later that such a last unshuffle pattern is not needed by properly modifying the self-routing rule in the USN. The shuffle pattern preceding the first stage in SN has been removed as in the other shuffle-based switches (Shuffleout and Shuffle Self-Routing), as it is useless from the routing standpoint.

Packets are switched stage by stage on each network by reading a specific bit of the self-routing tag. When a deflection occurs, the packet is routed for one stage through the other network, so that the distance of the packet from the addressed local outlets in general increases only by two, without needing to be reset at $n - 1$, as in the Shuffle Self-Routing switch. In order to better understand such a routing procedure, let us consider the $2N \times 2N$ network resulting from the structure of Figure 9.8 when the two 2×2 *core* SEs (the local outlets are disregarded) with the same index in the two networks can be seen as a single 4×4 *core* SE. The two top (bottom) SE outlets, that is 00 and 01 (10 and 11) originate the unshuffle (shuffle) links, whereas the two top (bottom) SE inlets terminate the unshuffle (shuffle) links. It is convenient to label the interstage unshuffle and shuffle links as $(1a,0b)$ and $(0a,1b)$, respectively to better describe the routing procedure in case of deflections. Based on the topological properties of the shuffle and unshuffle connections the SE $a = a_{n-1}...a_1$ is connected to the SE $b = b_{n-1}...b_1$ of the following stage by a shuffle (unshuffle) link if

$$a_{n-2}...a_1 = b_{n-1}...b_2 \quad (a_{n-1}...a_2 = b_{n-2}...b_1)$$

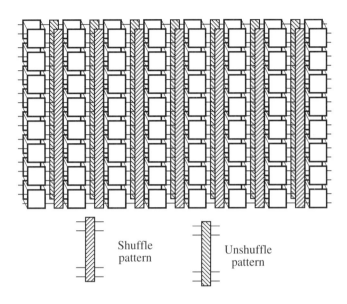

Figure 9.8. Interconnection network of Dual Shuffle

It is easy to verify that the shuffle and unshuffle links are labelled $(1b_1, 0a_{n-1})$ and $(0b_{n-1}, 1a_1)$, respectively. Such network is shown in Figure 9.9 for $N = 8$ and $K = 5$ (for the sake of readability the local outlets have been omitted).

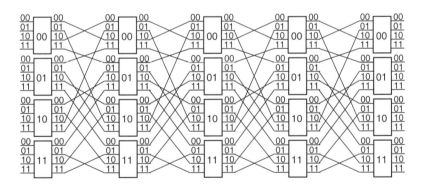

Figure 9.9. Interconnection network of Dual Shuffle with 4 × 4 core SEs

As in the two previous architectures based on the bit-by-bit packet self-routing, that is the Shuffle Self-Routing and the Rerouting switch, the destination tag of the cell includes two fields specifying the addressed network outlet $o = o_{n-1} o_{n-2} \ldots o_0$ and the output distance d. Owing to the size 4×4 of the core SE the output address includes now $2n$ bits whose arrangement depends on the network to cross, i.e., $1x_{n-1} 1x_{n-2} \ldots 1x_0$ with $x_{n-1} \ldots x_0 = o_{n-1} \ldots o_0$ for routing through SN and $0x_{n-1} 0x_{n-2} \ldots 0x_0$ with $x_{n-1} \ldots x_0 = o_0 \ldots o_{n-1}$ for routing through USN. The cell initial distance is always set to

$n-1$. Let us assume that a cell is sent through SN. If the cell can be routed without conflicts and $d > 0$, the SE routes the cell to its outlet $1x_d$, by decreasing its distance by one, whereas the proper local outlet is chosen according to bit x_0 if $d = 0$. In case of conflicts for the same interstage outlet, the winner is again the cell closer to the addressed outlet, that is the cell with the smaller distance d. If the cell switched through the SN with distance $d = k$ is the loser of a conflict for the outlet $1x_k$, it is deflected onto one of the three other core SE outlets $1\bar{x}_k, 0x_k, 0\bar{x}_k$. Assuming that the cell is switched to the core SE outlet $1\bar{x}_k$ onto the link $(1\bar{x}_k, 0a_{n-1})$ the distance is increased to $k+1$ and the new output address is set to $1x_{n-1}, ..., 1x_{k+1}, 0a_{n-1}, 1x_k, ..., 1x_0$. This operation corresponds to crossing first a shuffle link (upon deflection) and then an unshuffle link such that the cell reaches at stage $k+2$ the same row as at stage k. Therefore each deflection causes in general the cell to leave the network two stages later. Note that the same result would have been obtained by selecting any of the two unshuffle links $(0\bar{x}_k, 1a_1)$ or $(0x_k, 1a_1)$ and properly adjusting the output address (this time the unshuffle link is crossed before the shuffle link). The degree of freedom in choosing the SE outlet on which the packet is deflected can be exploited in case of more than one cell to be deflected in the same 4×4 core SE.

A packet reaches the addressed row in SN after crossing n switching stages, whereas also the unshuffle interstage pattern following the last switching stage must be crossed too in USN to reach the addressed row (this is because the shuffle pattern precedes the switching stage in the SN, whereas it follows it in the USN). Removing the shuffle pattern that precedes stage 1 in SN, as we always did, does not raise any problem, since it just corresponds to presenting the set of packets to be switched in a different order to the first switching stage of the SN. Also the final unshuffle pattern in USN can be removed by a suitable modification of the cell output address, which also results in a simpler design of the SE. In fact it is convenient that also the cell routed through the USN can exit the network at the local outlet of the n-th switching stage without needing to go through a final unshuffling; otherwise the SE should be designed in such a way that the SE local outlets would be accessed by cells received on shuffle links after the switching and on unshuffle links before the switching. This objective can be obtained by modifying the initial output address of a cell routed through the USN, that is $x_{n-1}...x_1x_0 = o_1...o_{n-1}o_0$, so that the least significant bit of the output address is used in the last switching stage in USN. It can be easily verified that this new addressing enables the cell to reach the addressed row just at the end of the n-th switching operation without deflections. Therefore also the last unshuffle pattern of USN can be removed. As with the Shuffle Self-Routing and Rerouting switches, also in this case the number of links entering each output queue interface reduces to $K \perp (n-1)$ since all cells must cross at least n stages.

The routing algorithm is also capable of dealing with consecutive deflections given that at each step the distance is increased and the output address is properly modified. However, if the distance is $d = n$ and the packet cannot be routed onto the proper SE outlet due to a conflict, it is convenient to reset its destination tag to its original configuration, either $1x_{n-1}1x_{n-2}...1x_0$ or $0x_{n-1}0x_{n-2}...0x_0$ so that an unbounded increase of the distance is prevented. Therefore, the self-routing tag should also carry the original output address since the actual output address may have been modified due to deflections.

From the above description it follows that each SE of the overall network has size 4×6 if we take also into account the two local outlets, that is the links to the output queues, shared between SN and USN. By considering that the implementation of the routing to the local outlets can have different solutions and can be seen as disjoint from the routing in the core SE, that is from the 4 inlets to the 4 interstage outlets, we can discuss how the core switching element with size 4×4 can be implemented. Two different solutions have been proposed [Lie94] to implement the core SE, either as a non-blocking crossbar network (Figure 9.10a) or as a two-stage banyan network built with 2×2 SEs (Figure 9.10b). In this latter solution, the first bit of the couple xo_d $(x = 0, 1)$, processed when the distance is d $(d > 0)$, is used in the first stage of the core SE to select either the shuffle network $(x = 1)$ or the unshuffle network $(x = 0)$. The second bit (o_d) routes the cell to the specific core SE outlet of the selected network. The banyan SE, which might be simpler to be implemented due to the smaller size of the basic 2×2 SE, is blocking since, unlike the crossbar SE, it does not set up all the 4! permutations.

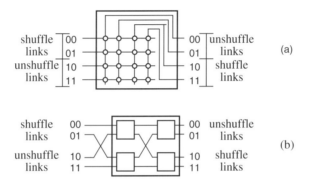

Figure 9.10. Implementation of the core 4 x 4 SE in Dual Shuffle

The routing example shown for the previous architectures based on deflection routing is repeated in Figure 9.11 for the Dual Shuffle switch with $K = 5$ stages. The four packets enter the network on the shuffle inlets of the SEs 10 and 11. A conflict occurs in both SEs and after a random choice both packets addressing network outlet $010 = 2$ are deflected. Note that the unshuffle link $(0o_{n-1}, 1a_1) = (00, 10)$ is selected for deflection in SE 10 and the shuffle link $(1\bar{o}_{n-1}, 0a_{n-1}) = (11, 01)$ in SE 11. These two packets restart their correct routing after two stages, that is at stage 3, where they find themselves in the original rows before deflections, that is 10 and 11, respectively. They reach the addressed row at stage 5, where only one of them can leave the network to enter the output queue and the other one is lost since the network includes only 5 stages. Both packets addressing outlet $000 = 0$ and routed correctly in stage 1 reach the addressed row at stage 3 and only one of them leaves the network at that stage, the other one being deflected onto the unshuffle link $(0\bar{o}_0, 1a_1) = (01, 10)$. This last packet crosses two more stages to compensate the deflection and then leaves the network at stage 5.

Unlike all the other architectures based on deflection routing, the interconnection network of an $N \times N$ Dual Shuffle switch has actually $2N$ inlets of which at most N can be busy in each slot. Two different operation modes can then be envisioned for the Dual Shuffle switch. The basic mode consists in offering all the packets to the shuffle (or unshuffle) network and

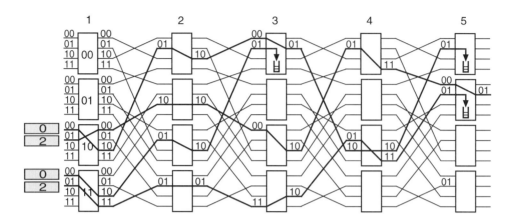

Figure 9.11. Routing example in Dual Shuffle

using the other only to take care of the cell deflections occurring in the main network. Alternatively *traffic sharing* can be operated, meaning that the generic inlet i of the switch $(i = 0, ..., N-1)$ is equipped with a 1×2 splitter which randomly routes the received cell to inlet i of SN or USN with even probability. If we look at the Dual Shuffle switch as a single network including all 4×4 SEs, such traffic sharing corresponds to randomly routing the cell received at the switch inlet $i_{n-1}...i_1 i_0$ onto inlet $1 i_0$ (routing on SN) and $0 i_0$ (routing on USN) of the SE $i_{n-1}, ..., i_1$ (recall that the shuffle preceding the SN first stage has been removed). Thus both networks are used to correct deflections when traffic sharing is accomplished.

9.2. Switch Architectures Based on Simpler SEs

Following the approach used in [Awd94] for the Shuffle Self-Routing switch, all the switch architectures described in Section 9.1 can be implemented in a slightly different version in which the basic switching element has size 2×2 rather than 2×4, meaning that the local outlets are removed from the switching element. Now also the cells that have reached the addressed row are routed onto an interstage link which this time feeds both the downstream inlets and the output queues. The general model of this switch architecture is shown in Figure 9.12 which now includes K switching blocks of size $N \times N$. Owing to the absence of dedicated local outlets, now each switching block inlet and each output queue is provided with a cell filter that accepts or discards a cell based on its destination tag. Therefore the destination tag of the cells will be provided with an additional field D, which is set to 0 by the switching block if the cell is being routed onto its outlet i $(0 \le i \le N-1)$ and the cell destination address is i. The field D is set to 1 otherwise. So the cell filter of the output queue (or of the switching block inlet) discards the cell if its field D is 1 (or 0).

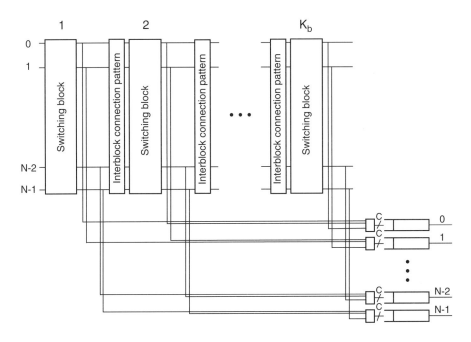

Figure 9.12. General model of ATM switch architecture based on deflection routing with simpler SEs

9.2.1. Previous architectures with 2×2 SEs

The architectures Shuffleout, Shuffle Self-Routing and Rerouting can be engineered easily to have simpler 2×2 SEs. For all of them each switching block of Figure 9.12 becomes a single switching stage, so that their ATM switch model becomes the one shown in Figure 9.13. Adopting simpler SEs in the Dual Shuffle switch means that the two local outlets of the original 4×6 SE must now be merged with two of the four interstage outlets, either the shuffle or the unshuffle links outgoing from the SE. More complex solutions with local outlets originating from each of the four interstage links will be examined in Section 9.5.

9.2.2. The Tandem Banyan switch

Unlike all previous architectures, the Tandem Banyan switching fabric (TBSF) [Tob91] does not fit into the model of Figure 9.13, since each switching block of Figure 9.12 is now a full n-stage banyan network $(n = \log_2 N)$ with 2×2 switching elements. Therefore a Tandem Banyan switch includes $K = nK_b$ stages (in this case $n_s = n$ since each block is a single I/O path network). Now the interstage block is simply given by the identity permutation (a set of N straight connections), so that the architecture of the $N \times N$ Tandem Banyan switch is shown in Figure 9.14. The cell destination tag now includes two fields, the network outlet address $o = o_{n-1}, ..., o_0$ and the flag D specifying if the cell transmitted by the last stage of a banyan network requires to enter the output queue or the downstream banyan network.

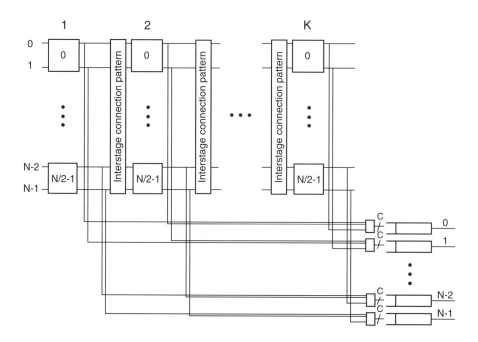

Figure 9.13. Model of deflection routing ATM switch with single-stage switching block and simpler SEs

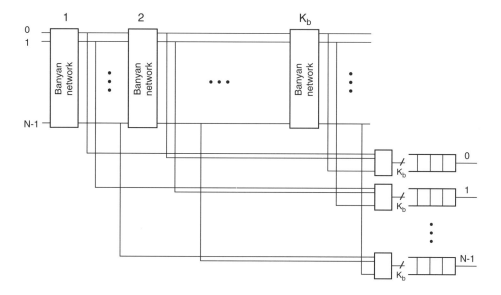

Figure 9.14. Architecture of the Tandem Banyan switch

The first banyan network routes the received packets according to the simplest bit-by-bit self-routing (the bit to be used depends on the specific topology of the banyan network). In case of a conflict for the same SE outlet, the winner, which is chosen randomly, is routed correctly; the loser is deflected onto the other SE outlet, by also setting to 1 the field D of the destination tag which is initially set to 0. Since each banyan network is a single I/O-path network, after the first deflection the packet cannot reach its addressed outlet. In order to avoid that a deflected packet causes the deflection of an undeflected packet at a later stage of the same network, the SE always routes correctly the undeflected packet when two packets with different values of D are received.

Different topologies can be chosen for the banyan networks of the Tandem Banyan switch. Here we will consider three of the most common structures in the technical literature, that is the Omega, the Baseline and the SW-banyan network (see Section 2.3.1.1 for a network description). All these topologies are n-stage networks with single I/O path providing full accessibility between inlets and outlets. They just differ in the interstage connection patterns. Another well-known topology, that is the n-cube network, has not been considered here as it is functionally equivalent to the Omega network (one is obtained from the other by simply exchanging the positions of some SEs). An example of routing in a 16×16 SW-banyan network of six packets with three deflections occurring (white squares) is shown in Figure 9.15.

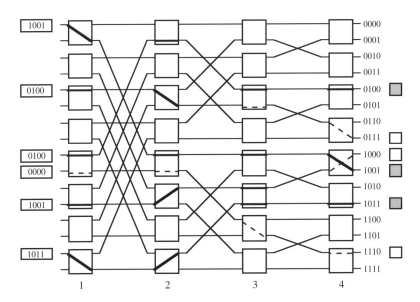

Figure 9.15. Routing example in Tandem Banyan

The output queue i is fed by the outlet i of each of the K_b banyan networks, through proper packet filters that select for acceptance only those packets carrying $D = 0$, whose destination tag matches the output queue address. The k-th banyan network $(k = 2, ..., K)$ behaves accordingly in handling the packets received form the upstream network $k - 1$: it filters out all the undeflected packets and accepts only the deflected packets $(D = 1)$.

Analogously to all the previous architectures, packets that emerge from the network K with $D = 1$ are lost.

Unlike all the previous switch architectures where a switching block means a switching stage, here a much smaller number of links enter each output queue (one per banyan network), so that the output queue in general need not be equipped with a concentrator. On the other hand, in general cells cross here a larger number of stages since the routing of a deflected cell is not started just after a deflection, but only when the cell enters the next banyan network.

In order to limit the number of networks to be cascaded so as to provide a given cell loss performance, all the links in the interconnection network can be "dilated" by a factor K_d, thus resulting in the Dilated Tandem Banyan switch (DTBSF) [Wid91]. Now each of the K_b banyan networks becomes a dilated banyan network (DBN) with dilation factor K_d (see Section 4.2.3), the pattern between banyan networks includes NK_d links and each output queue is fed by $K_b K_d$ links. In this configuration each network is able to switch up to K_d cells to each output queue, so that in general less networks are required compared to the basic TBSF to obtain a given cell loss performance. An example of a Dilated Tandem Banyan switch with $K_d = 2$ is given in Figure 9.16. Note that the network complexity of two networks with different dilation factor and equal product $K_b K_d$ is not the same. In fact even if they have the same total number of links to the output queues, the SE complexity grows with the square of the dilation factor.

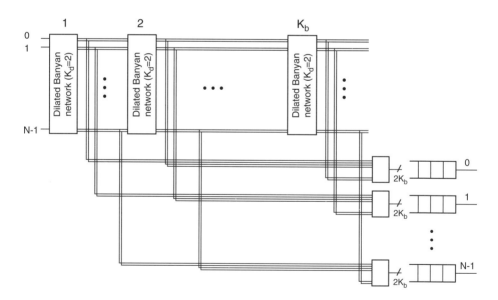

Figure 9.16. Architecture of the Dilated Tandem Banyan switch

9.3. Architecture Enhancements

The ATM switch architectures based on the deflection routing that have been described in the previous sections can be enhanced by adopting some changes in the interconnection network that basically affect the structure of the basic switching element. These two approaches are extended routing and interstage bridging.

9.3.1. Extended routing

Let us consider those switch architectures based on deflection routing in which each switching block of Figure 9.1 is just a switching stage and the routing does not adopt the shortest path algorithm, that is Shuffle Self-routing, Rerouting and Dual Shuffle. It has already been mentioned that in these architectures all cells exit the interconnection network to enter the proper output queue not earlier than stage n. Therefore 2×2 SEs could be adopted in stage 1 through $n - 1$ rather than 2×4, although this option is not viable in practice since the VLSI implementation of the SEs suggests the building of all identical SEs.

It has been proposed for the Rerouting switch [Uru91] how to improve the switch performance by allowing cells to enter the output queue in any of the first $n - 1$ stages if during its bit-by-bit routing in the first n stages the cell reaches the addressed row. This event occurs when the cell, which is addressing the network outlet $x_{n-1}, x_{n-2}, ..., x_1, x_0$, enters the SE with index $x_{n-1}, x_{n-2}, ..., x_1$. A better traffic performance is expected since some cells exit the network earlier and thus reduce the number of conflicts that could occur at later stages. The drawback, rather than having all identical 2×4 SEs, is a number of links feeding each output queue interface identical to the number of network stages, as in Shuffleout. This architecture enhancement, referred to as *extended routing*, will be evaluated here also for the Shuffle Self-Routing switch as its complexity is comparable to the Rerouting switch. Analogous modification, although applicable in principle, will not be considered for the Dual Shuffle as the complexity of its SEs is already higher than the other two architectures. Therefore the performance of the two architectures Extended Rerouting and Extended Shuffle Self-Routing will be compared to the others in Section 9.4.

The routing example of Figure 9.7 for the Rerouting switch is redrawn in Figure 9.17 for the Extended Rerouting switch. Now the cell received on inlet 4 can enter the local outlet at stage 2 rather than 3. It is interesting to observe that applying the extended routing to the example of Figure 9.5 of the Shuffle Self-Routing switch would provide the result shown in Figure 9.4. Therefore, at least for the specific example, Extended Shuffle Self-Routing and Shuffleout route packets in the same way.

9.3.2. Interstage bridging

We already noticed in Section 9.1.3 that in the shuffle-based switches (i.e., the Shuffleout and the Shuffle Self-Routing switch) the shortest path to the addressed network outlet of a cell that is deflected depends on the SE index and on the SE outgoing link for the cell. An interesting idea concerning the upgrade of these architectures consists in providing more interstage links, called *bridges*, so that the cell distance is increased as less as possible upon a deflection [Mas92],

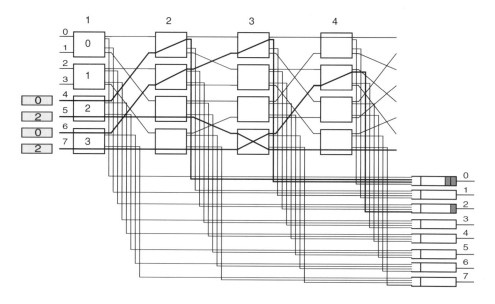

Figure 9.17. Routing example in Extended Rerouting

[Zar93a]. We will show how it is possible to design a network where the new distance of the deflected cell does not increase if it is routed through an interstage bridge, so that the cost of a deflection is just an additional stage to be crossed, rather than a variable value ranging up to n.

The most simple case of bridged shuffle-based network is a structure where each SE is provided with an additional "straight" outgoing link connecting it to the SE of the same row in the following stage. It is clear that if a cell in row i of stage k has to be deflected from its path requiring to enter the SE j at stage $k+1$ due to a conflict, its distance can be kept at the previous value d if it is routed onto the bridge to the SE of the same row i in stage $k+1$. Thus row j can be reached at stage $k+2$. Therefore in this case the network is built out of SEs with size 3×5, which accounts for the local outlets. An example is shown in Figure 9.18 for $N = 16$, $K = 5$.

The distance can thus be maintained at the same value and the internal path is delayed by only one stage, given that only one cell is deflected in the SE. If three cells enter the SE and require the same interstage SE outlet, then two of them must be deflected. Therefore two bridges per SE could be equipped, therefore designing a network with 4×6 SEs. Rather than having two "straight" bridges between SEs in adjacent columns it is more efficient to exploit the shuffle interstage pattern to distribute the deflected cells onto different downstream SEs [Zar93a]. In fact assume that a cell with distance $d \geq 2$ is crossing SE i at stage k and reaches without deflections SE j at stage $k+2$. If this cell is deflected at SE i in stage k, it can reach the requested row j at stage $k+3$ by entering in stage $k+1$ either SE i (through a "straight" bridge) or SE $(i + N/4) \bmod (N/2)$ (through a "cross" bridge). An example of such an interconnection network with two bridges is shown in Figure 9.19 for $N = 16$, $K = 5$. Note that the internal path is increased by only one stage by using either of the two bridges.

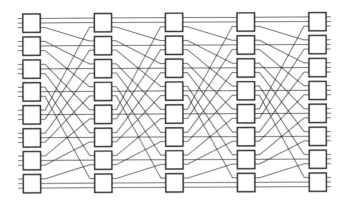

Figure 9.18. Example of 16 x 16 network with one interstage bridge

Nevertheless, the straight bridge is more "powerful" than the cross bridge, since the cells with $d = 1$ (that is those requiring a local SE outlet) do not increase their distance only if routed onto the straight bridge. So in general the cell with smaller distance should be preferably deflected onto the straight bridge. By following the same approach, network configurations with 2^k bridges $(k \geq 2)$ could be designed, but the resulting SE, $(2^k + 2) \times (2^k + 4)$, would become too large.

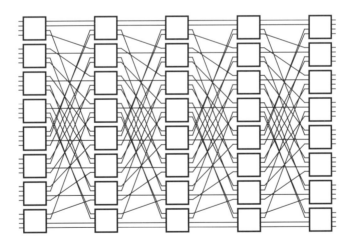

Figure 9.19. Example of 16 x 16 network with two interstage bridges

9.4. Performance Evaluation and Comparison

The performance of ATM switches with deflection routing is now evaluated under the random and uniform traffic assumption with average value p $(0 \leq p \leq 1)$. That is each switch inlet receives a cell in a slot with probability p and all cells, whose arrival events at the same or different inlet(s) are all mutually independent, are equally likely to be addressed to any switch outlet with the same probability $1/N$.

Unlike ATM switches with input queueing where the HOL blocking limits the throughput performance, the switch capacity $\rho_{max} = 1$ can well be achieved. Therefore our attention will be focused on the cell loss probability and delay performance. Cells can be lost in different parts of the switch, that is in the interconnection network (a cell reaches the last stage without entering the addressed local outlet), the concentrator, the output queue. If the cell loss probabilities corresponding to these events are indicated by π_n, π_c, π_q, respectively, then the overall packet loss probability is given by

$$\pi = 1 - (1 - \pi_n)(1 - \pi_c)(1 - \pi_q) \tag{9.1}$$

Since we are interested in obtaining very low cell loss probabilities, π in Equation 9.1 can be well approximated by $\pi = \pi_n + \pi_c + \pi_q$. The loss probability π_n depends on the specific network characteristics and will be evaluated in the following Sections 9.4.1–9.4.5 for the different switch architectures, with a comparative performance provided in Section 9.4.6. The overall switch performance, which takes into account also the behavior of the concentrator and of the output queue, will be analyzed in Section 9.4.7, which also provides a numerical example of the switch overall loss performance.

9.4.1. The Shuffleout switch

The performance analysis of the basic Shuffleout has been carried out developing a recursive analytical model: the load of stage $s + 1$ is computed using the load of stage s [Dec91a]. Therefore, starting from the first stage whose offered load is known, the load of all the network stages can be computed. Owing to the hypothesis of a purely random traffic offered to the switch, cell arrivals on SE inlets are mutually independent and each cell is likely to require any remote SE outlet (or local SE outlet if the cell distance is 0) with equal probability. The analysis will be first developed for a network built out of 2×4 SEs [Dec91a] and then extended to the case of 2×2 SEs.

Although this model could be in principle extended for the analysis of the architecture with interstage bridges, only computer simulation has been used to evaluate the performance of the bridged Shuffleout structure.

9.4.1.1. Network with 2×4 SEs

Let p_s be the load received by the network stage s, whereas $p_1 = p$ denotes the average external offered traffic. Moreover, let $q_{s,d}$ be the load of stage s, due to cells d hops away from the required switch outlet. The load p_s and the distribution $q_{s,d}$ are related by the equation

$$p_s = \sum_{d=0}^{n-1} q_{s,d} \tag{9.2}$$

The recursive equation relating p_{s+1} to p_s is obtained as follows. A *tagged* cell can exit the network from stage s only if it is 0 hops away from the required switch outlet. Furthermore one of the following conditions must be true:

- the cell is the only one to be routed by the SE, therefore no conflicts occur,
- the SE receives another cell with distance $d > 0$,
- the SE receives another cell with $d = 0$ and
  - no conflicts occur for the same local switch outlet or
  - a conflict occurs, but it is won by the tagged cell, due to the random choice algorithm.

These conditions are formally described by the equation

$$p_{s+1} = p_s - q_{s,0}\left[(1-p_s) + \sum_{d=1}^{n-1} q_{s,d} + \frac{3}{4}q_{s,0}\right]$$

(the result of a sum is set to 0, if the lower index is greater than the upper index in the sum).

To proceed with the iteration from the first stage to stage, K we also need a set of equations relating the distribution $q_{s+1,d}$ to the distribution $q_{s,d}$. The former distribution is computed as

$$q_{s+1,d} = \sum_{\delta=0}^{n-1} Pr[q_{s+1,d} \mid q_{s,\delta}] q_{s,\delta} \tag{9.3}$$

where we need to express the conditional probability $Pr[q_{s+1,d} \mid q_{s,\delta}]$ for all possible cases.

Case of $\delta = d+1$. A tagged cell passing from stage s to stage $s+1$ reduces (by one hop) its distance d if it is routed along the shortest path. This happens if one of these conditions is true:

- the tagged cell is the only one to be routed by the SE, therefore no conflicts can occur or
- the SE receives another cell with larger distance or
- the SE receives another cell with a smaller distance $d > 0$ and no contention occurs or
- the SE receives another cell with equal distance $d + 1$ and no contention occurs or
- the SE receives another cell with equal distance $d + 1$, a contention occurs and the tagged cell wins the conflict or
- the SE receives another cell with distance $d = 0$,

which implies

$$Pr[q_{s+1,d} \mid q_{s,d+1}] = 1 - p_s + \sum_{\delta=d+2}^{n-1} q_{s,\delta} + \frac{1}{2}\sum_{\delta=1}^{d} q_{s,\delta} + \frac{3}{4}q_{s,d+1} + q_{s,0} \tag{9.4}$$

Case of δ = d. When $d \neq n - 1$ the computation proceeds as follows. A cell passing from stage s to stage $s + 1$ can retain its original distance only if it is deflected. Cells following the shortest path reduce their distance, while deflected cells generally increase it. Due to the topological properties of the shuffle pattern, a deflected cell retains its distance only if it is routed along interstage links connecting to outlets 0 or $N-1$ of the SE stage (the *extreme* interstage links). Therefore, a tagged cell is deflected and retains its distance if:

- the SE receives another cell with a smaller distance $d > 0$ and a conflict occurs or
- the SE receives another cell with equal distance d, a conflict occurs and the tagged cell is deflected

and

- it is routed along an extreme interstage link,

which means that

$$\Pr[q_{s+1, d} \mid q_{s, d}] = \frac{2}{N}\left[\frac{1}{2}\sum_{\delta = 1}^{d-1} q_{s, \delta} + \frac{1}{4}q_{s, d}\right] \tag{9.5}$$

When a cell has maximum distance $(d = n - 1)$ to the required switch outlet, every deflection keeps it at distance $n - 1$, irrespective of the interstage link it follows (extreme or non-extreme). Therefore, if $d = n - 1$, Equation 9.5 applies with the factor $2/N$ removed.

Case of δ < d. A cell passing from stage s to stage $s + 1$ increases its distance from δ to d only if it is deflected. A tagged cell is deflected if:

- the SE receives another cell with a smaller distance $d > 0$ and a conflict for an interstage link occurs or
- the SE receives another cell with equal distance, a contention for an interstage link (or for a local switch outlet) occurs and the tagged cell is deflected.

Therefore we have

$$\Pr[q_{s+1, d} \mid q_{s, \delta}] = \Gamma(d, \delta)\left[\frac{1}{2}\sum_{k = 1}^{\delta-1} q_{s, k} + \frac{1}{4}q_{s, \delta}\right] \tag{9.6}$$

The factor $\Gamma(d, \delta)$ is the probability that a cell gets a new distance $d > \delta$ after being deflected when it was δ hops distant from the required switch outlet. A rather good approximate expressions for $\Gamma(d, \delta)$ is obtained by exhaustively counting, for all the possible couples i,j, the new distance d that a cell on link i that needs outlet j gets once deflected.

9.4.1.2. Network with 2×2 SEs

The previous model developed for 2×4 SEs has to be slightly modified to take into account that a cell with distance 0 can cause the deflection of cells with distance $d > 0$ if the two cells require the same SE outlet. Therefore the probability that a cell reduces its distance to the required outlet, given by Equation 9.4, is so modified

$$\Pr[q_{s+1,d} \mid q_{s,d+1}] = 1 - p_s + \sum_{\delta=d+2}^{n-1} q_{s,\delta} + \frac{1}{2}\sum_{\delta=0}^{d} q_{s,\delta} + \frac{3}{4}q_{s,d+1}$$

Analogously a cell retains its distance and is thus deflected onto one of the two extreme inter-stage links with probability

$$\Pr[q_{s+1,d} \mid q_{s,d}] = \frac{2}{N}\left[\frac{1}{2}\sum_{\delta=0}^{d-1} q_{s,\delta} + \frac{1}{4}q_{s,d}\right]$$

whereas Equation 9.5 applies to the case of 2×4 SEs. The event of a cell increasing its distance, previously expressed by Equation 9.6, occurs now with probability

$$\Pr[q_{s+1,d} \mid q_{s,\delta}] = \Gamma(d,\delta)\left[\frac{1}{2}\sum_{k=0}^{\delta-1} q_{s,k} + \frac{1}{4}q_{s,\delta}\right]$$

9.4.1.3. Network performance

The iterative model can be run starting from the probability $q_{1,d}$ that a cell entering the first stage of the network is d hops distant from the required switch outlet. The probability distribution $q_{1,d}$ is computed measuring the distance from the switch inlet i to the switch outlet j, for all the couples i,j. This procedure is correct under the hypothesis that a cell enters the switch from a specific inlet chosen with probability $1/N$ and is destined for a specific outlet with the same probability.

Once the iterative computation of the offered load stage by stage has ended, the cell loss probability for Shuffleout is immediately given by the ratio between the load that would be offered to stage $K + 1$ and the load received by stage 1:

$$\pi_n = \frac{p_{K+1}}{p} \tag{9.7}$$

The accuracy of the analytical models for the Shuffleout switch is rather good, as is shown in Figure 9.20 for both types of SEs and for various switch sizes under maximum input load $(p = 1.0)$. As a general rule, the number of stages K needed to guarantee a given loss performance increases with the switch size: for example $K = 16$ and $K = 46$ stages are necessary to guarantee a loss on the order of 10^{-8} for a switch size $N = 16$ and $N = 1024$ with 2×4 SEs, respectively. About 10% more stages are required when 2×2 SEs are used instead; in the previous example the two networks require now 16 and 51 stages, respectively.

The network packet loss probability given by the model with a variable level of the offered load for a switch with size $N = 1024$ equipped with 2×4 SEs is compared in Figure 9.21 with data from computer simulation. The graph shows that the model accuracy is roughly independent of the load level. Moreover the saving in number of stages by limiting the offered load and still providing the same loss performance is rather limited: compared to the case of maximum load $(p = 1.0)$, only 1/3 of the stages is saved when the input load is limited to

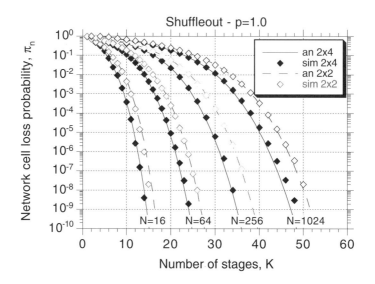

Figure 9.20. Model accuracy for the Shuffleout switch

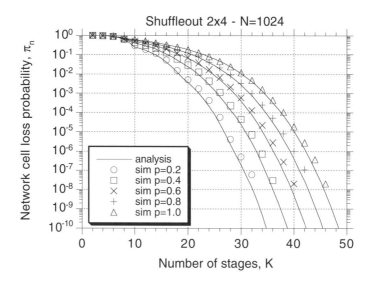

Figure 9.21. Performance of Shuffleout under a variable load level

$p = 0.2$ with a target loss probability $\pi_n = 10^{-8}$. Analogous comments apply to networks with different sizes or 2×2 SEs.

When interstage bridges are added to the basic Shuffleout switch, the loss performance improves significantly, as can be seen in Figure 9.22 for $N = 16, 1024$. Nevertheless, using only one bridge gives most of the performance enhancements provided by the bridging feature, especially for small switch sizes. With reference to the previous example with $N = 1024$, the 46 stages reduce now to 30 stages with one bridge, whereas 29 stages are needed using two bridges. In the case of the small 16×16 switch there is virtually no advantage in using two bridges rather than one: in both cases 11 stages are required instead of the 14 needed in the architecture without bridges.

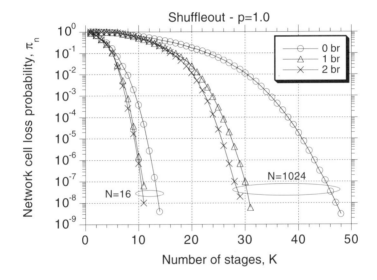

Figure 9.22. Performance of Shuffleout with interstage bridges

9.4.2. The Shuffle Self-Routing switch

The analysis of the Shuffle Self-Routing switch can be developed as a simplified version of the analytical model developed for Shuffleout considering the different definition of output distance that here makes the load p_s to stage s given by

$$p_s = \sum_{d=0}^{n-1} q_{s, d} \tag{9.8}$$

However, we will use here a different notation [Awd94] that is based on the definition of the probability $w_{s, d}$ that a cell with output distance d received on an SE inlet of stage s is deflected. The following analysis considers first the basic Shuffle Self-Routing network built out of 2×4 SEs and of 2×2 SEs. Afterwards the model is extended to account for the extended routing with both types of SEs.

The extension of this model to the Shuffle Self-Routing architecture with bridges when 2×4 SEs are used without extended routing is described in [Zar93a]. It can be seen in this reference that the accuracy of the model with one bridge is quite good, whereas it becomes rather poor when a network with two bridges is studied. Here only computer simulation has been used to evaluate the performance of the bridged Shuffle Self-Routing architecture.

9.4.2.1. Basic network with 2×4 SEs

Equation 9.3 specifying how the distribution $q_{s+1, d}$ relates to the distribution $q_{s, d}$ ($q_{s, d}$ is the load to stage s due to a cell with distance d) now simplifies since the distance can either decrease by one unit (no deflection) or be reset to $n - 1$ (deflection) after crossing one stage without entering the output queue. It follows that

$$q_{s+1, d} = \begin{cases} q_{s, d+1} (1 - w_{s, d+1}) & (0 \leq d \leq n - 2) \\ \displaystyle\sum_{\delta = 0}^{n-1} q_{s, \delta} w_{s, \delta} & (d = n - 1) \end{cases} \tag{9.9}$$

Now the distribution $q_{1, d}$ needed to start the iteration is straightforwardly obtained since all cells have output distance n, that is

$$q_{1, d} = \begin{cases} 0 & (0 \leq d \leq n - 2) \\ p & (d = n - 1) \end{cases} \tag{9.10}$$

The network packet loss probability is then given as in Shuffleout by Equation 9.7.

In order to compute the probability of cell deflection $w_{s, d}$ we have to specify the type of routing operated by the network. In this case of the basic architecture (n stages without deflections must always be crossed before entering the output queue), a cell, the *tagged* cell, is deflected if:

- the SE receives another cell with a smaller distance $d > 0$ and a conflict for an interstage link occurs or
- the SE receives another cell with equal distance d, a contention for an interstage link (or for a local switch outlet) occurs and the tagged cell loses

so that the deflection probability is given by

$$w_{s, d} = \frac{1}{2} \sum_{\delta = 1}^{d-1} q_{s, \delta} + \frac{1}{4} q_{s, d} \qquad (0 \leq d \leq n - 1) \tag{9.11}$$

(again the result of a sum is set to 0, if the lower index is greater than the upper index in the sum). Recall that in a network with 2×4 SEs a cell with distance $d = 0$ does not affect the routing of cells with a larger distance.

9.4.2.2. Basic network with 2 × 2 SEs

The analysis of the Shuffle Self-Routing switch with 2×2 SEs is substantially the same developed for the case of 2×4 SEs developed in Section 9.4.2.1. The only difference lies in the larger number of deflections occurring with switching elements without the local outlets, so that a cell with distance $d > 0$ can be deflected also when the SE receives another cell with $d = 0$. Therefore the previous analysis for the basic Shuffle Self-Routing switch applies now too given that Equation 9.11 is replaced by

$$w_{s, d} = \frac{1}{2} \sum_{\delta = 0}^{d - 1} q_{s, \delta} + \frac{1}{4} q_{s, d} \qquad (0 \le d \le n - 1) \qquad (9.12)$$

9.4.2.3. Basic network performance

The model accuracy in the analysis of the basic Shuffle Self-Routing (without extended routing) is displayed in Figure 9.23 for various network sizes for both types of SEs. The accuracy looks rather good independent of the switch size: only a very small underestimation of the real cell loss probability is observed. As with the Shuffleout switch, adopting the simpler 2×2 SEs rather than 2×4 SEs implies that a given cell loss performance is provided with about 10% more stages.

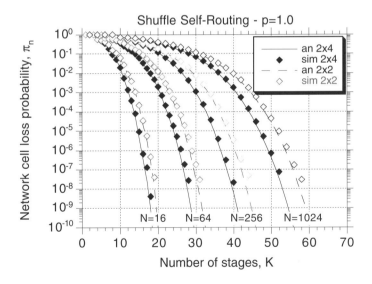

Figure 9.23. Model accuracy for the basic Shuffle Self-Routing switch

Figure 9.24, which compares analysis and simulation data for a 1024×1024 switch with 2×2 SEs, shows that the model is accurate for all load levels. An important phenomenon occurs for low load levels, which is evidenced both by analysis and by simulation: the loss curve as a function of the number of stages K shows irregular decreases around $K = k \log_2 N$

with k integer. In particular, if $\pi_n(K)$ denotes the network loss probability with K stages, $\pi_n(k\log_2 N - 1) / \pi_n(k\log_2 N) \gg \pi_n(k\log_2 N) / \pi_n(k\log_2 N + 1)$. In other words the loss probability decreases much more when K grows, say, from 19 to 20 than from 20 to 21. This phenomenon is related to the routing mechanism in the network which causes packets to leave the network exactly at stage $n = \log_2 N$ in absence of deflections, which is likely to occur at low load levels (e.g., $p = 0.2$). Nevertheless, in general more packets address the same network outlet, so that only one can leave the network at stage n and the other conflicting for the local outlet will try to exit at stage $2n$. Another conflict at stage $2n$ with another packet deflected twice or more will again increase the probability of having outputs at stage $3n$ and so on. This explains the irregularities of the loss curve due to larger sets of packets leaving the network at stages with indices that are multiples of n. Analogous results are obtained for networks with different sizes or with 2×4 SEs.

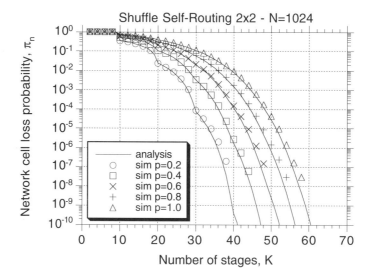

Figure 9.24. Performance of Shuffle Self-Routing under a variable load level

9.4.2.4. Network with extended routing and 2×4 SEs

Equation 9.9 specifying the distribution $q_{s+1, d}$ of the cell distance in the basic network must now be modified to take into account that a cell can enter the output queue before crossing n network stages if it reaches the addressed row along its routing. Thus the new distribution $u_{s, d}$ is introduced denoting the probability that a cell with distance d ($1 \leq d \leq n-1$) leaves the interconnection network at stage s to enter the output queue before completing its n steps of routing. Therefore we have now

$$q_{s+1, d} = \begin{cases} q_{s, d+1}(1 - w_{s, d+1} - u_{s, d+1}) & (0 \leq d \leq n-2) \\ \displaystyle\sum_{\delta=0}^{n-1} q_{s, \delta} w_{s, \delta} & (d = n-1) \end{cases} \tag{9.13}$$

The deflection probability $w_{w, d}$ expressed by Equation 9.11 for the basic Shuffle Self-Routing network must now be modified to take into account that a cell deflection with distance $d > 0$ can occur only if both contending cells are not in the addressed row. The previous expression still applies when $d = 0$: in fact it is assumed that a tagged cell that has completed its routing can be deflected only when it contends with another cell with same distance that addresses the same SE local outlet and the tagged cell is the contention loser. Therefore the probability of deflection becomes

$$w_{s, d} = \begin{cases} \dfrac{1}{4} q_{s, d} & (d = 0) \\ (1 - r)^2 \left[\dfrac{1}{2} \displaystyle\sum_{\delta=1}^{d-1} q_{s, \delta} + \dfrac{1}{4} q_{s, d} \right] & (1 \leq d \leq n-1) \end{cases} \tag{9.14}$$

Computing the distribution $u_{s, d}$ requires the identification of all the situations in which a cell can leave the interconnection network earlier. This event occurs when the tagged cell is crossing an SE of the addressed row, and this event is said to occur with probability r, and

- the other SE inlet is idle in the slot or
- the SE receives another cell that has completed its routing $(d = 0)$ and no conflict occurs with the tagged cell for the local outlet or
- the SE receives another cell that has not completed its routing $(d \geq 1)$ and is not in its addressed row or
- the SE receives another cell that has not completed its routing $(d \geq 1)$ but is in its addressed row and
  — requires a local outlet different from the tagged cell or
  — requires the same local outlet as the tagged cell and the tagged cell wins the contention.

It follows that

$$u_{s, d} = r \left[(1 - p_s) + \frac{1}{2} q_{s, 0} + (1 - r) \sum_{\delta=1}^{n-1} q_{s, \delta} + \left(\frac{1}{2} + \frac{1}{4} \right) r \sum_{\delta=1}^{n-1} q_{s, \delta} \right] \quad (1 \leq d \leq n-1) \tag{9.15}$$

Note that our model assumes that the cell distance affects only the deflection process, whereas it is irrelevant with respect to earlier exits. In fact a cell with lower distance wins the contention, as shown in Equation 9.14, while a random winner is selected upon a contention for the same SE local outlet between two cells with arbitrary distance $d > 0$, as shown in Equation 9.15. It is interesting to observe how the expression of the probability of earlier exit

$u_{s,d}$ given by Equation 9.15 becomes by using Equation 9.8 after some algebraic manipulations:

$$u_{s,d} = r\left[1 - \frac{1}{2}q_{s,0} - \frac{1}{4}r\sum_{\delta=1}^{n-1} q_{s,\delta}\right]$$

In other words the tagged cell can leave earlier the interconnection network if it is in the addressed row and two events do not occur: another cell is received by the SE which addresses the same SE local outlet as the tagged cell and whose distance is $d > 0$, with the tagged cell being the contention loser, or $d = 0$.

As far as the probability r that a packet is crossing an SE of the addressed row, the model assumes $r = 2/N$ independent of the stage index based on the fact that each row interfaces two out of N outlets. This is clearly an approximation since the real probability, which depends on the interconnection network topology, increases in general stage by stage as it approaches the final routing step.

The distribution $q_{1,d}$ needed to start the iterative computation of all the probability distributions stage by stage is still given by Equation 9.10, since the feature of extended routing has apparently no effect on the external condition that all packets entering the network have distance n from the addresses network outlet.

9.4.2.5. Network with extended routing and 2×2 SEs

The analysis of a network with 2×2 SEs adopting extended routing is rather close to that developed for 2×4 SEs in the previous Section 9.4.2.4. The probability distributions $q_{s+1,d}$ and $u_{s,d}$ are still given by Equations 9.13 and 9.15, whereas the deflection probability becomes much more complex now compared to the case of 2×4 SEs expressed by Equation 9.14. In fact, due to the link sharing between local and interstage SE outlets, now deflections can occur also when a contention involves a cell that is ready to exit the network because it has either completed its n steps of routing $(d = 0)$ or it can actually perform an earlier exit $(d > 0)$.

A tagged cell with $d = 0$ is deflected when the SE receives another cell with the same distance, a conflict occurs and the tagged cell is the loser. A tagged cell with $d > 0$ is deflected when

- the SE receives another cell with distance $d = 0$ and a conflict occurs for the same SE outlet or
- the SE receives another cell with distance $d > 0$, both cells are in the addressed row, a conflict occurs for the same SE outlet and the tagged cell is the loser or
- the tagged cell is not in the addressed row and
  — the SE receives another cell with distance $d > 0$ which is in the addressed row and a conflict occurs or
  — the SE receives another cell with lower distance which is not in the addressed row and a conflict occurs or
  — the SE receives another cell with the same distance d which is not in the addressed row, a conflict occurs and the tagged cell is the loser.

It follows that

$$
w_{s,d} = \begin{cases}
\dfrac{1}{4} q_{s,d} & (d = 0) \\[2ex]
\dfrac{1}{2} q_{s,0} + \dfrac{1}{4} r^2 \displaystyle\sum_{\delta=1}^{n-1} q_{s,\delta} + (1-r) \left(\dfrac{1}{2} r \displaystyle\sum_{\delta=1}^{n-1} q_{s,\delta} + \dfrac{1}{2}(1-r) \displaystyle\sum_{\delta=1}^{d-1} q_{s,\delta} + \dfrac{1}{4}(1-r) q_{s,d} \right) \\
& (1 \le d \le n-1)
\end{cases}
$$

9.4.2.6. Network performance with extended routing

The network performance of the Shuffle Self-Routing switch with extended routing given by the analytical model for maximum load are plotted in Figure 9.25 for different switch sizes and are compared with results given by computer simulation. Again the model is rather accurate: it slightly underestimates the real network loss probability. The loss performance with network size $N = 1024$ and 2×2 SEs under different input load levels is shown in Figure 9.26. The discussion for the basic Shuffle Self-Routing architecture concerning the performance with different SEs and the irregularities of the loss curve at low loads (see Section 9.4.2.3) apply here too when extended routing is accomplished.

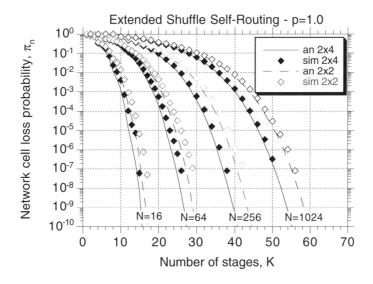

Figure 9.25. Model accuracy for the Extended Shuffle Self-Routing switch

For the two network sizes $N = 16$ and $N = 1024$ Figure 9.27 provides the cell loss performance of the Shuffle Self-Routing switch when using interstage bridges and/or extended routing. The adoption of extended routing, in which packets can exit the interconnection network to enter the addressed output queue before crossing $n = \log_2 N$ stages, gives a little performance improvement only for a small network when interstage bridging is not used at the same time. The effect of using bridges on the cell loss performance is substantially similar

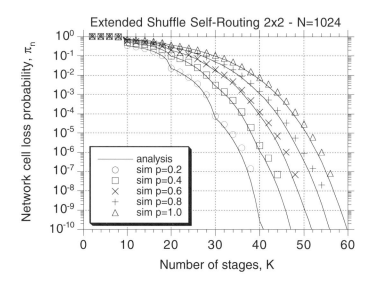

Figure 9.26. Performance of Shuffle Self-Routing with extended routing under a variable load level

to Shuffleout: one bridge greatly reduce the number of stages required to provide a loss performance target, especially for large switches. Such reduction ranges from 25% for small switches ($N = 16$) up to 40% for large switches ($N = 1024$) for loss probability around 10^{-8}. The further stage reduction enabled by adding one more bridge is very small, that is less than 10% in all the cases. Therefore, given that the SE complexity grows significantly with the number of bridges, the best cost-performance trade-off seems to be provided by networks built out of SEs with one bridge, especially for large size switches.

9.4.3. The Rerouting switch

The analytical model described in Section 9.4.2 for the basic Shuffle Self-routing switch does not take into account the particular interstage pattern, which in that case is always the shuffle connection. Therefore that model applies to the Rerouting switch too, whose interconnection network differs from the Shuffle Self-Routing one only in the interstage connection patterns. In particular Equations 9.8–9.11 apply with 2×4 SEs, whereas Equation 9.12 replaces Equation 9.11 if 2×2 SEs are used instead.

The analysis of the Extended Rerouting switch is carried out considering that an earlier exit can occur only before the first deflection, based on the relative position of the cell network inlet and the addressed network outlet. Recall that in the Rerouting switch earlier exits after the first deflection according to the extended routing concept are not allowed by the network topology. In fact the shortest path to the addressed outlet after a deflection always requires $n = \log_2 N$ stages to be crossed.

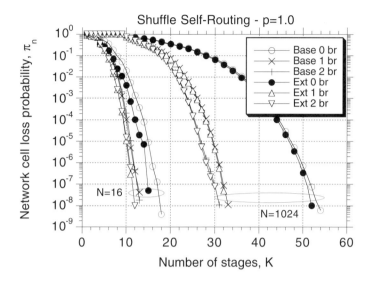

Figure 9.27. Performance of Shuffle Self-Routing with interstage bridges and extended routing

Earlier exits before the first deflection in the Rerouting switch can be taken into account by means of a proper initial distance distribution $q_{1,d}$. In fact, it is observed that a cell entering on an arbitrary row $i_{n-1}i_{n-2}\ldots i_1$ has distance 0 from the 2^1 output queues interfaced by row $i_{n-1}i_{n-2}\ldots i_1$, distance 1 from the 2^1 queues fed by row $\bar{i}_{n-1}i_{n-2}\ldots i_1$ (a butterfly permutation β_{n-1} follows stage 1), distance 2 from the 2^2 queues fed by the two rows $\bar{i}_{n-1}i_{n-2}\ldots i_1$ and $i_{n-1}\bar{i}_{n-2}\ldots i_1$ (a permutation β_{n-2} follows stage 2) and so on. Therefore

$$q_{1,d} = \begin{cases} \dfrac{2}{N}p & (d = 0) \\[3mm] \dfrac{2^{d-1}}{N}p & (1 \le d \le n-1) \end{cases}$$

Figures 9.28 and 9.29 show the model accuracy for the Rerouting switch in the basic structure and in the enhanced architecture with extended routing. In general the model turns out to be more accurate for smaller switch sizes. Furthermore, the analysis overestimates the real loss probability for 2×4 SEs, whereas underestimates it in the case of 2×2 SEs.

Using extended routing is more effective in the Rerouting switch than in Shuffle Self-Routing; in fact performance improvements, although limited, are obtained for any switch size, as shown in Figure 9.30 for 2×4 SEs. This result can be explained considering that each SE has a "straight" connection to the following stage, that is SEs with the same index in adjacent rows are always connected. Therefore the earlier exits from the interconnection network occur more frequently than in a network with all shuffle interstage patterns. An analogous improvement is given by the extended routing in the Rerouting switch with 2×2 SEs.

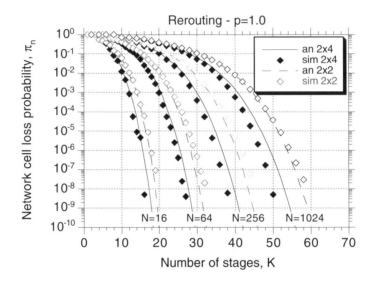

Figure 9.28. Model accuracy for the Rerouting switch

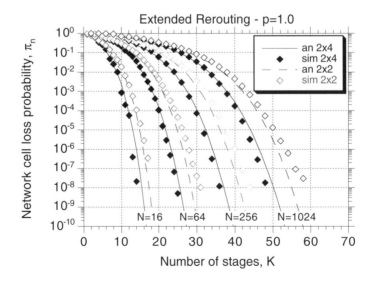

Figure 9.29. Model accuracy for the Rerouting switch with extended routing

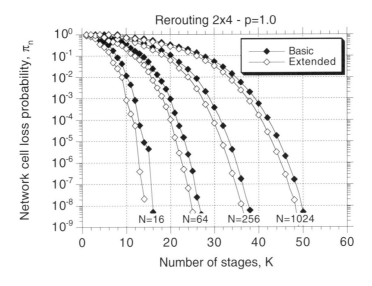

Figure 9.30. Performance of Rerouting with extended routing

9.4.4. The Dual Shuffle switch

The traffic performance of the Dual Shuffle switch has been evaluated by means of computer simulation: analytical models analogous to those developed for the previous architectures would become very complex. Figure 9.31 quantifies the performance improvement that a non-blocking crossbar SE causes compared to a blocking two-stage banyan SE for the basic Dual Shuffle switch without traffic sharing (the external traffic enters only N out of the $2N$ inlets of the interconnection network). A given loss performance is in general given by the crossbar SE using a couple of network stages less than using the banyan SE. It is interesting to note that the two SEs can be classified as having the same complexity in terms of crosspoints; in fact, by disregarding the local outlets, a crossbar 4×4 SE accounts for 16 crosspoints, the same total number required by the four 2×2 (crossbar) elements in the two-stage banyan structure. Therefore, the crossbar SE element has the best cost/performance ratio.

Using 4×4 rather than 4×6 SEs, slightly degrades the network loss performance, as shown in Figure 9.32 for different network sizes under maximum load: one or two more stages are needed in the former case to provide about the same loss probability as in the latter case. If the network load can be kept below the maximum value, e.g. $p = 0.2 - 0.8$ for the results shown in Figure 9.33, up to 30% of the stages needed at $p = 1.0$ can be saved by still providing the same cell loss performance.

The loss performance of traffic sharing of the Dual Shuffle with crossbar 4×6 SEs compared to the basic operation without sharing is shown in Figure 9.34. The saving in the number of stages given by the sharing feature increases with the switch size, it ranges from 10% for $N = 16$ up to about 25% for $N = 1024$. Therefore the traffic sharing is always convenient in Dual Shuffle, since the hardware is basically the same (we disregard the cost of the splitters).

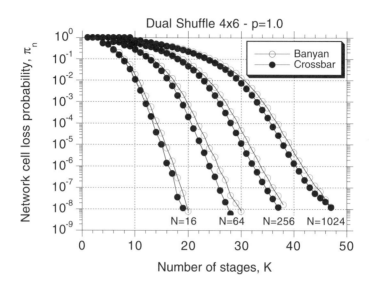

Figure 9.31. Performance of Dual Shuffle for two types of core SEs

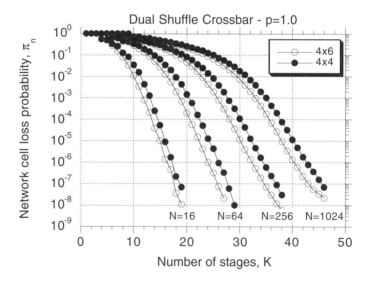

Figure 9.32. Performance of Dual Shuffle for two types of SEs

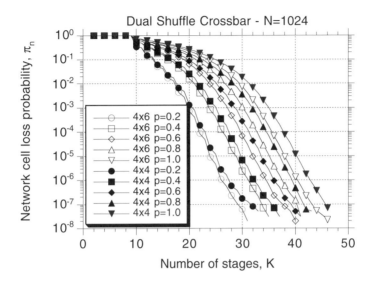

Figure 9.33. Performance of Dual Shuffle under a variable load level

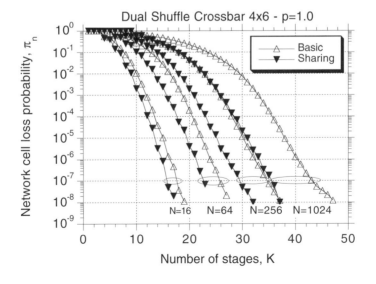

Figure 9.34. Performance of Dual Shuffle with traffic sharing

9.4.5. The Tandem Banyan switch

The performance of the Tandem Banyan switch for the two network sizes $N = 32, 1024$ is given in Figure 9.35 by considering three topologies of the basic banyan network, that is Omega, SW-banyan and Baseline network. Consistently with other previously known results [Tob91], the Omega topology gives the best performance and the Baseline performs much worse, especially for large switches. For very low loss figures, using the Baseline topology requires about 50% more networks than with the Omega topology. The SW-banyan behaves very close to the Omega topology. This is due to the different interstage patterns of the various banyan networks which results in presenting the deflected packets in a different configuration to the next banyan network. It is interesting to note that the large switch with $N = 1024$ requires only one network more ($K = 11$) than is required in a small switch with $N = 32$ ($K = 10$) to provide the same loss probability of 10^{-8}. Nevertheless the overall network with $N = 32$ includes $5 \cdot 10 = 50$ switching stages, whereas $11 \cdot 10 = 110$ stages are used with $N = 1024$.

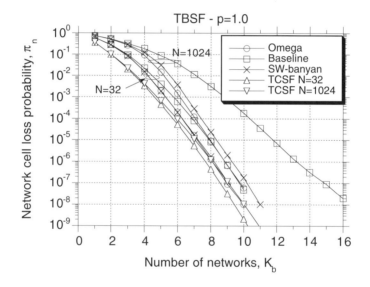

Figure 9.35. Performance of the Tandem Banyan switch

Figure 9.35 also shows for the two switch sizes the lower bound in the required number of networks when internal conflicts in the banyan networks do not occur. This structure is referred to as Tandem Crossbar switching fabric (TCSF) [Tob91], since each of the basic networks is now assumed to be a crossbar network. Therefore the loss performance of the TCSF gives the minimum number of networks required to cope with the unavoidable external conflicts, that is conflicts among packets addressing the same network outlet. The TCSF packet loss probability is given by Equation 9.16 expressing the loss probability in a concentrator where C outlets (corresponding to K networks in TCSF) are available to receive packets generated by K_b sources (N in TCSF) with all the outlets being addressed with the same probability. It is worth noting that a TBSF with Omega topology gives a performance rather close to the TCSF bound.

Interestingly enough, when the offered load is decreased, the network performance becomes less dependent on the switch size, as shown in Figure 9.36 for the Omega topology. In fact the cell loss probability for the two network sizes $N = 32, 1024$ is almost the same.

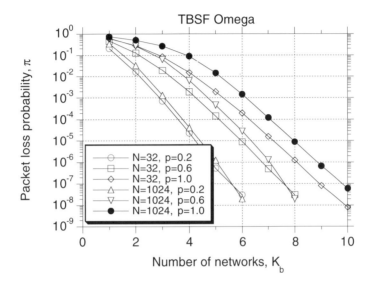

Figure 9.36. Performance of Tandem Banyan under a variable load level

Adoption of link dilation in each banyan network according to the architecture shown in Figure 9.16 improves the overall network performance given the same number K_b of banyan networks in the structure. In fact up to K_d packets can be successfully routed to the addressed banyan queue by each banyan network where each link is dilated by a factor K_d. Table 9.1 shows how the link dilation dramatically improves the cell loss performance for both switch sizes $N = 32, 1024$ under maximum load $p = 1.0$ (recall that $K_d = 1$ corresponds to the absence of link dilation). In fact less than half banyan networks are needed with the minimum dilation factor $K_d = 2$ to give a loss performance comparable with that of a non-dilated TBSF. A further improvement is obtained with a dilation factor $K_d = 3$. Apparently the price to be paid for such a performance improvement is a higher complexity of each SE in the banyan networks.

9.4.6. Interconnection network performance comparison

The loss performance provided by the various switch architectures presented in this section are now compared to each other in order to identify the structure providing the best performance for a given class of architecture complexity. Three basic parameters can be used to qualify the complexity of the interconnection network, that is the number of network stages, the number of links conveying cells to each output queue and the size of the basic SE.

Table 9.1. Packet loss probability of a dilated TBSF with Omega topology under maximum load

K_b	N=32			N=1024		
	K_d=1	K_d=2	K_d=3	K_d=1	K_d=2	K_d=3
1	$6.0 \cdot 10^{-1}$	$2.0 \cdot 10^{-2}$	$4.2 \cdot 10^{-2}$	$7.4 \cdot 10^{-1}$	$3.4 \cdot 10^{-1}$	$1.0 \cdot 10^{-1}$
2	$2.9 \cdot 10^{-1}$	$8.3 \cdot 10^{-3}$	$9.7 \cdot 10^{-5}$	$5.0 \cdot 10^{-1}$	$3.1 \cdot 10^{-2}$	$2.7 \cdot 10^{-4}$
3	$8.6 \cdot 10^{-2}$	$1.0 \cdot 10^{-4}$	$1.6 \cdot 10^{-8}$	$2.8 \cdot 10^{-1}$	$3.3 \cdot 10^{-4}$	$1.8 \cdot 10^{-7}$
4	$1.6 \cdot 10^{-2}$	$4.7 \cdot 10^{-7}$		$9.0 \cdot 10^{-2}$	$2.4 \cdot 10^{-6}$	
5	$2.0 \cdot 10^{-3}$			$1.5 \cdot 10^{-2}$		
6	$2.0 \cdot 10^{-4}$			$1.5 \cdot 10^{-3}$		
7	$1.6 \cdot 10^{-5}$			$1.2 \cdot 10^{-4}$		
8	$1.3 \cdot 10^{-6}$			$9.1 \cdot 10^{-6}$		
9	$9.0 \cdot 10^{-8}$			$6.7 \cdot 10^{-7}$		
10				$6.0 \cdot 10^{-8}$		

Given a number of network stages, the TBSF architecture is clearly the one with the least complexity in the connection pattern to the output queue. In fact it supports only one link every $\log_2 N$ stages per output queue, whereas all the other architectures based on deflection routing are equipped with one link every stage. Nevertheless, the TBSF switch is expected to provide a worse performance than the other architectures, since the packet self-routing can start only every N stages in TBSF and in general every one or two stages in the other architectures. This result is shown in Figure 9.37 for a network with $N = 1024$ under maximum load for a TBSF with Omega topology without ($K_d = 1$) or with ($K_d > 1$) link dilation compared to the Shuffle Self-Routing. The former network without dilation requires about twice the stages of the latter network to give a similar low cell loss performance. Adopting link dilation in the TBSF significantly improves the loss performance, even compared to the Shuffle Self-Routing network. Nevertheless, remember that the cost of a dilated banyan network grows quite fast with N (see Section 4.2.3). It is interesting also to compare the performance of the TBSF structure, where the K banyan networks are cascaded to each other with the BRBN architecture with random loading (see Section 6.3.1), where the K banyan networks are arranged in parallel and each of them receives $1/K$ of the total offered load. As is shown again in Figure 9.37, the serial structure (TBSF) outperforms the parallel structure (BRBN) by several orders of magnitude.

By excluding now the TBSF architecture, the other structures based on deflection routing are now compared in terms of packet loss probability for a given complexity of the basic SE. For the case of 2×4 SEs, the architectures Shuffle Self-Routing and Rerouting with extended routing, together with Shuffleout, have been considered. For the case of 4×6 SEs, the architectures Shuffle Self-Routing with extended routing and Shuffleout with two bridges, together with the Dual Shuffle with traffic sharing have been compared. Therefore for each architecture the version providing the best performance has been selected. The loss perfor-

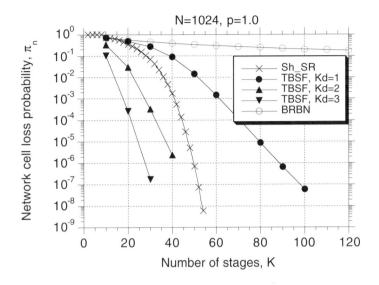

Figure 9.37. Loss performance comparison for different architectures

mance is shown in Figure 9.38 for $N = 16$ and in Figure 9.39 for $N = 1024$. With 2×4 SEs the Shuffleout switch gives the best performance owing to its shortest path routing which is more effective than the simple bit self-routing applied in the Shuffle Self-Routing and Rerouting switch. With the more complex 4×6 SEs the Shuffleout SEs still perform the best, with the Shuffle Self-Routing giving a rather close performance especially for large networks. The Dual Shuffle switch gives a significantly higher loss probability for the same network stages.

We would like to conclude this loss performance review, by comparing the number of stages K required for a variable network size to guarantee a loss performance of 10^{-6} for the six architectures based on deflection routing just considered: the results are given in Figure 9.40, respectively. All the curves grow almost linearly with the logarithmic switch size. Therefore we can conclude that minimum complexity of the order of $N\log_2 N$ characterizes not only the Dual Shuffle switch, as pointed out in [Lie94], but all the other architectures based on deflection routing. It can be shown that this property applies also to the Tandem Banyan switch. With this feature established, we need to look at the gradient and the stage number for a minimum size switch in order to identify the real optimum architecture. According to our previous considerations, Shuffleout requires the minimum amount of hardware in the interconnection network to attain a given loss figure, whereas the Shuffle Self-Routing (Dual Shuffle) needs the maximum amount of hardware with 2×4 (4×6) SEs. The conclusions to draw from these figures are that Shuffleout gives the best cost/performance ratio at the expense of implementing shortest-path routing. If bit-by-bit routing is preferred, Shuffle Self-Routing and Rerouting with extended routing provide the best solution with 2×4 and 4×6 SEs. Analogous results have been obtained for a different loss probability target.

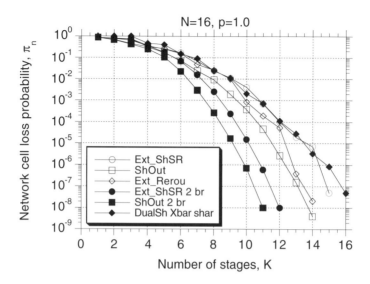

Figure 9.38. Loss performance comparison for different architectures

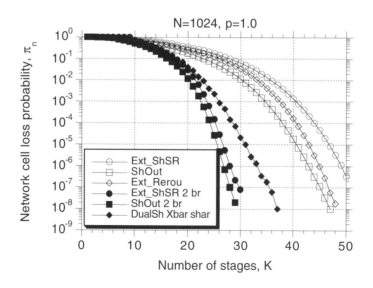

Figure 9.39. Loss performance comparison for different architectures

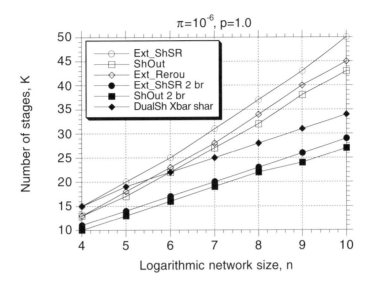

Figure 9.40. Network stages required to provide a given loss performance

In order to provide a completely fair comparison of the different architectures, also the *fairness* property has to be addressed. We refer here to the capability of the switching architecture to provide a similar loss performance for each virtual connection supported, that is the objective of a fair switching architecture is a cell loss probability that is independent of the specific inlet/outlet couple we look at. In general all the architectures in which the routing is bit-based, that is those in which the SE outlet is just indicated by one specific bit of the self-routing tag, are completely fair given that all the selections upon a conflict are just random. We refer to the selection of the packet to be deflected upon a conflict for an interstage link and to the selection of the SE interstage outlet upon conflict for the same SE local outlet. Therefore all the architectures based on deflection routing other than Shuffleout are completely fair.

In Shuffleout the cell loss probability is a function of the outlet index. In particular the packet flows addressing central switch outlets (those around addresses $N/2 - 1$ and $N/2$) are better served by the network than flows addressing the peripheral network outlets (those close to 0 and $N - 1$). The reason is that the shuffle interstage pattern makes available a larger number of paths to reach central outlets than peripheral outlets after crossing a given number of stages. If reasonable engineering criteria are used, the interconnection network must be dimensioned in order to guarantee a given cell loss probability independently from the switch outlet index. Therefore, unlike all the other deflection-based switches, Shuffleout must be engineered looking at the worst performing network outlets, that is the peripheral ones. Nevertheless the drawback of Shuffleout in terms of additional hardware required is rather limited. In fact, as is shown in [Bas92], [Dec92], engineering the switch based on the worst loss performance only requires an additional stage compared to a dimensioning based on average performance so as to provide a comparable performance figure.

9.4.7. Overall switch performance

After analyzing the loss performance occurring in the interconnection network due to the finite number of network stages, attention is now given to the other two components where cell loss can take place, that is the output concentrator and the output queue. Recall that in general concentrators are not equipped in the Tandem Banyan switch owing to the small number of lines feeding each output queue.

The average load p_c offered to the concentrator is given by $p_c = p(1 - \pi_n)$. Following [Yeh87], we assume that all the K_b local outlets feeding the concentrator carry the same load and are independent of one another, so that the packet loss probability in the concentrator is given by

$$\pi_c = \frac{1}{p_c} \sum_{i = C + 1}^{K_b} (i - C) \binom{K_b}{i} \left(\frac{p_c}{K_b} \right)^i \left(1 - \frac{p_c}{K_b} \right)^{K_b - i} \tag{9.16}$$

Figure 9.41 shows the loss performance of a concentrator with $C = 9$ output lines for a variable offered load level and three values of the number of its inlets, $K_b = 16, 32, 64$, which also represent the number of switching stages in the interconnection network. It is observed that the performance improves as the number of sources decreases for a given offered load as a larger set of sources with lower individual load always represents a statistically worse situation. In any case the number of concentrator outlets can be easily selected so as to provide the target loss performance once the stage number and the desired load level are known. Alternatively if K_b and C are preassigned, a suitable maximum load level can be suitably selected according to the performance target.

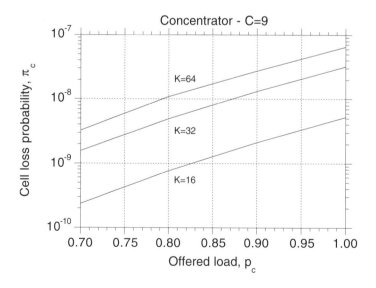

Figure 9.41. Loss performance in the concentrator

The average load p_q offered to the output queue is given by $p_q = p_c(1 - \pi_c)$; apparently $p_q = p_c$ in the Tandem Banyan switch which does not employ concentrators. Each output queue is fed by L links, where $L = C$ for all the architectures with output concentrators and $L = K_b$ for the Tandem Banyan switch. So the cell arrival process at the queue has a binomial distribution, that is the probability of i cells received in a slot is

$$\binom{L}{i} \left(\frac{p_q}{L}\right)^i \left(1 - \frac{p_q}{L}\right)^{L-i}$$

The server of the queue transmits one packet per slot, therefore the output queue can be classified as discrete-time $Geom(L)/D/1/B_o$, where B_o (cells) is the output queue capacity. By solving numerically this queue (see Appendix), the results in Figure 9.42 are obtained for different load levels ranging from $p_q = 0.5$ to $p_q = 0.99$ when the concentrator feeding the queue is equipped with $C = 9$ output lines. As one might intuitively expect, limiting the offered load level is mandatory to satisfy a given loss performance target.

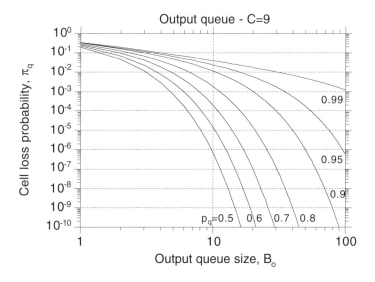

Figure 9.42. Loss performance in the output queue

Observe that the above evaluations of the cell loss in the concentrator and in the output queue are based on the assumption that all the K_b traffic sources generate the same traffic. This is clearly an approximation since the lines feeding each concentrator give a load that decreases from 1 to K_b (the load carried by the network stages decreases due to the earlier network exits). Nevertheless, our analysis is conservative since it gives pessimistic results compared to reality: assuming homogeneous arrivals always corresponds to the worst case compared to any type of heterogeneous arrivals with the same load [Yeh87].

The total packet loss probability π_t given by Equation 9.1 is plotted in Figure 9.43 for the Shuffleout switch with size $N = 256$, $C = 9$ output lines per concentrator and output queues with a capacity of $B_o = 64$ cells each under an offered load $p = 0.7, 0.8, 0.9$. The loss probability decreases as the number of stages increases since the interconnection network is capable of carrying more and more packets, until a saturation effect takes place that prevents the loss from decreasing again. The constant loss values that are reached for offered load $p = 0.7$ and $p = 0.8$ are due to the loss in the concentrator (see also Figure 9.41), whereas the saturation occurring at $p = 0.9$ is due to the overflow in the output queue (see also Figure 9.42).

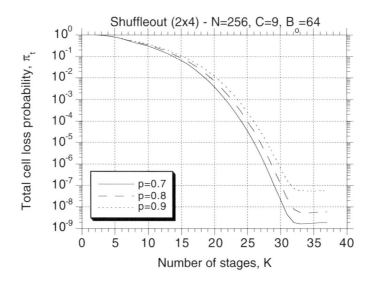

Figure 9.43. Total loss probability in Shuffleout

The packet delay is given by the time spent in the output queue (waiting time plus service time). The output queue delay is simply given by the analysis of the $Geom(L)/D/1/B_o$ queue reported in the Appendix. If we assume the output buffer to be large enough, then the average delay in the output queue is given by Equation 9.17, as explained in the Appendix.

$$T = 1 + \frac{C-1}{C}\frac{p_q}{2(1-p_q)} \tag{9.17}$$

9.5. Switch Architectures with Parallel Switching Planes

All the switch architectures based on deflection routing and described in Section 9.1 can be built with a smaller number of stages by still providing the same traffic performance if multiple switching planes, say n_k, can be equipped. We consider here the simple case in which 2 planes are available for switching and packets received at each switch inlet are routed randomly into one of the planes in order to be switched to its required destination. It follows that each switch inlet is equipped with a 1×2 splitter. The general structure of this switch architecture with 2 parallel planes is shown in Figure 9.44 where each plane includes K_b switching blocks. Therefore each output queue is fed now by $2K_b$ links. As with the basic architectures based on deflection routing, one switching block consists of one switching stage with the Shuffleout, Shuffle Self-Routing and Rerouting switch, whereas it includes a banyan network with $\log_2 N$ stages with the Tandem Banyan switch. Therefore the SEs have now the same size as in the respective basic architectures. In the case of the Dual Shuffle switch, two planes already exist in the basic version, which under the traffic sharing operation includes also the splitters. Considering the parallel architecture in the Dual Shuffle switch means that each SE in a plane has its own local outlets to the output queues. Therefore once we merge the SEs in the same stage and row into a single SE, whose internal "core" 4×4 structure is either crossbar or banyan, the basic SE of the single resulting plane now has the size 4×8.

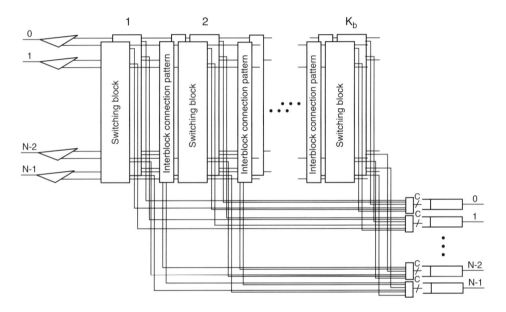

Figure 9.44. Model of ATM switch architecture with deflection routing and parallel planes

The performance of architectures using parallel planes is now evaluated using computer simulation. The loss performance of the parallel Shuffleout switch is compared in Figure 9.45 with the corresponding data for the basic architecture under maximum load. It is noted that using two planes reduces the number of stages compared to the single plane case by a factor in the range 10–20%; larger networks enable the largest saving. For example a loss target of 10^{-8} requires 47 stages in the basic architecture and 39 stages in the parallel one for $N = 1024$. Nevertheless, the parallel architecture is altogether more complex, since its total number of stages (and hence the number of links to each output queue) is $39 \cdot 2 = 78$. Similar comments apply using interstage bridging and extended routing, when applicable, in all the versions of the three architectures Shuffleout, Shuffle Self-Routing and Rerouting. Table 9.2 gives the number of stages required to guarantee a network packet loss probability smaller than

10^{-6} in a switch with size $N = 16, 64, 256, 1024$ with parallel planes $(n_k = 2)$ compared to the basic single-plane architecture $(n_k = 1)$.

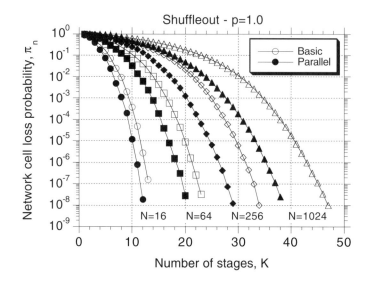

Figure 9.45. Loss performance of Shuffleout with parallel planes

Similar results have been obtained for the Tandem Banyan switch, whose performance comparison between the basic one-plane and the parallel two-plane architectures is given in Figure 9.46. Now the parallel structure requires about two banyan networks less than the basic structure. Also in this case the former architecture is much more complex since it needs in total $8 \cdot 2 = 16$ banyan networks versus 10 networks of the latter one.

Adopting the parallel architecture in the Dual Shuffle switch means making available twice the local links (to the output queues) in each SE compared to the basic structure: we have now 4×8 rather than 4×6 SEs. Figure 9.47 compares the loss performance of the basic structure adopting traffic sharing with the parallel one. It is observed that the additional local outlet available in the parallel switch carrying a very small performance advantage.

In conclusion, all the parallel architectures based on the deflection routing seem not to provide any advantage over their respective basic single-plane versions.

9.6. Additional Remarks

Other ATM switch architectures based on deflection routing have been proposed that are not reported here. For example "closed-loop" structures have been proposed in which recirculating lines are used instead of additional stages so as not to lose packets that have not reached their destination at the last network stage. Application of this concept to the Shuffleout switch

Table 9.2. Stages required to provide a packet loss probability of 10^{-6}

	n_k	$N=16$	$N=64$	$N=256$	$N=1024$
S_Out	1	13	22	32	43
	2	11	19	27	36
S_Out 2br	1	10	16	22	27
	2	9	14	18	22
Sh_SR	1	16	27	38	50
	2	15	23	31	41
Sh_SR 2 br	1	12	18	24	29
	2	11	16	20	24
Ext_Sh_SR	1	15	25	37	50
	2	13	22	31	41
Ext_Sh_SR 2 br	1	11	18	24	31
	2	10	15	20	26
Rerou	1	16	25	35	47
	2	13	21	30	38
Ext_Rerou	1	13	23	34	45
	2	12	20	28	37

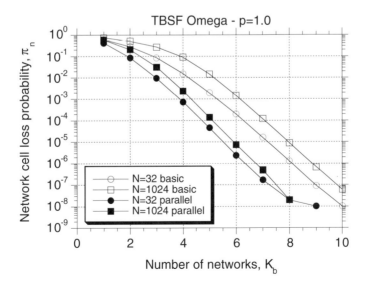

Figure 9.46. Loss performance of Tandem Banyan with parallel planes

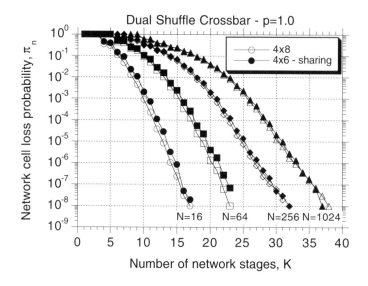

Figure 9.47. Loss performance of Dual Shuffle with parallel planes

and to the Shuffle-Self-Routing switch have been described in [Dec91b], Zar93b, respectively. The analysis of Shuffleout with SEs of generic size $b \times 2b$ is described in [Bas94].

The traffic performance of deflection-based architectures has been evaluated based on the random and uniform traffic assumption. It is worth pointing out that traffic correlation has no impact on the dimensioning of the number of stages in ATM switches with arbitrary-depth networks. In fact the interconnection network is internally unbuffered and the only queueing takes place at the output interfaces. Therefore the correlation of traffic patterns just affects the dimensioning of the output queues of the structure. This kind of engineering problem has been studied for example in [Hou89]. Non-uniformity in the traffic pattern, that is unbalanced in the output side, has been studied in [Bas92] for the Shuffleout architecture. Non-uniform traffic patterns in the Shuffleout switch have been studied in [Gia96].

9.7. References

[Awd94]   R.Y. Awdeh, H.T. Mouftah, "Design and performance analysis of an output-buffering ATM switch with complexity of O(Nlog2N)", *Proc. of ICC 94*, New Orleans, LA, May 1994, pp. 420-424.

[Bas92]   S. Bassi, M. Decina, A. Pattavina, "Performance analysis of the ATM Shuffleout switching architecture under non-uniform traffic patterns", *Proc. of INFOCOM 92*, Florence, Italy, May 1992, pp. 734-742.

[Bas94]   S. Bassi, M. Decina, P. Giacomazzi, A. Pattavina, "Multistage shuffle networks with shortest path and deflection routing: the open-loop Shuffleout", *IEEE Trans. on Commun.*, Vol. 42, No. 10, Oct. 1994, pp. 2881-2889.

[Dec91a]   M. Decina, P. Giacomazzi, A. Pattavina, "Shuffle interconnection networks with deflection routing for ATM switching: the Open-Loop Shuffleout", *Proc. of 13th Int. Teletraffic Congress*, Copenhagen, Denmark, June 1991, pp. 27-34.

[Dec91b]   M. Decina, P. Giacomazzi, A. Pattavina, "Shuffle interconnection networks with deflection routing for ATM switching: the Closed-Loop Shuffleout", *Proc. of INFOCOM 91*, Bal Harbour, FL, Apr. 1991, pp. 1254-1263.

[Dec92]   M. Decina, F. Masetti, A. Pattavina, C. Sironi, "Shuffleout architectures for ATM switching", *Proc. of Int. Switching Symp.*, Yokohama, Japan, Oct. 1992, Vol. 2, pp. 176-180.

[Gia96]   P. Giacomazzi, A. Pattavina, "Performance analysis of the ATM Shuffleout switch under arbitrary non-uniform traffic patterns", *IEEE Trans. on Commun.*, Vol. 44, No.11, Nov. 1996.

[Hou89]   T.-C. Hou, "Buffer sizing for synchronous self-routing broadband packet switches with bursty traffic", *Int. J. of Digital and Analog Commun. Systems*, Vol. 2, Oct.-Dec. 1989, pp. 253-260.

[Lie94]   S.C. Liew, T.T. Lee, "N log N dual shuffle-exchange network with error correcting routing", *IEEE Trans. on Commun.*, Vol. 42, No. 2-4, Feb.-Apr. 1994, pp. 754-766.

[Mas92]   F. Masetti, A. Pattavina, C. Sironi, "The ATM Shuffleout switching fabric: design and implementation issues", *European Trans. on Telecommun. and Related Technol.*, Vol. 3, No. 2, Mar.-Apr. 1992, pp. 157-165.

[Tob91]   F.A. Tobagi, T. Kwok, F.M. Chiussi, "Architecture, performance and implementation of the tandem banyan fast packet switch", *IEEE J. on Selected Areas in Commun.*, Vol. 9, No. 8, Oct. 1991, pp. 1173-1193.

[Uru91]   S. Urushidani, "A high performance self-routing switch for broadband ISDN", *IEEE J. on Selected Areas in Commun.*, Vol. 9, No. 8, Oct. 1991, pp. 1194-1204.

[Wid91]   I. Widjaja, "Tandem banyan switching fabric with dilation", *Electronics Letters*, Vol. 27, No. 19, Sept. 1991, pp.1770-1772.

[Yeh87]   Y.S. Yeh, M.G. Hluchyj, A.S. Acampora, "The knockout switch: a simple, modular architecture for high-performance packet switching", IEEE J. on Selected Areas in Commun., Vol. SAC-5, No. 8, Oct. 1987, pp. 1274-1283.

[Zar93a]   R. Zarour, H.T. Mouftah, "Bridged shuffle-exchange network: a high performance self-routing ATM switch", *Proc. of ICC 93*, Geneva, CH, June 1993, pp. 696-700.

[Zar93b]   R. Zarour, H.T. Mouftah, "The closed bridged shuffle-exchange network: a high performance self-routing ATM switch", *Proc. of GLOBECOM 93*, Houston, TX, Nov. 1993, pp. 1164-1168.

9.8. Problems

9.1   Find the routing algorithm for an ATM switch using deflection routing derived from the Rerouting switch. The interconnection network of this switch is built cascading Baseline networks rather than reverse SW-banyan networks with last and first stages of two cascaded networks merged, as in Rerouting.

9.2   Derive the modification to be applied to the analytical models developed in Section 9.4 to account for the availability of multiple planes.

9.3   Use the models derived in Problem 9.2 to compute the packet loss probability of different switch architectures and verify how the analytical results match those given by computer simulation reported in Figure 9.45 and Table 9.2.

9.4   Extend the analytical model of Shuffleout to account for the availability of bridges.

9.5   Extend the analytical model of Shuffle Self-Routing to account for the availability of bridges.

9.6   Verify that Equations 9.3, 9.4 and 9.6 of Shuffleout simplify into Equations 9.9 and 9.11 of Shuffle Self-Routing.

9.7   Find the minimum number of stages in the Shuffleout switch 16×16 that makes null the packet loss probability in the interconnection network.

9.8   Repeat Problem 9.7 for the Rerouting switch.

9.9   Repeat Problem 9.7 for the Tandem Banyan switch.

Appendix   *Synchronous Queues*

A characteristic common to all the queues considered here is their *synchronous* behavior, meaning that customers join and leave a queue only at discrete epochs nt ($n = 1, 2, \ldots$) that are integer multiples of the basic time interval t that we call a *slot*. Furthermore, starts and ends of service can only occur at slot boundaries. For the sake of convenience we adopt the slot duration as the basic time unit by expressing all the time measures in slots, so that state transitions can take place only at times n ($n = 1, 2, \ldots$). Unless specified otherwise the queue discipline is assumed to be first-in–first-out (FIFO).

The main random variables characterizing the queue are:

- A: *arrival size*, that is number of service requests offered to the queue in a given time period;
- W: *waiting line size*, that is number of customers in the queue waiting for the server availability;
- Q: *queue size*, that is total number of customers in the queue;
- θ: *service time*, that is amount of service requested to the queue by the customer;
- η: *waiting time*, that is time spent in the queue by a customer before its service starts;
- δ: *queueing time*, that is the total time spent by a customer in the queue (waiting time + service time).

In the following λ will denote the average arrival rate to the queue, which for a synchronous queue will be indicated by p expressing the probability that a service request is received in a slot by the queue. Two other parameters specifying the behavior of a queue other than the moments of the above random variables can be defined:

- ρ: *server utilization factor*, that is the time fraction in which each server in the queue is busy;
- π: *loss probability*, that is probability that a new customer is not accepted by the queue.

A.1. Synchronous Single-server Queues

In this section synchronous queues with one server are studied. Queues with deterministic service time are examined first, in the case of both Poisson and geometric arrival processes. Then queues with general distribution of the service time are considered with infinite or finite waiting line.

A.1.1. The *M/D*/1 queue

We are interested here in the evaluation of the *synchronous M/D/*1 queue in which the customer arrivals occur only at discrete time epochs and the arrival process is Poisson. We are going to show that this queueing system has a close relationship with the (*asynchronous*) *M/D/*1 in which arrivals and starts of service can take place at any epoch of a continuous time axis. We assume that a queue is asynchronous if not specified otherwise. Since the study of the *M/D/*1 queue relies on the analysis of an *M/G/*1 queue (see, e.g. [Gro85]), we will briefly examine this latter queue.

A.1.1.1. The asynchronous M/G/1 queue

In an *M/G/*1 queue the customers join the queue according to a Poisson process with average arrival rate λ and the service times are independent and identically distributed (IID) with an arbitrary probability distribution. Although in general such a queue is not a Markov process, it is possible to identify a set of "renewal" epochs in which the Markov property of the system holds. Through this technique an imbedded Markov chain can be identified that enables us to study the properties characterizing the system at these epochs. These renewal epochs are the times at which customers complete the service in the queue and thus leave the system. The number of customers Q_n left behind by the *n*-th departing customer who leaves the system at epoch n $(n > 0)$ is then described by

$$Q_n = \max\{0, Q_{n-1} - 1\} + A_n \tag{A.1}$$

in which A_n is the number of arrivals during the service time of the *n*-th customer.

The evolution of the system can be described by defining:
- the transition probability matrix $\boldsymbol{p}^n = [p_{i,j}^n]$ in which $p_{i,j}^n = \Pr[Q_n = j | Q_{n-1} = i]$;
- the distribution of arrivals in a service time $a_i^n = \Pr[A_n = i]$

We assume that a steady state is achievable (the necessary conditions are specified in any queueing theory book, see e.g. [Gro85]) and thus omit the index *n*, so that

$$p = \begin{bmatrix} a_0 & a_1 & a_2 & \cdots \\ a_0 & a_1 & a_2 & \cdots \\ 0 & a_0 & a_1 & \cdots \\ 0 & 0 & a_0 & \cdots \\ \cdot & \cdot & \cdot & \cdots \end{bmatrix}$$

The state probability vector at departure epochs $q = [q_i]$, in which $q_i = \Pr[Q = i]$, is then given by

$$q = qp$$

or equivalently

$$q_i = q_0 a_i + \sum_{j=1}^{i+1} q_j a_{i-j+1} \qquad (i \geq 0) \tag{A.2}$$

It can be shown that the steady-state probability distribution of the queue size q obtained through Equation A.2 together with the boundary condition $\Sigma_i q_i = 1$ at the departure epochs is the same at any arbitrary time epoch (see [Gro85]). Thus, in particular, q also gives the probability distribution we are interested in, that is at customer arrival epochs.

Let us define now the probability generating function (PGF) $\Pi(z)$ of the queue size Q

$$\Pi(z) = \sum_{i=0}^{\infty} q_i z^i \qquad (|z| \leq 1)$$

and of the customer arrivals A

$$A(z) = \sum_{i=0}^{\infty} a_i z^i \qquad (|z| \leq 1)$$

After some algebraic manipulation, including application of L'Hôpital's rule, and recalling that $\Pi(1) = 1$, $A(1) = 1$, $A'(1) = \rho$ where $\rho = \lambda E[\theta]$, we finally obtain the Pollaczek–Khinchin (P–K) transform equation

$$\Pi(z) = \frac{(1-\rho)(1-z)A(z)}{A(z) - z} \tag{A.3}$$

The average number of customers in the system $E[Q]$ is then given by $\Pi'(1)$, which after some algebraic computations results in the well-known Pollaczek–Khinchin (P–K) mean-value formula

$$E[Q] = \rho + \frac{\lambda^2 E[\theta^2]}{2(1-\rho)}$$

However, such a P–K mean value formula can also be obtained through a simpler mean–value argument that only uses Equation A.1 (see, e.g., [Kle75] and [Gro85]). The mean value of the queueing delay is now obtained via Little's formula

$$E[\delta] = \frac{E[Q]}{\lambda} = E[\theta] + \frac{\lambda E[\theta^2]}{2(1-\rho)}$$

The average number of customers $E[W]$ waiting in the queue for the server availability is obtained from $E[Q]$ considering that the average number of customers in the server is ρ and the corresponding waiting time $E[\eta]$ is given by Little's formula, that is

$$E[W] = E[Q] - \rho = \frac{\lambda^2 E[\theta^2]}{2(1-\rho)} \tag{A.4}$$

$$E[\eta] = E[\delta] - E[\theta] = \frac{E[W]}{\lambda} = \frac{\lambda E[\theta^2]}{2(1-\rho)} \tag{A.5}$$

Equations A.4–A.5 are independent of the specific service order of the customers in the queue. A simple recurrence formula giving all the moments of waiting time that applies only to a FIFO queue has been found by Takacs [Tak62]

$$E[\eta^k] = \frac{\lambda}{1-\rho} \sum_{i=1}^{k} \binom{k}{i} \frac{E[\theta^{i+1}]}{i+1} E[\eta^{k-i}] \qquad (k \geq 1) \tag{A.6}$$

A.1.1.2. The asynchronous M/D/1 queue

The asynchronous $M/D/1$ queue is just a particular case of the above $M/G/1$ queue in which the service time is deterministic and equal to θ. We assume θ as the basic time unit, so that the service time is 1 unit of time and the server utilization factor is $\rho = \lambda = p$ where p denotes the mean arrival rate of the Poisson distribution. In this case the PGF of the number of arrivals is

$$A(z) = \sum_{i=0}^{\infty} \frac{e^{-p} p^i}{i!} z^i = e^{-p(1-z)}$$

so that the queue size PGF (Equation A.3) becomes

$$\Pi(z) = \frac{(1-p)(1-z)}{1 - ze^{p(1-z)}}$$

Expansion of this equation into a Maclaurin series [Gro85] gives the queue size probability distribution

$$
\begin{cases}
q_0 = 1 - p \\[2mm]
q_1 = (1 - p)(e^p - 1) \\[4mm]
q_n = (1 - p) \displaystyle\sum_{k=1}^{n} (-1)^{n-k} e^{kp} \left[\frac{(kp)^{n-k}}{(n-k)!} + \frac{(kp)^{n-k-1}}{(n-k-1)!} \right] \qquad (n \geq 2)
\end{cases}
\tag{A.7}
$$

where the second factor in q_n is ignored for $k = n$.

The above Equations A.4–A.6 expressing the average queue size and delay for the $M/G/1$ provide the analogous measures for the $M/D/1$ queue by simply putting $\theta = 1$ and $\lambda = \rho = p$. We just report here for the waiting time of the $M/D/1$ queue the first moment given by Equation A.5:

$$
E[\eta] = E[\delta] - 1 = \frac{p}{2(1-p)}
\tag{A.8}
$$

as well as all the other moments for a FIFO service provided by Equation A.6:

$$
E[\eta^k] = \frac{p}{1-p} \sum_{i=1}^{k} \binom{k}{i} \frac{E[\eta^{k-i}]}{i+1} \qquad (k \geq 1)
\tag{A.9}
$$

A.1.1.3. The synchronous M/D/1 queue

Let us study a synchronous $M/D/1$ queue by showing its similarities with the asynchronous $M/D/1$ queue. In the synchronous $M/D/1$ queue the time axis is slotted, each slot lasting the (deterministic) service time of a customer, so that all arrivals and departures take place at slot boundaries n ($n = 1, 2, \ldots$). This corresponds to moving all the arrivals that would take place in the asynchronous $M/D/1$ queue during a time period lasting one slot at the end of the slot itself. So the number of arrivals taking place at each slot boundary has a Poisson distribution. In particular, at the slot boundary the queue *first* removes a customer (if any) from the queue (its service time is complete) and *then* stores the new customer requests. Since each customer spends in the queue a minimum time of 1 slot (the service time), this queue will be referred to as a *synchronous $M/D/1$ queue with internal server*, or S-IS $M/D/1$ queue. The queueing process of customers observed at discrete-time epochs (the slot boundaries) is a discrete-time Markov chain described by the same Equation A.1 in which Q_n and A_n represent now the customers in the queue and the new customers entering the queue, respectively, at the beginning of slot n. Thus the probability distribution q_i ($i = 0, 1, \ldots$) found for the asynchronous $M/D/1$ queue (Equation A.7) also applies to an S-IS $M/D/1$ queue for which the moments of the waiting time are thus expressed by Equation A.9. Therefore the first two moments for a FIFO service are given by

$$
E[\eta] = \frac{p}{2(1-p)}
\tag{A.10}
$$

$$E[\eta^2] = \frac{p^2}{2(1-p)^2} + \frac{p}{3(1-p)} = \frac{p(2+p)}{6(1-p)^2} \qquad (A.11)$$

Let us consider now a random order (RO) service, which means that the server upon becoming idle chooses randomly the next customer to be served. In this case the first moment of the waiting time is still given by Equation A.10 (the service order of the customers does not affect the number of waiting users and hence the average waiting time), whereas the second moment is given by [Tak63]

$$E[\eta^2] = \frac{p(2+p)}{3(1-p)^2(2-p)} \qquad (A.12)$$

In this queue the carried load ρ, which equals the mean arrival rate p, since the queue capacity is infinite, is immediately related to the probability q_0 of an empty queue $(Q = 0)$, since in any other state the server is busy. So,

$$\rho = 1 - q_0$$

which is consistent with Equation A.7.

We examine now a different synchronous $M/D/1$ queue in which at the slot boundary the queue *first* stores the new customer requests and *then* removes a customer request (if any) in order for that request to be served. In this case the customer is supposed to start its service time as soon as it is removed from the queue. In other words the queue only acts as a waiting line for customers, the server being located "outside the queue". Now a customer can be removed from the queue even if it has just entered the queue, so that the minimum time spent in the queue is 0 slot. For this reason this queue will be referred to as a *synchronous $M/D/1$ queue with external server*, or an S-ES $M/D/1$ queue. The queue then evolves according to the following equation:

$$Q_n = \max\{0, Q_{n-1} + A_n - 1\} \qquad (A.13)$$

in which Q_n and A_n have the same meaning as in the S-IS $M/D/1$ queue.

Note that in this case the queue size equals the waiting list size, since a customer leaves the queue in order to start its service. The S-ES $M/D/1$ queue can be studied analogously to the asynchronous $M/D/1$ queue. The steady-state number of customers in the queue $q_i = \Pr[Q = i]$ is obtained solving again the equation $\mathbf{q} = \mathbf{qp}$ in which

$$\mathbf{p} = \begin{bmatrix} a_0 + a_1 & a_2 & a_3 & \cdots \\ a_0 & a_1 & a_2 & \cdots \\ 0 & a_0 & a_1 & \cdots \\ 0 & 0 & a_0 & \cdots \\ \cdot & \cdot & \cdot & \cdots \end{bmatrix}$$

or equivalently

$$
\begin{cases}
q_0 = q_0 (a_0 + a_1) + q_1 a_0 \\
q_i = \displaystyle\sum_{j=0}^{i+1} q_j a_{i-j+1} \qquad (i \geq 1)
\end{cases}
$$

together with the boundary condition $\sum_i q_i = 1$. The PGF of the queue size is thus obtained

$$
\Pi(z) = \frac{(1-p)(1-z)}{A(z) - z} = \frac{(1-p)(1-z)}{e^{-p(1-z)} - z}
$$

Expansion of this equation into a Maclaurin series [Kar87] gives the probability distribution of the number of customers in the queue $q_i = \Pr[Q = i]$ waiting for the server availability:

$$
\begin{cases}
q_0 = (1-p) e^p \\
q_1 = (1-p) e^p (e^p - 1 - p) \\
q_n = (1-p) \displaystyle\sum_{k=1}^{n} (-1)^{n+1-k} e^{kp} \left[\frac{(kp)^{n+1-k}}{(n+1-k)!} + \frac{(kp)^{n-k}}{(n-k)!} \right] \qquad (n \geq 2)
\end{cases}
\tag{A.14}
$$

where the second factor in q_n is ignored for $k = n$.

Also in this queue the carried load ρ can be expressed as a function of the probability of an empty system. However now, unlike in the S-IS $M/D/1$ queue, the server is idle in a slot if the queue holds no customers and a new customer does not arrive. Thus

$$
\rho = 1 - q_0 a_0
$$

which is consistent with Equation A.14.

All the performance measures of the S-ES $M/D/1$ queue can be so obtained using the probability distribution q or the correspondent PGF $\Pi(z)$. In particular, by differentiating $\Pi(z)$ with respect to z and taking the limit as $z \to 1$, we obtain the average number of customers in the queue, which immediately gives the average time spent in the queue $E[\delta]$, that is

$$
E[\delta] = \frac{E[Q]}{p} = \frac{p}{2(1-p)}
$$

It is very interesting but not surprising that the average queueing time in the S-ES $M/D/1$ queue equals the average waiting time in the S-IS $M/D/1$ queue. In fact, the only difference between the two queues is that in the former system a customer receives service while sitting in the queue, whereas in the latter system the server is entered by a customer just after leaving the queue.

A.1.2. The *Geom(N)/D/*1 queue

A synchronous $Geom(N)/D/1$ queue is a system very similar to a synchronous $M/D/1$ queue with the only difference lying in the probability distribution of customer arrivals at each slot. Now the service requests are offered by a set of N mutually independent customers each offering a request with probability p/N $(0 \le p \le 1)$. Then the probability distribution of request arrivals A to the queue in a generic slot is binomial:

$$a_i = \Pr[A = i] = \binom{N}{i}\left(\frac{p}{N}\right)^i\left(1 - \frac{p}{N}\right)^{N-i} \qquad (0 \le i \le N) \qquad (A.15)$$

with mean value p and PGF

$$A(z) = \sum_{i=0}^{\infty} a_i z^i = \left(1 - \frac{p}{N} + \frac{p}{N}z\right)^N$$

A synchronous $Geom(N)/D/1$ queue with internal server, or S-IS $Geom(N)/D/1$ queue (defined analogously to a S-IS $M/D/1$ queue) is characterized by the same evolution equation (Equation A.1) as the S-IS $M/D/1$ queue, provided that the different customer arrival distribution is properly taken into account. Then the PGF of the queue occupancy is

$$\Pi(z) = \frac{(1-p)(1-z)A(z)}{A(z) - z} = \frac{(1-p)(1-z)\left(1 - \frac{p}{N} + \frac{p}{N}z\right)^N}{\left(1 - \frac{p}{N} + \frac{p}{N}z\right)^N - z}$$

Following the usual method, the average queue occupancy is obtained:

$$E[Q_{S-IS\ Geom(N)/D/1}] = p + \frac{N-1}{N}\frac{p^2}{2(1-p)} = p + \frac{N-1}{N}E[W_{S-IS\ M/D/1}]$$

Hence the average queue occupancy of an S-IS $Geom(N)/D/1$ queue approaches the average queue occupancy of an S-IS $M/D/1$ queue as $N \to \infty$. This result is consistent with the fact that the binomial distribution of the customer arrivals (see Equation A.15) approaches a Poisson distribution with the same rate p as $N \to \infty$.

The same reasoning applies to a synchronous $Geom(N)/D/1$ with external server, or S-ES $Geom(N)/D/1$, so that Equation A.13 holds in this case by giving rise to the queue occupancy PGF:

$$\Pi(z) = \frac{(1-p)(1-z)}{A(z) - z} = \frac{(1-p)(1-z)}{\left(1 - \frac{p}{N} + \frac{p}{N}z\right)^N - z}$$

The average queue occupancy, computed as usual through derivation and limit, gives, after using Little's formula,

$$E[\delta_{S-ES\ Geom(N)/D/1}] = \frac{N-1}{N}\frac{p}{2(1-p)} = \frac{N-1}{N}E[\delta_{S-ES\ M/D/1}]$$

A.1.3. The *Geom*/*G*/1 queue

A (synchronous) $Geom/G/1$ queue is a discrete-time system in which customer arrivals and departures take place at discrete-time epochs evenly spaced, the slot boundaries. The arrival process is Bernoulli with mean p $(0 \leq p \leq 1)$, which represents the probability that a customer requests service at a generic slot boundary. The service time θ is a discrete random variable with general probability distribution $\theta_j = \Pr[\theta = j]$ assuming values that are integer multiples of the basic time unit, the slot interval $(\theta = 1, 2, \ldots)$. The analysis of this queue, first developed in [Mei58], is again based on the observation of the number of customers left behind by a departing customer. So the balance equations of an $M/G/1$ queue (Equation A.2) apply also in this case considering that now the distribution a_i of customer arrivals in a number of slots equal to the service time has a different expression. Such a distribution is obtained computing the number of arrivals $a_{i,m}$ when the service time lasts m slots and applying the theorem of total probability while considering that the number of arrivals in m slots is not larger than m:

$$a_i = \sum_{m=0}^{\infty} a_{i,m} \theta_m = \sum_{m=i}^{\infty} \binom{m}{i} p^i (1-p)^{m-i} \theta_m \qquad (i \geq 0)$$

Apparently the PGF of the number of customer in the system is again given by Equation A.3 in which now

$$A(z) = \sum_{i=0}^{\infty} (1-p+pz)^i \theta_i$$

By applying twice L'Hôpital's rule and considering that

$$A'(1) = pE[\theta] = \rho$$

$$A''(1) = p^2 E[\theta(\theta-1)]$$

we finally obtain

$$E[Q] = pE[\theta] + \frac{p^2 E[\theta(\theta-1)]}{2(1-pE[\theta])}$$

The average queueing time and waiting time are obtained applying Little's formula:

$$E[\delta] = \frac{E[Q]}{p} = E[\theta] + \frac{pE[\theta(\theta-1)]}{2(1-pE[\theta])}$$

$$E[\eta] = \frac{E[W]}{p} = E[\delta] - E[\theta] = \frac{pE[\theta(\theta-1)]}{2(1-pE[\theta])}$$

A.1.4. The *Geom*/*G*/1/*B* queue

A (synchronous) $Geom/G/1/B$ queue is defined as the previous $Geom/G/1$ queue in which the queue capacity, B, is now finite rather being infinite. This queue is studied observing in each slot the process (Q, θ_r) where Q is the number of customers in the queue and θ_r is the residual number of slots for the service completion of the customer being served. Note that the service time for a new customer entering the queue is given by the probability distribution $\theta_{r,j} = \Pr[\theta_r = j]$. The process (Q, θ_r) observed at the end of each slot represents a Markov Chain for which the state transitions from the generic state (n, j) and the relevant probabilities can be easily found in a non-empty queue $(Q \geq 1)$ according to the residual value of the service time of the customer being served, p $(0 \leq p \leq 1)$ representing the probability of a new arrival in a slot:

1. Last slot of the service time $(\theta_r = 1)$. The system enters state (n, j) with probability $p\theta_j$ (a new customer is received) or $(n - 1, j)$ with probability $(1 - p)\theta_j$ (a new customer is not received).

2. Residual service time greater than one slot $(\theta_r > 1)$. The system enters state $(B, j - 1)$ if the queue is full $(Q = B)$, state $(n, j - 1)$ with probability $1 - p$ or state $(n + 1, j - 1)$ with probability p if the queue is not full.

Thus, if $q_{n,j}$ indicates the probability of the state $(n, \hat{j})$, the balance equations for the queue are:

$$
\begin{cases}
q_{0,0} = q_{0,0}(1 - p) + q_{1,1}(1 - p) \\
q_{1,j} = q_{1,j+1}(1 - p) + q_{2,1}(1 - p)\theta_{r,j} + q_{1,1}p\theta_{r,j} + q_{0,0}p\theta_{r,j} \\
q_{n,j} = q_{n,j+1}(1 - p) + q_{n-1,j+1}p + q_{n+1,1}(1 - p)\theta_{r,j} + q_{n,1}p\theta_{r,j} \quad (1 < n < B, \ j \geq 1) \\
q_{B,j} = q_{B,j+1} + q_{B,1}p\theta_{r,j} + q_{B-1,j+1}p
\end{cases}
$$

The numerical solution of this system provides the probability $q_{n,j}$ that gives us all the performance measures of interest for the queue. In particular, the load carried by the queue ρ, that is, the activity factor of the server, is given by the probability that the server is busy, that is

$$
\rho = 1 - q_{0,0}
$$

and the loss probability, π, is computed considering that customer departures take place at the end of the slot in which the residual service time is 1 slot, that is

$$
\pi = 1 - \frac{\sum_{n=1}^{B} q_{n,1}}{p}
$$

The average queueing and waiting time are then obtained applying Little's formula

$$
E[\delta] = \frac{E[Q]}{\rho} = \frac{\sum_{n=1}^{B} n \sum_{j=1}^{\infty} q_{n,j}}{p(1 - \pi)}
$$

$$E[\eta] = \frac{E[W]}{\rho} = \frac{\displaystyle\sum_{n=2}^{B} (n-1) \sum_{j=1}^{\infty} q_{n,j}}{p(1-\pi)}$$

A.2. Synchronous Multiple-server Queues

Synchronous queues with multiple servers are now studied in which the arrival process is either Poisson or geometric.

A.2.1. The $M/D/C$ queue

In an asynchronous $M/D/C$ queue customers join the queue according to a Poisson distribution with mean rate $\lambda = pC$ and a customer immediately starts its service lasting 1 unit of time (while sitting in the queue) if at the moment of its arrival less than C servers are busy. Such a queue can be studied analogously to the $M/D/1$ queue, so that the system evolution is described by the equation

$$Q_n = \max\{0, Q_{n-1} - C\} + A_n \tag{A.16}$$

in which Q_n is the queue size left behind by the n-th departing customer and A_n denotes the number of customer arrivals during the service time of that departing customer. As in the single-server case, a steady state is assumed to be reached and the queue size distribution $q_i = \Pr[Q = i]$ that we are going to find for the customer departure epochs also applies to an arbitrary epoch [Gro85]. The balance equations $\boldsymbol{q} = \boldsymbol{qp}$ for the queue are now given by

$$\begin{cases} q_0 = \displaystyle\sum_{i=0}^{C} q_i a_0 \\ q_n = \displaystyle\sum_{i=0}^{C} q_i a_n + \sum_{i=0}^{n-1} q_{n+C-i} a_i \qquad (n \geq 1) \end{cases} \tag{A.17}$$

where a_i is given by the Poisson distribution. Using these equations the queue size PGF is thus obtained:

$$\Pi(z) = \frac{\displaystyle\sum_{n=0}^{C} q_n z^n - z^C \sum_{n=0}^{C} q_n}{1 - z^C e^{\lambda}(1-z)} \tag{A.18}$$

The $C+1$ unknowns can be removed through non-trivial arguments including the solution of the transcendental equation at the denominator of Equation A.18 (see [Gro85]). However, the final expression for $\Pi(z)$ does not enable us to get a closed-form expression either for the queue size probabilities $q_i = \Pr[Q = i]$, or for the moments of the queue size or waiting line size.

An exact expression for the first moment of the waiting time has been computed in [Cro32] using a different approach, that is

$$E[\eta] = \sum_{i=1}^{\infty} e^{-ipC} \left[\sum_{n=iC}^{\infty} \frac{(ipC)^n}{n!} + \frac{1}{p} \sum_{n=iC+1}^{\infty} \frac{(ipC)^n}{n!} \right]$$

Analogously to the approach followed for single server queues, we can show that a synchronous $M/D/C$ queue, in which all customer arrivals and departures take place at slot boundaries, is described by the same Equations A.16 and A.17. Thus the results obtained for the asynchronous $M/D/C$ queue apply to the analogous synchronous $M/D/C$ queue as well.

A.2.2. The Geom(N)/D/C/B queue

In a $Geom(N)/D/C/B$ queue C servers are available to N mutually independent customers, each requesting service with probability pC/N $(0 \le p \le 1)$, and B positions (including the C places for the customers in service) are available to hold the customers in the system. The probability distribution of customer requests A in a generic slot is binomial:

$$a_i = \Pr[A = i] = \binom{N}{i} \left(\frac{pC}{N}\right)^i \left(1 - \frac{pC}{N}\right)^{N-i} \qquad (0 \le i \le N)$$

with mean value pC.

Consistently with the previously described operations of a queue with internal servers, a $Geom(N)/D/C/B$ queue with internal servers, or S-IS $Geom(N)/D/C/B$ queue, *first* removes the customers in service (at most C) and *then* stores the new customers in the currently idle positions of the queue. The system evolution is described by

$$Q_n = \min\{\max\{0, Q_{n-1} - C\} + A_n, B\}$$

in which Q_n and A_n represent the customers in the queue and the new customers requesting service, respectively, at the beginning of slot n. Thus the balance equations for the queue are

$$
\left\{
\begin{array}{l}
q_0 = \displaystyle\sum_{i=0}^{C} q_i a_0 \\[3ex]
q_n = \displaystyle\sum_{i=0}^{C} q_i a_n + \sum_{i=0}^{n-1} q_{n+C-i} a_i \qquad (0 < n \leq N) \\[3ex]
q_n = \displaystyle\sum_{i=0}^{N} q_{n+C-i} a_i \qquad (N < n \leq B-C) \\[3ex]
q_n = \displaystyle\sum_{i=n-B+C}^{N} q_{n+C-i} a_i \qquad (B-C < n \leq B-1) \\[3ex]
q_B = 1 - \displaystyle\sum_{i=0}^{B-1} q_i
\end{array}
\right.
$$

The average load carried by the queue, that is the average number of customers served in a slot, can be expressed as C times the utilization factor ρ of each server. The average carried load is readily obtained once we know the probability distribution q through

$$
\rho C = \sum_{i=0}^{C} i q_i + C \sum_{i=C+1}^{B} q_i
$$

The loss probability π is then computed by considering that the average load offered to the queue is pC:

$$
\pi = 1 - \frac{\rho C}{pC} \tag{A.19}
$$

The average time spent in the system and in the queue waiting for the service are given by Little's formula:

$$
E[\delta] = \frac{E[Q]}{\rho C} = \frac{\displaystyle\sum_{i=0}^{B} i q_i}{\rho C} \tag{A.20}
$$

$$
E[\eta] = \frac{E[W]}{\rho C} = \frac{\displaystyle\sum_{i=C+1}^{B} (i-C) q_i}{\rho C}
$$

In order to compute the waiting time distribution with a FIFO service [Des90], let us observe a tagged customer who is requesting service. Let v_j be the probability that the service request by the tagged customer arrives in a group of $j+1$ requests (the tagged request plus j other service requests). The probability v_j is proportional to the group size, $j+1$, and to the probability, a_{j+1}, that such group size occurs, that is

$$v_j = \frac{(j+1) a_{j+1}}{pC}$$

Note that pC is the normalizing constant such that $\Sigma_j v_j = 1$. The probability, η_j, that the tagged customer waits j slots before being served is a function of the queue location, and is given when it enters the queue. The event of receiving a location k $(k = 1, ..., B)$ given that the tagged customer is not lost, whose probability is indicated by t_k, means that $k-1$ customers will be served before the tagged customer. So,

$$\eta_j = \delta_{j+1} = \sum_{k=jC+1}^{(j+1)C} t_k \qquad (j \geq 0)$$

Since all the $j+1$ customers in the group have the same probability of being assigned to the $j+1$ idle queue locations with lower index, $1/(j+1)$ is the probability that the tagged customer is given location k. Since the probability of entering the queue is $1 - \pi$, we obtain

$$t_k = \frac{1}{1-\pi} \sum_{i=0}^{\min\{k+C-1, B\}} q_i \sum_{j = k-\max\{0, i-C\}-1}^{N-1} \frac{v_j}{j+1}$$

In fact, in order for the tagged customer to enter location k, the size $j+1$ of the group must at least equal the number of idle locations $k - \max\{0, i-C\}$ (C or i customers leave the queue in the slot if $i \geq C$ or $i < C$, respectively). Moreover the maximum number of customers in the queue compatible with the value k as the tagged customer location is $k + C - 1$, given that the capacity B of the queue is not exceeded.

A different approach must be used to compute the queueing time distribution with the random order (RO) service, in which a server becoming idle chooses randomly the next customer to be served among those waiting for service. We follow the approach described in [Bur59] with reference to an asynchronous $M/D/1$ queue. The probability that the tagged customer spends j slots in the queue can be expressed as a function of the conditional probability $\delta_{j,k}$ that a customer spends j slots given that it competes for the servers together with other $k-1$ customers at the time of its arrival, whose probability is s_k, that is

$$\delta_j = \sum_{k=1}^{B} \delta_{j,k} s_k \qquad (j \geq 1)$$

The event of spending j slots given that the tagged customer competes for the servers in a group of k customers occurs when the customer is not selected for service in the slot and then spends $j-1$ slots whatever is the number of new customers arriving in the following slot. Then

$$\delta_{j,k} = \frac{k-C}{k} \sum_{n=0}^{N} \delta_{j-1, \min\{B, k-C+n\}} a_n \qquad (j \geq 2, C < k \leq B)$$

whereas the boundary conditions are

$$\begin{cases} \delta_{1,k} = 1 & k \leq C \\ \delta_{j,k} = 0 & j > 1, k \leq C \\ \delta_{1,k} = \dfrac{C}{k} & k > C \end{cases}$$

The probability s_k that the tagged customer arrives in the queue so that k customers in total compete for the servers is obtained considering the current queue status and all the possible combination of arrivals resulting in k customers in the system waiting for service. That is

$$\begin{cases} s_k = \dfrac{1}{1-\pi} \displaystyle\sum_{i=0}^{\min\{k+C-1,\,B\}} q_i \Gamma & (1 \leq k < B) \\[3mm] s_B = \dfrac{1}{1-\pi} \displaystyle\sum_{i=0}^{B} q_i \sum_{j=B-\max\{0,\,i-C\}-1}^{N-1} v_j \end{cases}$$

with

$$\Gamma = \begin{cases} v_{k-\max\{0,\,i-C\}-1} & \text{if} \quad k-\max\{0,\,i-C\} \leq N-1 \\ 0 & k-\max\{0,\,i-C\} > N-1 \end{cases}$$

Using the same approach of the tagged customer, we can compute π as the probability that the tagged customer is received in a group whose size exceeds the current idle locations $B - \max\{0, i-C\}$ and it is not selected to occupy any of these idle locations

$$\pi = \sum_{i=0}^{B} q_i \sum_{j=B-\max\{0,\,i-C\}}^{N-1} v_j \frac{j+1-[B-\max\{0,\,i-C\}]}{j+1}$$

Let us now examine a $Geom\,(N)\,/\,D\,/\,C\,/\,B$ queue with external servers, or S-ES $Geom\,(N)\,/\,D\,/\,C\,/\,B$ queue, in which the queue *first* accepts as many customer as possible, so as to occupy the B locations in the waiting line and the C server positions, *then* moves C of the customers (if available) to the servers and stores the remaining customers (up to B) in the queue. The system evolution is described by

$$Q_n = \min\{\max\{0, Q_{n-1} - C + A_n\}, B\}$$

Thus the balance equations for the queue are

$$
\begin{cases}
q_0 = \displaystyle\sum_{j=0}^{C} q_j \sum_{i=0}^{C-j} a_i \\[2ex]
q_n = \displaystyle\sum_{i=0}^{C+n} q_{n+C-i} a_i & (0 < n \le N - C) \\[2ex]
q_n = \displaystyle\sum_{i=0}^{N} q_{n+C-i} a_i & (N - C < n \le B - C) \\[2ex]
q_n = \displaystyle\sum_{i=n-B+C}^{N} q_{n+C-i} a_i & (B - C < n \le B - 1) \\[2ex]
q_B = 1 - \displaystyle\sum_{i=0}^{B-1} q_i
\end{cases}
$$

The average carried load is computed as

$$
\rho C = \sum_{j=0}^{C} q_j \left(\sum_{i=0}^{C-j} (i+j)\, a_i + C \sum_{i=C-j+1}^{N} a_i \right) + C \sum_{j=C+1}^{B} q_j
$$

whereas the loss probability and the average queueing time are given by Equations A.19 and A.20, respectively.

Let us first compute the delay distribution with a FIFO service. In the S-ES $Geom\,(N)/DC/B$ queue the servers are external so that the only meaningful time distribution is the queueing time, δ_i, that is the time spent in the queue waiting to access a server. The probability that a customer accesses a server directly is given by

$$
\delta_0 = 1 - \sum_{k=1}^{B} t_k
$$

whereas the probability that a customer spends a non-zero time in the queue is obtained analogously to the queue with internal servers, that is

$$
\delta_j = \sum_{k=(j-1)C+1}^{jC} t_k \qquad (j \ge 1)
$$

The probability t_k of receiving a location k $(k = 1, \ldots, B)$ given that the tagged customer is not lost is now given by

$$
t_k = \frac{1}{1-\pi} \sum_{i=0}^{\min\{k+C-1,\,B\}} q_i \sum_{j=k-\max\{0,\,k-(i-C)-1\}}^{N-1} \frac{v_j}{j+1}
$$

In the case of RO service, the probability that the tagged customer spends j slots in the queue is again expressed as a function of the conditional probability $\delta_{j,k}$ that a customer spends j slots given that it competes for the servers together with another $k-1$ customers at the time of its arrival, whose probability is s_k, that is

$$\delta_j = \sum_{k=1}^{B+C} \delta_{j,k} s_k \qquad (j \geq 0)$$

in which

$$\delta_{j,k} = \frac{k-C}{k} \sum_{n=0}^{N} \delta_{j-1,\min\{B+C,\,k-C+n\}} a_n \qquad (j \geq 1,\, C < k \leq B+C)$$

with the boundary conditions

$$\begin{cases} \delta_{0,k} = 1 & k \leq C \\ \delta_{j,k} = 0 & j > 0,\, k \leq C \\ \delta_{0,k} = \dfrac{C}{k} & k > C \end{cases}$$

The probability s_k that the tagged customer arrives in the queue so that k customers in total compete for the servers is obtained now as

$$\begin{cases} s_k = \dfrac{1}{1-\pi} \displaystyle\sum_{i=0}^{\min\{k-1,\,B\}} q_i \Gamma & (1 \leq k < B+C) \\[4mm] s_{B+C} = \dfrac{1}{1-\pi} \displaystyle\sum_{i=0}^{B} q_i \displaystyle\sum_{j=B+C-i-1}^{N-1} v_j \end{cases}$$

with

$$\Gamma = \begin{cases} v_{k-i+1} & \text{if } \begin{aligned} k-i+1 &\leq N-1 \\ k-i+1 &> N-1 \end{aligned} \\ 0 & \end{cases}$$

As in the S-IS $Geom(N)/D/C/B$ queue, we can compute π as the probability that the tagged customer is received in a group whose size exceeds the current idle locations $B+C-i$ and it is not selected to occupy any of these idle locations:

$$\pi = \sum_{i=0}^{B} q_i \sum_{j=B+C-i}^{N-1} v_j \frac{j+1-[B+C-i]}{j+1}$$

A.3. References

[Bur59]   P.J. Burke, "Equilibrium delay distribution for one channel with constant holding time, Poisson input and random service", *Bell System Tech. J.*, Vol. 38, July 1959, pp. 1021-1031.

[Cro32]   C.D. Crommelin, "Delay probability formulae when the holding times are constant", *P.O. Elect. Engrs. Journal*, Vol. 25, 1932, pp. 41-50.

[Des90]   E. Desmet, G.H. Petit, "Performance analysis of the discrete time multiserver queueing system Geo(N)/D/c/K", *Proc. of BLNT RACE Workshop*, Munich, Germany, July 1990, pp. 1-20.

[Kar87]   M.J. Karol, M.G. Hluchyj, S.P. Morgan, "Input versus output queueing on a space-division packet switch", *IEEE Trans. on Commun.*, Vol. COM-35, No. 12, Dec. 1987, pp. 1347-1356.

[Gro85]   D. Gross, C.M. Harris, *Fundamentals of Queueing Theory*, Second Edition, John Wiley & Sons, New York, 1985.

[Kle75]   L. Kleinrock, *Queueing Systems, Volume I: Theory*, John Wiley & Sons, New York, 1975.

[Mei58]   T. Meisling, "Discrete-time queueing theory", *Operations Research*, Vol. 6, No. 1, Jan.-Feb. 1958, pp. 96-105.

[Tak62]   L. Takacs, "A single server queue with Poisson input", *Operations research*, Vol. 10, 1962, pp. 388-397.

[Tak63]   L. Takacs, "Delay distributions for one line with Poisson input, general holding times, and various service orders", *Bell System Tech. J.*, Vol. 42, March 1963, pp. 487-503.

Index